SEISMIC EXPLORATION

Volume 33

SEISMIC AMPLITUDE INVERSION IN REFLECTION TOMOGRAPHY

by

Yanghua WANG
Head of Research and Development
Robertson Research International
Swanley, Kent, UK

2003
PERGAMON
An Imprint of Elsevier Science
Amsterdam – Boston – London – New York – Oxford – Paris
San Diego – San Francisco – Singapore – Sydney – Tokyo

ELSEVIER SCIENCE Ltd
The Boulevard, Langford Lane
Kidlington, Oxford OX5 1GB, UK

First edition 2003

Library of Congress Cataloging in Publication Data
A catalog record from the Library of Congress has been applied for.

British Library Cataloguing in Publication Data
A catalogue record from the British Library has been applied for.

ISBN: 0-08-044243-9
ISSN: 0950-1401 (Series)

∞ The paper used in this publication meets the requirements of ANSI/NISO Z39.48-1992 (Permanence of Paper).
Printed in The Netherlands.

*This book is dedicated to
my wife Guo-ling, and
my two children Brian
and Claire.*

Preface

This is the first book of its kind on *seismic amplitude inversion* in the context of reflection tomography. The aim of this monograph is to advocate the use of ray-amplitude data, separately or jointly with traveltime data, in reflection seismic tomography.

One of the fundamental problems in conventional reflection seismic tomography using only traveltime data is the possible ambiguity between the velocity variation and the reflector depth. The inclusion of amplitude data in the inversion may help to resolve this problem because the amplitudes and traveltimes are sensitive to different features of the subsurface model, and thereby provide more accurate information about the subsurface structure and the velocity distribution.

The book consists of eleven chapters. The introductory chapter presents a prototype proof that the velocity-depth ambiguity in traveltime inversion may be resolved by using the additional amplitude information. It places a special emphasis on the important contribution of the interface geometry to reflection seismic amplitudes. Chapter 2 introduces a stable ray tracing method necessary for ray theoretical tomography and the dynamic ray system, which allows the estimation of the geometrical spreading of ray amplitude in heterogeneous media. Amplitude changes due to the model perturbation are predicted using ray perturbation theory.

Using the quadratic approximations to the Zöppritz equations of amplitude coefficients, derived in Chapter 3, the sensitivity matrix of ray amplitudes with respect to the elastic parameters can be computed analytically, although not necessarily numerically. Thus the amplitude inversion procedure may be speeded up. The approximations also provide an instructive insight into pore-fluid content in conventional amplitude variation versus offset (AVO) analysis.

The reflection seismic amplitude is determined partly by the reflection coefficient and partly by the curvature of the reflector. Chapter 4 shows that information contained in amplitude versus offset data suffices to constrain the geometry of an arbitrary 2-D reflector separating constant velocity layers in an accurate way. Chapter 5 investigates amplitude inversion for velocity variations, and demonstrates that amplitude data contain information that can constrain unknown velocity variation.

In reflection seismic, traveltime and amplitude data do contain complementary information, as is revealed systematically in Chapter 6. Traveltimes are sensitive to the model components with small wavenumbers and the amplitudes are more sensitive to the components with high wavenumbers. A joint inversion can then provide a better solution than the inversion using traveltime data alone.

Chapter 7 focuses on the ability of amplitude inversion to constrain the interface geometry of a multi-layered structure, for which a so-called multi-stage damped, subspace inversion scheme is developed. An essential goal of this monograph is to make the amplitude inversion method work with real reflection seismic data. Chapter 8 presents a practical approach to the application of real seismic amplitude data in the context of reflection seismic tomography. The application example represents a valuable contribution to the discussion of how the combined effort of imaging and inversion of seismic data should be organized when dealing with field data.

In Chapter 9, traveltimes and amplitudes in reflection seismic are used jointly in a simultaneous inversion for the interface geometry and the elastic parameters at the reflector. Practical approaches are proposed for handling the inverse problem which has different physical dimensions in both the data and model spaces. Chapter 10 develops a conciliatory approach to decomposing the structural amplitude effect and the AVO attributes. By proper accounting of the structural effects associated with focusing and defocusing due to the reflector curvature, the traditional AVO analysis may also be improved.

Most of the material presented here represents a pioneering and preliminary investigation. The work is by no means complete but requires further research and refinement. Chapter 11 delineates the basic improvement needed, rather than the road ahead, in order to enable seismic amplitude inversion to be used in practical applications.

Much of the research was carried out when the author was at Monash University, Australia, and Imperial College of Science, Technology and Medicine, London University, UK. Insightful suggestions from and stimulating discussions with Professors Greg Houseman, Gerhard Pratt, Roy White and Michael Worthington are very much appreciated. The author is also grateful to Birkhäuser Publishers, Blackwell Publishing and Society of Exploration Geophysicists for their permission to reproduce in this monograph the material from research articles published in Pure and Applied Geophysics, Geophysical Journal International and Geophysics, respectively.

Dr. Yanghua Wang

Kent, England
July 2002

Introduction

Seismic exploration methods have developed a steadily increasing power and complexity during the past 50 years. With the principal industrial application in the high-cost activity of hydrocarbon exploration, the value of better seismic imaging systems soon makes an impact on exploration cost-effectiveness. The development of fast programmable digital analysis systems in the 1960's made possible the first major development of the seismic technique: multifold acquisition and processing using the common mid-point method allowed a dramatic increase in signal-to-noise ratio relative to the earlier split-spread analyses. Subsequent development of digital processing techniques, such as filtering, deconvolution and migration, allowed seismic images of startling clarity and accuracy to be developed. The seismic section could (with some caution) be interpreted directly as a geological section, especially if it could be tied directly to well data.

The exponential growth in capacity of electronic data recording and processing systems during the past 40 years has encouraged increasingly ambitious strategies for the imaging of subsurface structures. The next major development in the technique arrived in the 1990's with the increasingly widespread use of 3-D surveys. 3-D images remove the uncertainty about structural variation in the direction perpendicular to the section. Furthermore, because the 3-D wavefield can also be sectioned in the horizontal plane, it is now possible to produce seismic maps at different depths that may be interpreted (again with some caution) as geological maps. Advances in computerized visualization and display techniques have actually moved the business of seismic interpretation beyond what is possible simply using maps and sections. With current visualization systems the interpreter can view the 3-D wavefield from within, changing perspective at will so that complex geological structure can be easily visualized.

Development of the seismic technique to this time has indeed focused on imaging. To the extent that quantitative measurement of seismic properties has mattered, the emphasis certainly has been on accurate estimation of depth to horizon. It is much harder to get accurate information about the variation of elastic properties within strata, or even to measure the effective impedance contrast between layers. Yet now that the imaging problem is more or less solved, the next major development in industrial seismology will surely be to focus on the problem of precise localized measurement of the elastic properties of the subsurface strata. There are good reasons to devote resources to solve what I refer to here as the measurement problem. Firstly, good measurements of the seismic velocity distribution are required in order to get accurate images. If the variation of seismic wavespeed is not accurately measured and used in migration of the recorded waveforms, imaged structures will be wrongly mapped, or may be entirely lost in noise. Secondly, accurate physical parameter measurements are a primary requirement for accurate identification of lithology. Finally, of course, such measurements are needed to identify (and potentially monitor) pore fluid pressure and composition.

Recognizing the motivation to develop such methods, one rapidly developing field is based on the use of triaxial ground motion sensors. By using all three components of the ground motion record, the shear-wave field can be measured in addition to the compressional-wave field. The shear-wave field is an independent dataset that, in principle, provides a separate constraint on the independent variation of the elastic moduli and of the density. Moreover, with such data it is possible to attempt to measure anisotropy in the response of the medium to elastic waves. Methods that use triaxial data and analyse for anisotropic parameters are, however, still in the early stages of development. In contrast it is perhaps ironic that, in decades of industrial seismology, one major aspect of waveform data that potentially is easier to measure and analyse has generally been ignored. I refer here, of course, to the information content of seismic wave amplitudes, even as measured using acoustic pressures or single component data. Amplitudes, of course, are influenced by a number of distinct physical processes: geometrical spreading, reflection and transmission coefficients, attenuation, mode conversion and sensor noise, to name the more important ones. Perhaps this potential complexity has deterred most researchers from a more thorough investigation of the practical use of seismic amplitude data. But of course the emphasis of seismic data processing was on imaging, and for that purpose amplitude variation was something to be got rid of, so that coherent phase arrivals could be seen more easily throughout the section.

The author of this volume, Dr Yanghua Wang, here presents an authoritative and detailed study of amplitude data, as used in conjunction with traveltime data, to provide better constraints on the variation of seismic wave speed in the subsurface. He demonstrates by the use of synthetic model calculations that traveltimes and amplitudes of seismic arrivals provide complementary information. Applied as joint constraints in an inversion context, including both types of data can result in more accurate tomographic solutions. His inversion methods are based firmly in the framework of geophysical inverse theory, in particular making use of subspace methods and singular value decomposition techniques that are well established. Geophysical inverse theory generally relies on iterative improvement of a model from which data can be predicted by means of a forward calculation. In this case the forward calculation is based on the high-frequency ray-theory approach.

Ray theory has significant advantages here, not least because it is possible to get some physical intuition as to how and why seismic amplitude data are affected by subsurface structure. A key aspect of this is the recognition that seismic wave speed variation must generally be represented at two extreme scales: the gradual variation that may occur with facies variation within strata, and the discontinuous change that occurs at stratigraphic boundaries. Tomographic inversion based only on traveltimes may be unable to discriminate between layer velocity variation and layer thickness variation. Amplitude data bring new information to bear on this problem because reflected wave amplitude is potentially very sensitive to curvature of the reflecting interface. While the method, in principle, is relatively straightforward, the implementation is complex and multi-step. As is appropriate for this kind of study, the methods are first tested on synthetic datasets, in order to find the theoretical limitations. In later chapters the methods are applied to actual datasets from an oil exploration context.

It remains to be seen whether such methods will find wide application in practice. Dr Wang has made, however, a fairly compelling case for why these methods deserve further investigation and development. Tomographic inversions based solely on traveltime data, could be rendered significantly more accurate and reliable by incorporating amplitude data. The data are collected anyway, and the incremental processing cost would be minor, once the procedures are clearly established. Better tomographic solutions will result in better seismic images and thus better interpretations.

Greg A. Houseman

Professor of Geophysics
University of Leeds
July 2002

Contents

Chapter 1

Introduction to amplitude inversion

Abstract   Reflection seismic tomography using only traveltime data may be unable to resolve the ambiguity caused by trade-off between reflector position and velocity anomaly. The inclusion of amplitude data in the inversion may help to resolve the velocity-depth ambiguity because the amplitudes and traveltimes are sensitive to different features of the model, thereby providing more accurate information about underground structures and velocity distribution. The amplitude of a reflected seismic wave is determined partly by the reflection coefficients and partly by the curvature of the reflector. The latter causes the spherical divergence of the seismic rays to be modified, focused or defocused, at the reflection point. We represent this modified geometrical spreading function using a simplified analytical expression, which clearly indicates that the interface geometry has an important effect on reflection amplitudes, and then the amplitude inversion may, in particular, extract the information about reflector geometry.

1.1.   Introduction

The concept of tomography is based on the idea that an observed dataset consists of integrals along lines or rays (i.e. projections) of certain physical quantities. The purpose of tomography is to reconstruct a model of the physical quantities, such that the projected data agree approximately with the measurements (e.g. Herman, 1980; Worthington, 1984; and references therein). Much research on the application of tomographic techniques to seismic data has been done in the reflection configuration as this has immediate relevance to much exploration work. To imitate tomographic inversion closely (i.e. line integration along rays), we may assume that the reflector is completely defined by *a priori* information and use tomography only to infer the velocity field (Neumann, 1981; Nercessian *et al.*, 1984; Fawcett and Clayton, 1984). Migration can also be exploited to elicit the reflector structure. Bording *et al.* (1987), Dyer and Worthington (1988), Stork (1992a) and Grau and Lailly (1993) proposed using depth migration in order to position the reflectors and then using tomography of traveltime inversion for the imaging of the velocity field.

The alternative approach is, of course, to use inversion methods to invert for the velocity distribution simultaneously with reflector positions. The unknown reflectors are parameterized in some suitable fashion and included in the inversion. Bishop *et al.* (1985) showed a test of tomographic reconstruction of seismic reflection data, in an attempt to map permafrost thickness to aid in the determination of static corrections. The tomographic depth estimates at the line intersection were significantly more accurate than the corresponding depth estimates derived from conventional processing. Similar approaches have also been taken by Farra and Madariaga (1988) and Williamson (1990). Since the extra parameters are not really tomographic, i.e. they are not line integrated in the same way as slowness (reciprocal velocity) is for traveltimes, a general optimization approach must be used if we wish to keep the two kinds of variables on a similar footing. This inversion then is very nonlinear, because small changes in the interface position produce large changes in ray trajectories. Thus all source-receiver rays have to be retraced at each iteration of the inverse problem.

Where non-planar reflection surfaces and spatially variable velocity variations are solved, there may be an ambiguity in the tomographic solutions in the form of a trade-off between reflector depth and velocity anomaly (Bishop *et al.*, 1985; Farra and Madariaga, 1988; Williamson, 1990; Blundell, 1992; Stork and Clayton, 1992; Bube *et al.*, 1995; Bube and Meadows, 1999). Bishop *et al.* (1985) investigated the trade-off in these parameters and pointed out that poorly determined quantities included linear slowness variations (with zero vertical mean) and the reflector tilt (with a compensating linear slowness variation). Ivansson (1986) offered a complement to the work of Bishop *et al.* (1985), and further provided some basic insight into the theoretical possibilities and limitations of reflection tomography for determining the subsurface velocity structure. Farra and Madariaga (1988) demonstrated, by calculation of the relevant components of a singular value decomposition (SVD), that pure vertical variation of slowness across the whole width of the model is essentially unrecoverable from rays reflecting off the designated reflector only. Williamson (1990) suggested that the short wavelength velocity-depth trade-offs at the reflector are unresolvable without further information.

Using only traveltime data in reflection seismic tomography, it may not be possible to resolve this ambiguity, particularly if the velocity anomaly is close to the reflector (Williamson, 1990). The inclusion of amplitude data in the inversion may help to resolve this ambiguity. One approach is the waveform inversion (e.g. Nolet, 1987; Tarantola, 1987a). Information about velocity anomalies in an inhomogeneous medium is concealed in the waveform data but is not used in standard traveltime tomographic techniques. Waveform inversion overcomes the limitations imposed by the high-frequency approximation of traveltime inversion and the weak scattering approximation of Born methods, by perturbing the velocity model until the synthetic seismograms match the observed seismograms.

However, the computational demands of waveform inversion render it an impractical choice for routine velocity inversion in exploration seismology. Although a 2.5-D modelling, exploiting the configuration symmetry to achieve an economical solution of the full 3-D problem, can be produced by modelling the ordinary 2-D acoustic wave equation and then applying the relevant filter (e.g. Liner, 1991), the forward model usually yields the wrong amplitudes (Williamson and Pratt, 1995).

A compromise between the two approaches, traveltime inversion and waveform inversion, is to use simplified amplitude data and traveltime data in tomographic inversions. In this book, we advocate the use of seismic amplitude data, separately or simultaneously with the traveltime data, in the reflection tomographic inversion.

1.2.   Velocity-depth ambiguity in traveltime inversion

To discuss the possible ambiguity between slowness and reflector depths, let us consider a model with constant slowness above a flat horizontal reflector, except for a small area above the reflector with lower slowness (higher velocity) than the surrounding region. If the computer code assumes that the slowness above the reflector is constant and the reflector is not necessarily flat, then attempts to match the traveltimes measured from the original model will yield a computed model whose traveltimes fit the measured data reasonably well but not exactly. The reflector of the computed model will be shallower under the slowness anomaly, a feature which may lead a structural geologist to suspect a trap for hydrocarbons. Thus inappropriate modelling can lead to incorrect conclusions concerning the model features which are of interest.

Conventional stacking velocity analysis yields a similar incorrect model (e.g. Bishop *et al.*, 1985). Bickel (1990) explored and explained some of the ambiguities in the conversion of time to depth in seismic reflection data. Rapid changes in stacking velocity are a symptom of a long-wavelength ambiguity between interval velocity and interface depth, even for the case of one laterally inhomogeneous layer overlying a half-space. Stork (1992b) reported that certain patterns of velocity variations could produce the same traveltime patterns as some reflector depth variations. One of the reasons for applying reflection tomography in the first place is to be able to distinguish between these models.

The ambiguity is not caused by the particular inversion algorithm being used in the tomography but is a feature of the recording geometry and the geometry of the subsurface, and thus it is a pervasive problem in reflection seismology. The factors that control how an anomaly affects seismic traveltimes (whether the anomaly correlates or anticorrelates with traveltimes or causes ambiguities) are its wavelength, its thickness and its height above a reflector. Other factors such as the source-receiver offset and slowness anomaly magnitude affect the magnitude of the traveltimes, but not their behaviour. Bickel (1990) showed that time structures that have a wavelength of about 2.7 times the average depth cannot be resolved into velocity and depth components using either travel or stacking velocity information. At this wavenumber, large changes in the sub-surface give rise to small changes in observed traveltimes. Tieman (1994) showed that thin anomalies induce velocity-depth ambiguities at a wavelength equal to $4.44h$ (where h is the height of the anomaly above a reflector), and that thick anomalies spanning the entire space from surface to reflector have ambiguities at wavelengths of approximately $2.57d$, where d is the thickness of the anomaly.

Since the wavelength of the ambiguity changes with source-receiver offset, a complete description of the velocity and depth fields can, in theory, be extracted from a combination of multiple-offset traveltime measurements (Lines, 1993). However, the wavelength of the ambiguity is such a weak function of source-receiver separation that multiple offset processing in practice does little to resolve the ambiguity. Kosloff and Sudman (2002) showed that when the ratio of the maximum-offset to the layer-depth is about one, the tomographic inversion can reconstruct the velocity and interface depth variations of wavelengths less than 2.4 times the layer thickness. The resolution is the lowest when $\lambda/H = 2.4$, where λ is the wavelength and H is the layer depth. Increasing the ratio of the maximum-offset to the layer-depth only slightly improves the resolution.

In the following section, however, a simple proof is given based on the ray theory assumption, that the use of reflection amplitude may help to combat the ambiguity problem in the traveltime inversion.

1.3.   Resolving ambiguity by using additional amplitude information

It is well known that in the reflection seismic experiment an ambiguity exists between velocity and reflector depth in the presence of velocity variations. To analyse the ambiguity problem in reflection tomography, let us consider a model with a planar reflecting interface at depth z and slowness u. A reflection for the source-receiver offset y is recorded at the surface. We wish to recover z and u from the data. This is an underdetermined problem that will show us the key ambiguity factor in the velocity analysis. This simple model was used by Stork (1992b), Lines (1993) and Kosloff and Sudman (2002) for the analysis of the ambiguity in traveltime inversion. In this section, we provide a different version of the traveltime variation, and compare it with the amplitude variation. We see that if we use both traveltime and amplitude data in the inversion the ambiguity between velocity variation and reflector depth can be reduced.

1.3.1   Ambiguity in traveltime inversion

Consider the problem of inverting for the slowness u and the reflector depth z, using traveltime information. The reflection traveltime equation for the source-receiver distance y is given by

$$T(u,\ z;\ y) = u\sqrt{y^2 + (2z)^2}\ . \tag{1.1}$$

The fractional variation in traveltime is then given in terms of the fractional changes in slowness and reflector depth,

$$\frac{dT}{T} = \frac{du}{u} + \frac{dz}{z}\cos^2\theta\ , \tag{1.2}$$

where θ is the angle of incidence.

Note that when θ is zero, i.e. the offset is zero, there is no change in traveltime whenever the fractional depth change equals the fractional velocity change. That is, the model is completely ambiguous (Lines, 1993). With a zero-offset section, it is impossible to invert for the velocity and depth separately. Zero-offset sections do not carry any information about velocity if the location of the reflection interface is unknown. Therefore, the angle of incidence plays an important role in the reconstruction of the velocity and depth functions, and controls the degree of ambiguity between velocity variation and reflector depth.

In this situation, considering a model with constant slowness above a flat horizontal reflector, there is no real slowness-depth ambiguity. Just two reflection traveltimes corresponding to rays with different θ's are sufficient to solve the problem. When the slowness field and the reflector depths can have spatial variation, the situation is not as simple. However, the extent of the slowness-depth ambiguity when using traveltimes alone is still characterized by $\cos^2\theta$ and therefore, equation (1.2) can be taken as a characteristic equation for a general case.

1.3.2   *Ambiguity in amplitude inversion*

In the case of one planar reflector model, a ray amplitude can be estimated as

$$A(u, z; y) \propto \frac{C}{D} , \qquad (1.3)$$

in terms of the reflection amplitude coefficient C and the geometrical spreading function D. The reflection amplitude coefficient can be approximated by (Wang and Houseman, 1995)

$$C(u, u_b; y) = \eta \frac{u - u_b}{u + u_b} , \qquad (1.4)$$

where u_b is the wave slowness in the underlying layer and η is a factor relating to the incident angle, $1.0 \geq \eta \geq 0.5$. The geometrical spreading function is

$$D(z; y) = \sqrt{y^2 + (2z)^2} . \qquad (1.5)$$

An evaluation of the fractional change in amplitude can be obtained as

$$\frac{dA}{A} = \frac{1}{2C_0} \frac{du}{u} - \frac{dz}{z} \cos^2\theta , \qquad (1.6)$$

where C_0 is the reflection amplitude coefficient at zero incident angle, given by

$$C_0 \approx \frac{u - u_b}{u + u_b} , \qquad (1.7)$$

and $dA/A = d(\ln A)$ is the perturbation of logarithmic amplitude. We can see that the amplitude inversion also shares, with the traveltime inversion, the problem of ambiguity between velocity variation and reflector depths. Equation (1.6) can be taken as a characteristic equation in the case of amplitude inversion.

1.3.3   *Joint inversion*

From equation (1.2) and equation (1.6) we see that the characteristic parameter which indicates how well we can distinguish in practice between slowness perturbations and depth perturbations near a section of a reflector is the range of values of the cosine of the angle of incidence of rays upon that section of the reflector. Using one type of data (traveltimes or amplitudes) alone cannot solve the problem. However, if we simply make an algebraic addition of equation

(1.2) and equation (1.6), the ambiguity term, $\cos^2\theta$, can be cancelled. This is the joint inversion.

To understand this situation in physical terms, let us consider again the model with constant slowness above a flat horizontal reflector. If the reflector were slightly shallower, the slowness above it would be slightly greater so that the reflection traveltimes in this perturbed model would almost equal those of the original model, but if we want perturbed amplitudes equal to those of the original model, the slowness would be slightly smaller. Constrained by both types of data, a compromised slowness would be produced, subject to the forms adopted for the data variations and model parameterization. From equations (1.2) and (1.6) we see that the fractional changes in velocity and reflector depth can be determined uniquely if we try a joint inversion using both traveltime and amplitude data simultaneously.

The aim of this monograph is, therefore, to investigate the use of simplified amplitude data in order to improve on the results of traveltime inversion. This method is an intermediate step between traveltime inversion and waveform inversion. As a practical application, it uses more information than traveltime inversion to model subsurface structure, and we expect that it will provide better velocity resolution than is possible with traveltime data alone, without excessive consumption of computational time.

To illustrate the use of amplitude data and to assess the information content in the context of tomographic inversion, we may initially exclude all traveltime information from the inversion. The inversion procedure using only amplitude data is hereafter referred to as the amplitude inversion. The procedure using both traveltime and amplitude data is referred to as the joint inversion.

1.4.   Overview of amplitude inversion

1.4.1   *Tomographic amplitude inversion*

Some examples of amplitude data used in tomographic inversion, both in 2-D and 3-D cases (including crosshole) have been given by, e.g. Menke (1984), Wong *et al.* (1987), Bregman *et al.* (1989), Zelt and Ellis (1990), and Brzostowski and McMechan (1992). In these studies amplitude data are used to estimate attenuation or "acoustic transparency". In this case the inversion becomes a pseudo-linear problem, similar to the problem of traveltime inversion for slowness variation. However a reliable velocity distribution is required *a priori*, so traveltime data are also used separately, for the determination of this velocity distribution.

There are few published examples of tomographic inversion of seismic amplitude data for the reconstruction of velocity structure. Thomson (1983) performed an iterative linearized inversion for 3-D structure under the NORSAR array (the Norwegian Seismic Array). The results obtained by the amplitude inversion did not compare satisfactorily with results from traveltime inversion because of the nonlinearity. Nowack and Lyslo (1989) showed, however, that it was possible to invert for velocity variation using reflection seismic amplitudes, using a slightly perturbed model in which the velocity of two smoothly splined velocity heterogeneities was increased by 1% above a constant background.

Nowack and Lutter (1988) used slightly perturbed velocity models (1.7%) to show that linearized inversions based on traveltime and ray-amplitude were complementary, being sensitive to different features of the model. Neele *et al.* (1993a) reported that traveltimes showed a more linear dependence on slowness perturbation. The greater stability of traveltimes suggested that traveltimes should be used in the first (nonlinear) iterations and that amplitude data should be included once the ray positions had become close enough to the true ray paths. In this way nonlinear behaviour of the amplitude equations was minimized. The method clearly showed that the amplitude data were sensitive to the curvature of the slowness distribution and therefore added independent information to traveltime data. Neele *et al.* (1993b) also reported the use of *P*-wave amplitude data in a joint tomographic inversion with traveltimes for upper-mantle velocity structure.

In 1992, I started an investigation into the efficacy of ray-amplitude inversion for reflection interface geometry and velocity variations, and co-authored a series of articles in geophysical journals. The investigation into the interface inversion suggested that the reflection amplitude data were sufficient to determine the geometry of the reflection interface accurately (Wang and Houseman, 1994). The study on velocity inversion suggested that the amplitude of a reflected wave was more sensitive to the slowness perturbation in the vicinity of the reflection point than to a comparable perturbation at any other point on the ray path (Wang and Houseman, 1995). Therefore, because of this property, it is difficult to reconstruct interval velocity variation from the seismic amplitudes of *reflected arrivals*, even though some quite good results have been obtained from inversion of the amplitude data of *direct arrivals* (e.g. Nowack and Lutter, 1988). However, the investigation indicated that seismic amplitudes were most sensitive to the location of the velocity anomalies.

During the investigation, we parameterized the reflection interface geometry and the velocity distribution using a 1-D and a 2-D discrete Fourier series, respectively (Wang and Pratt, 1997). Using this parameterization, we could conveniently represent the data (traveltime and amplitude) dependence on the model variation (velocity and interface) with different wavelengths. Sensitivity analysis in the reflection tomography reported that traveltime and amplitude were basically sensitive to different features of the subsurface model. Thus, joint inversion using both data sets would produce a better result than that using only one data set (Wang *et al.*, 2000; Wang *et al.*, 2002).

In the inversion, we used the subspace gradient inversion method favourably. The subspace method was proposed originally by Kennett *et al.* (1988). We developed several algorithms of subspace partitioning for problems such as single physical dimensional with different magnitudes (Wang and Houseman, 1994, 1995), multi-dimensional (Wang and Pratt, 1997; Wang, 1999a) in both data and model spaces, and inversion for a multi-layered structure (Wang and Pratt, 2000; Wang *et al.*, 2000) etc.

In this monograph, we collectively and systematically summarize the research on this aspect. One of the essential goals of this study is to make the amplitude inversion method work with real data.

1.4.2   *AVO inversion for lithology identification*

Apart from reflection tomography, there is another important aspect, mainly in the field of exploration seismic, of the use of amplitude variation with offset (AVO) information for the lithology identification. In AVO inversion, the formation mechanical properties, in particular Poisson's ratio, are extracted from the reflection amplitude data, based on the exact or approximated Zöppritz equations.

In the previous section, we reviewed the amplitude inversion for the subsurface velocity structure. In that case, we seek to extract the *P*-wave velocity and the depth of reflector. The density and the *S*-wave velocity used in the calculation of reflection and transmission coefficients (*via* the Zöppritz equations) are usually obtained from empirical relationships with the *P*-wave velocity. In contrast, AVO analysis uses reflection amplitude data to extract information about elastic parameters, such as Poisson's ratio and the shear modulus. The principal modelling equations used in AVO analysis are also the Zöppritz equations.

The Zöppritz equations, while exact, do not give a feeling for how amplitudes depend on the various factors involved. Several approximations have been made in an effort to achieve an equation form that gives more insight into the changes expected for various situations. When the contrasts in elastic properties at an interface are small, the reflection amplitude coefficient for the *P*-wave incident on a solid-solid interface can be expanded to first order with respect to the relative contrast of bulk density, the *P*-wave velocity and the *S*-wave velocity (Bortfeld, 1961; Richards and Frasier, 1976; Aki and Richards, 1980). The first order expression would allow us to see the separate effects of changes in density and *P*-wave or *S*-wave velocities, which are difficult to distinguish from the exact expression.

Shuey (1985) suggested that Poisson's ratio, σ, was the elastic constant most directly related to the variation in reflection amplitude with incident angle θ, following the investigation of Koefoed (1955). Koefoed analysed the angle variations in the *P-P* wave reflection amplitude coefficient for the interface separating media with different Poisson's ratios. A decrease in σ gives a decreasing reflection coefficient with reflection angle, while an increase in σ gives an increasing reflection coefficient with reflection angle. Ostrander (1984) extended Koefoed's calculations for a variety of models and showed how *P*-wave AVO analysis could be used as a hydrocarbon indicator. The work of Ostrander (1984) was the basis for the subsequent development of AVO technology.

The approximation for the AVO analysis is more commonly expressed as amplitude variation with the angle of incidence (AVA). A different approximation formula can also be obtained by using the ray-parameter p as the dependent variable (Mallick, 1993; Wang, 1999b). An advantage of using AVO analysis in the p domain is that a dynamic ray tracing system can be used in the forward calculation and explicit ray tracing is not necessary. The asymptotic formula given in terms of the ray-parameter p depends mainly on two elastic parameters: R_f, the fluid-fluid reflection coefficient (i.e. the reflection coefficient when the *S*-wave velocities in both media are set to zero), and $\Delta\mu/\rho$, the ratio of the contrast in shear modulus to the average bulk density across the reflecting interface. Using a least-squares inversion of theoretical values for the reflection coefficients, Mallick (1993) demonstrated that, in a linear inversion of amplitude versus offset data, $\Delta\mu/\rho$ is better estimated than the contrast in Poisson's ratio $\Delta\sigma$. AVO processing directly in the offset domain was also demonstrated by Ursin and Dahl (1992) and Ursin and Ekren (1995). Since the data are recorded as a function of offset, it is useful to model seismic amplitudes directly as a polynomial in the offset coordinate.

In AVO inversion, using either operator-based or model-based inversion algorithms (e.g. Hampson, 1991), the reflection amplitude data are, after some preprocessing, fitted to the approximation to the Zöppritz equations. This may be performed as detailed studies of a given event or it may be included as a step in the seismic processing, where the output is in the form of AVO attribute sections (e.g. Smith and Gidlow, 1987). A common application of an AVO analysis is to detect areas with anomalous amplitude behaviour, e.g. increase in absolute amplitude variation with offset, which can be used as an indication of gas sands in some regions (e.g. Ostrander, 1984; Rutherford and Williams, 1989; Fatti *et al.*, 1994).

An assumption is usually made in AVO analysis that the offset dependent amplitude variation is caused by the reflection coefficient, while the variations with offset of the geometrical spreading and the inelastic attenuation are often neglected or removed by preprocessing. We hope that, in the course of AVO analysis, the consideration of geometrical amplitude effects due to velocity structures may improve the accuracy with which elastic or lithologic parameters are extracted.

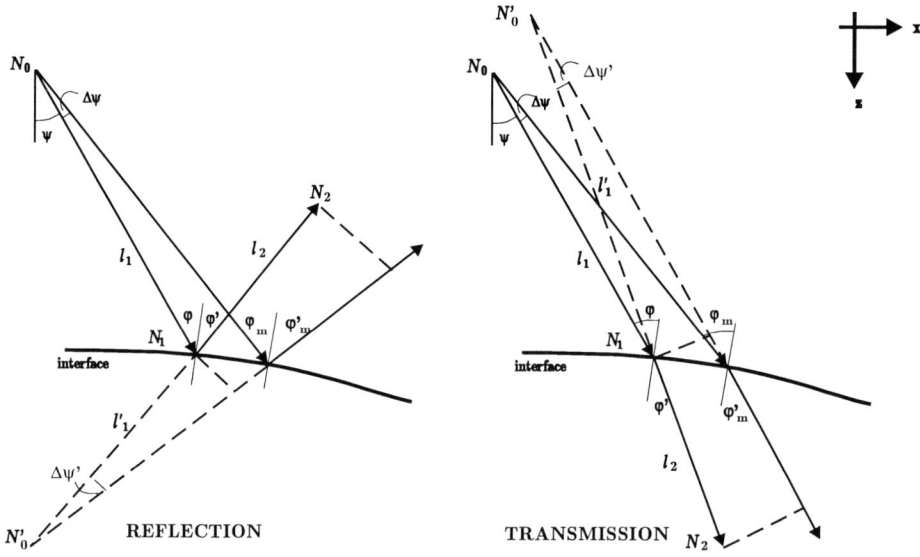

Figure 1.1. Geometry of incidence and reflection (or refraction). φ and φ' are the angles of incidence and reflection (or refraction) for a ray of take-off angle ψ, φ_m and φ'_m represent the modified angles where the ray take-off angle is $\psi + \Delta\psi$. N_1 is the incident point and N_2 is the initial observation point. The distance between N_0 and N_1 is ℓ_1 and between N_1 and N_2 it is ℓ_2.

1.5. Analytical expression for the geometrical spreading function for layered structures

The structural amplitude effect is included in the ray geometrical spreading function. Wavefronts spread during passage through a medium. The pattern of "spherical divergence" depends on the way that the velocity distribution varies with depth and is also affected by any velocity anisotropy. There are two stages of such spreading, the first from the source to the reflector and the second from the reflector (which can be viewed as a virtual source) to the receiver (Figure 1.1). The character of the observed reflections also depends strongly on the shape of the reflector. When the curvature of the reflector is greater than that of the incident wavefront, we have the possibility of the formation of subsurface focusing, with possible phase distortion of the observed reflection.

For a 2-D stratified medium consisting of constant velocity layers, a 2.5-D geometrical spreading is summarized here and the detailed derivation is given in Appendix A.1. When the ray geometrical spreading function is $L(\ell)$, in which ℓ is the distance from a source point measured along the ray path, the amplitude variation due to geometrical spreading is given by

$$A(\ell) = A_0 C \frac{\ell_0}{L(\ell)} \,, \tag{1.8}$$

where A_0 is taken here to be the amplitude of the wave at some distance ℓ_0 sufficiently close to the source that there are no intervening interfaces, but sufficiently far that near-source effects can be neglected and the wave front is spherical, and C is related to the changes due to acoustic impedance contrasts across interfaces using the Zöppritz relations.

Let N_i, for $i = 1, \cdots, K$, where K is the number of intersection points of a specified ray with successive interfaces, let N_0 and N_{K+1} represent the source and the receiver points, and let ℓ_i represent the length of the ray segment between two points N_{i-1} and N_i. To evaluate the geometrical spreading function $L(\ell)$ at ray distance ℓ, let us relate the spherical divergence of the ray beam to a virtual image of the source, as shown in Figure 1.1.

Because of interface curvature and impedance contrast, the distance between the incidence point N_i and the virtual image of the source N_0', in the ray plane, is ℓ_i', while the provisional observation position is at N_{i+1} . In the perpendicular plane the interface curvature is zero because of the assumption of 2-D structure and the distance between N_i and N_0' is ℓ_i'' . The geometrical spreading function can then be expressed as follows:

$$L(\ell) = \ell_1 \prod_{i=1}^{K} \left[\left(1 + \frac{\ell_{i+1}}{\ell_i'} \right) \left(1 + \frac{\ell_{i+1}}{\ell_i''} \right) \right]^{1/2} , \quad \text{for} \quad \ell_{K+1} = \ell - \sum_{i=1}^{K} \ell_i \,. \tag{1.9}$$

In the ray plane, the virtual ray distance ℓ_i' is given by

$$\frac{1}{\ell_i'} = \frac{1}{(\ell_{i-1}' + \ell_i)} \frac{v_{i+1}}{v_i} \frac{\cos^2 \varphi_i}{\cos^2 \varphi_i'} + \frac{1}{\cos \varphi_i'} \left(\frac{v_{i+1}}{v_i} \frac{\cos \varphi_i}{\cos \varphi_i'} \pm 1 \right) \Theta_i \,, \tag{1.10}$$

where the "+" sign refers to the reflection case and the "−" sign refers to the refraction case, v_i is the local velocity (assumed constant) along ray segment ℓ_i, φ_i and φ_i' are incident and reflection or refraction angles at the point N_i, and Θ_i is the factor describing the effect of local curvature of the ith interface, defined by

$$\Theta_i(x) = \pm \frac{\dfrac{d^2 z_i}{dx^2}}{\left[1 + \left(\dfrac{dz_i}{dx} \right)^2 \right]^{3/2}} \,, \tag{1.11}$$

where here the $\pm$ signs refer to the incident ray with an acute or an obtuse angle with the z-axis (or to the down- and up-going rays, respectively). The interface is here represented by a single-valued function $z_i(x)$ in which x is the horizontal coordinate and z is depth below some reference level at coordinate x.

In the perpendicular direction equation (1.10) with $\Theta_i = 0$ and $\varphi_i = 0$ can be used to define the apparent distance ℓ_i'' to the virtual image of the source:

$$\frac{1}{l_i''} = \frac{v_{i+1}}{\sum\limits_{j=1}^{i} \ell_j v_j} \; .$$ (1.12)

The explicit, analytical expression above clearly indicates that reflection seismic amplitude data do contain the information on reflector geometry. The tomographic amplitude inversion will, in particular, extract the information on reflector geometry which causes the focusing and defocusing of the ray amplitude.

Chapter 2

Traveltime and ray-amplitude in heterogeneous media

Abstract   In this chapter, we describe the forward calculation of the reflection seismic pulse propagating in heterogeneous media in the following sequence. (1) Within a variable velocity structure, a bending ray tracing method, based on Fermat's principle, is reduced to an iterative solution of a linearized tridiagonal equation system. (2) Traveltime derivation derives the Euler-Lagrange ray-tracing equation, which can be transformed to a Hamiltonian formula. (3) The first-order perturbation to the Hamiltonian ray equation forms a linearized ray-equation system which describes the paraxial rays. (4) The solution of the paraxial ray system is analytically expressed in terms of the propagator. (5) The propagator of paraxial rays is used to calculate the geometrical spreading which partly determines the ray amplitude. (6) Amplitude perturbations due to the model perturbations are then determined based on the ray perturbation theory.

2.1.   Introduction

With a high-frequency approximation, classical ray theory (Červený, 2001) can be used to obtain an approximate solution to the wave propagation problem in the far field. In this book, we focus on the ray theory-based inverse problem in the reflection configuration, and consider the stratified structure, consisting of continuously varying interfaces separating heterogeneous velocity layers. In the calculation, we assume that the principal radii of curvature of the wavefront along the interfaces are continuous in the neighbourhood of points of reflection or refraction on any interfaces, and that the radii of curvature of the interfaces are large in comparison with the acoustic wavelength so that diffraction effects can be neglected. Inhomogeneous scattering is also neglected.

The material in this chapter is organized as in the following flow chart:

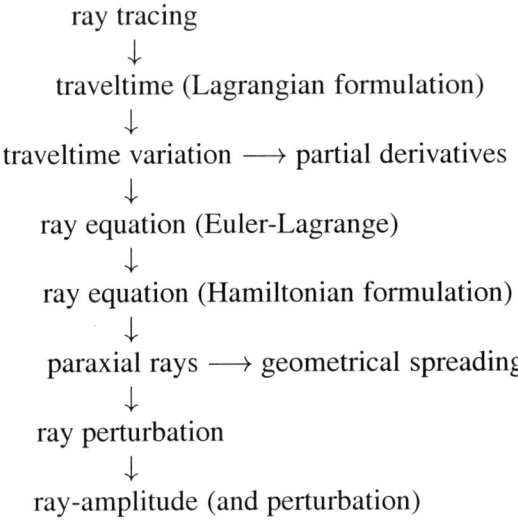

In a ray theory-based tomographic inversion it is necessary to have a robust ray tracing routine. In section 2.2, we present a bending method for the two-point ray tracing within a 3-D heterogeneous media (*cf.* Moser *et al.*, 1992). The bending method, based on Fermat's principle, is reduced to an iterative solution of a linearized tridiagonal equation system.

For the forward calculation of traveltimes and their partial derivatives with respect to the model perturbations, we review in section 2.3 the Lagrangian formulation of traveltime and its variation due to a model perturbation. The Euler-Lagrange equation of ray tracing can be transformed, *via* the Legendre transform (Kline and Kay, 1965), to the Hamiltonian formulation of the ray system.

The first-order perturbation to the Hamiltonian formulation, forming a linearized ray-equation system, can be applied to describe paraxial rays (Thomson and Chapman, 1985; Farra and Madariaga, 1987; Nowack and Lutter, 1988). The solution to this linearized ray-equation system can be expressed analytically in terms of the propagator matrix as a function of ray parameter along the central ray (Gilbert and Backus, 1966; Aki and Richards, 1980). The propagator matrix in turn is used to estimate the geometrical spreading of seismic arrivals. It is presented in section 2.4.

In section 2.5, we see how to use the ray perturbation theory (*cf.* Thomson and Chapman, 1985; Farra and Madariaga, 1987) for the determination of the "two-point" perturbation to the central ray and paraxial rays caused by the slowness perturbation and the interface perturbation.

The ray amplitude, in principle, includes the influences of the geometrical spreading, reflection-transmission coefficients and the intrinsic attenuation (section 2.6). The effects, such as reflection, refraction and diffraction caused by discontinuities in the impedance, can be included in the category of elastic effects. For real seismic amplitude data, the intrinsic inelastic attenuation is also a crucial factor. The attenuation formula may be expressed in terms of the absorption coefficient $\eta(\mathbf{x})$, which can be easily included in the inversion problem by using the partial derivatives calculated for the traveltime inversion.

2.2.   Bending ray tracing method

For ray amplitude calculations, the model slowness distribution must vary smoothly within a layer. In a medium with smoothly varying slowness, any ray is composed of an arc with continuously varying curvature. The traveltime and its derivatives with respect to the model parameters can be calculated and, if the ray-tube around the reference ray smoothly diverges, calculation of the ray-amplitude is also stable. At an interface between layers, the assumption of a smooth interface (i.e. the existence of its partial derivatives of first order and second order) is necessary in order to calculate the transformation of the ray and the paraxial rays.

According to Fermat's principle, the ray path is the path γ which minimizes the traveltime T, given by

$$T(\gamma) = \int_{\gamma} \frac{ds}{c} \rightarrow \text{min} , \qquad (2.1)$$

where c is velocity and s is the ray arc-length. For computational purposes, the path under consideration can be discretized into a polygonal path,

$$\gamma = \{\mathbf{x}_0, \mathbf{x}_1, \cdots \mathbf{x}_K, \cdots \mathbf{x}_{2K}\} , \qquad (2.2)$$

consisting of $2K + 1$ points in three-dimensional space, numbered from 0 to $2K$, and connected by straight line segments or circular arcs. The traveltime can then be expressed explicitly as

$$T = \frac{1}{2} \sum_{i=1}^{2K} (u_i + u_{i-1}) \, ds_i , \qquad (2.3)$$

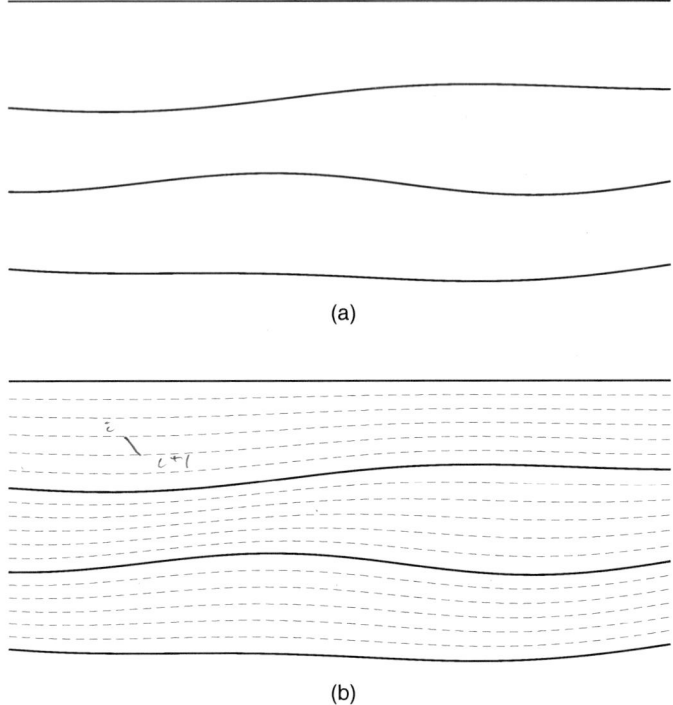

(a)

(b)

Figure 2.1.   A layered earth structure (a) and the interpolation between interfaces (b). A reflected ray trajectory intersects all the interfaces and interpolation levels.

where $u_i = 1/c_i$ and ds_i is the length of the ray segment between points $\mathbf{x}_i$ and $\mathbf{x}_{i-1}$. For ray tracing, we may consider $\mathbf{x}_K$ as the reflection point, and the end points, $\mathbf{x}_0$ and $\mathbf{x}_{2K}$, of the ray path are fixed. Fermat's principle can then be expressed as

$$\nabla_\gamma (\gamma) = 0 \, , \tag{2.4}$$

which states that seismic energy travels along a path "for which the first-order variation with all neighbouring paths is zero" (Sheriff, 1991).

   Let us consider a layered structure (Figure 2.1a) with M interfaces defined by

$$f_k(\mathbf{x}) = 0 \, , \qquad \text{for} \quad k = 1, \cdots, M.$$

We may divide each layer into N_k thin layers and insert N_k-1 interpolation levels between each pair of interfaces (Figure 2.1b). The jth interpolation level is specified by a smooth surface defined by means of cubic spline interpolation among a set of discrete depth values $\mathbf{z}_j(x, y)$ which are obtained by a linear interpolation in the vertical direction between adjacent interfaces:

$$\mathbf{z}_j = \mathbf{z}_k + \frac{j}{N_k}(\mathbf{z}_{k+1} - \mathbf{z}_k), \qquad \text{for} \quad j = 0, \cdots, N_k, \qquad (2.5)$$

where $\mathbf{z}_k$ and $\mathbf{z}_{k+1}$ are data sets consisting of depths of the kth and $(k+1)$th interfaces $f_k(\mathbf{x}) = 0$ and $f_{k+1}(\mathbf{x}) = 0$. Assume a reflected ray trajectory intersects a total of K of interfaces and interpolation levels, then

$$K = M + \sum_{k=1}^{M}(N_k - 1),$$

and the intersection points are ordered as $1, 2, \cdots, 2K-1$.

In the 3-D problem, the number of free parameters is thus reduced to $2(2K-1)$ with

$$\xi = \{\xi_1, \cdots, \xi_K, \cdots, \xi_{2K-1}\}, \qquad (2.6)$$

where ξ includes x or y components of the intersection points of interfaces and interpolation levels. The corresponding gradient of traveltime $\nabla_\xi T(\gamma)$ in equation (2.4) can be written explicitly as

$$\frac{\partial T}{\partial \xi_i} = \frac{1}{2}\left\{ \frac{(u_i + u_{i+1})}{ds_{i+1}}\left[(\xi_i - \xi_{i+1}) + (z_i - z_{i+1})\frac{\partial z_i}{\partial \xi_i}\right] \right.$$

$$+ \frac{(u_i + u_{i-1})}{ds_{i-1}}\left[(\xi_i - \xi_{i-1}) + (z_i - z_{i-1})\frac{\partial z_i}{\partial \xi_i}\right]$$

$$\left. + (ds_i + ds_{i+1})\left(\frac{\partial u_i}{\partial \xi_i} + \frac{\partial u_i}{\partial z_i}\frac{\partial z_i}{\partial \xi_i}\right)\right\}, \qquad (2.7)$$

$$\text{for} \quad i = 1, \cdots, 2K - 1.$$

Using a second order accurate representation of the derivatives, equation (2.7) gives a tridiagonal linear equation system in the set of unknowns $\{\xi_i\}$, which can be solved iteratively.

As an example, Figure 2.2 shows the ray paths through a 2-D example with arbitrary slowness and interface variation. This algorithm is robust for rays whose angle of inclination relative to the interpolation surfaces is not too shallow. Because the ray path is approximated by straight line segments, the algorithm should not be expected to give accurate results in the vicinity of turning rays.

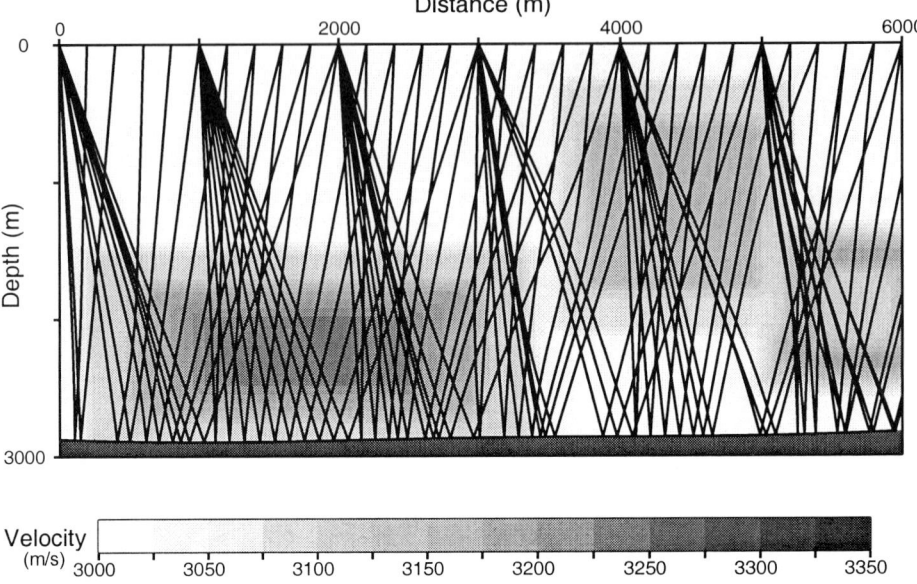

Figure 2.2.   The geometry of seismic reflection rays within a variable velocity medium, reflected from a curved interface.

2.3.   Traveltime and its perturbations

2.3.1   *Traveltime and its variation*

Introducing the Lagrangian L as

$$L(\mathbf{x}, \dot{\mathbf{x}}, \sigma) = \frac{1}{2}[\dot{\mathbf{x}}^2 + u^2(\mathbf{x})] \,, \tag{2.8}$$

where σ is an independent variable,

$$d\sigma = \frac{ds}{u} \tag{2.9}$$

defined in terms of the slowness $u(\mathbf{x})$ along the ray curve and the arc-length s, and differentiation with respect to σ is denoted with a dot, the traveltime integral along the ray can be expressed as

$$T = \int_{\sigma_0}^{\sigma} L(\mathbf{x}, \dot{\mathbf{x}}, \sigma) \, d\sigma \,. \tag{2.10}$$

Its first variation is

$$\delta T = \int_{\sigma_0}^{\sigma} \left[\frac{\partial L}{\partial \dot{\mathbf{x}}} \cdot \delta \dot{\mathbf{x}} + \frac{\partial L}{\partial \mathbf{x}} \cdot \delta \mathbf{x} + \frac{\partial L}{\partial u} \delta u \right] d\sigma .$$

Integrating the first term by parts this expression can be written as (Snieder and Spencer, 1993)

$$\delta T = \left[\frac{\partial L}{\partial \dot{\mathbf{x}}} \cdot \delta \mathbf{x} \right]_{\sigma_0}^{\sigma} + \int_{\sigma_0}^{\sigma} \left[\frac{\partial L}{\partial \mathbf{x}} - \frac{\partial}{\partial \sigma} \left(\frac{\partial L}{\partial \dot{\mathbf{x}}} \right) \right] \cdot \delta \mathbf{x} \, d\sigma + \int_{\sigma_0}^{\sigma} \frac{\partial L}{\partial u} \delta u \, d\sigma ,$$

$$(2.11)$$

where $\partial L / \partial u = u(\mathbf{x})$, from equation (2.8), in the third integral term.

2.3.2 *Ray tracing equations*

From the Fermat condition that the traveltime is stationary for perturbations in the ray position, we have $\delta T = 0$. For fixed endpoints and assuming no model (slowness and boundary) perturbations, the first and third terms on the right hand side of equation (2.11) are zero. The following ray equation is then obtained,

$$\frac{\partial L}{\partial \mathbf{x}} - \frac{\partial}{\partial \sigma} \left(\frac{\partial L}{\partial \dot{\mathbf{x}}} \right) = 0 , \tag{2.12}$$

known as the Euler-Lagrange equation. Describing the ray trajectory by the canonical vector

$$\mathbf{y}(\sigma) = \begin{bmatrix} \mathbf{x}(\sigma) \\ \mathbf{p}(\sigma) \end{bmatrix}$$

in the position $\mathbf{x}$ and momentum $\mathbf{p}$, the ray equation (2.12) can also be expressed in a Hamiltonian form (Chapman and Drummond, 1982),

$$\dot{\mathbf{x}} = \nabla_{\mathbf{p}} H ,$$
$$\dot{\mathbf{p}} = -\nabla_{\mathbf{x}} H , \tag{2.13}$$

where the Hamiltonian function is defined as

$$H(\mathbf{x}, \mathbf{p}, \sigma) = \frac{1}{2} [\mathbf{p}^2 - u^2(\mathbf{x})] , \tag{2.14}$$

corresponding to the definition of the independent variable σ used in this text. This expression for the Hamiltonian was proposed by Burridge (1976). Thomson and Chapman (1985), Červený (1989) and Kendall and Thomson (1989) discussed various expressions for the Hamiltonian. Transformation between equation (2.12) and equation (2.13) can be bridged by means of the Legendre transformation (Kline and Kay, 1965),

$$L(\mathbf{x}, \dot{\mathbf{x}}, \sigma) = -H(\mathbf{x}, \mathbf{p}, \sigma) + \mathbf{p} \cdot \dot{\mathbf{x}} . \qquad (2.15)$$

We use the first-order perturbation to equation (2.13) later to describe paraxial rays and their perturbation, for the calculation of ray-amplitude and variation.

2.3.3   Traveltime perturbations

Let us now consider traveltime perturbations due to the perturbations of slowness and interface variations, respectively.

For a ray with fixed endpoints, the perturbation of traveltime for a variation in material slowness,

$$u(\mathbf{x}) = u_0(\mathbf{x}) + \delta u(\mathbf{x}) ,$$

is given by the third term on the right-hand side of equation (2.11),

$$\delta T = \int_{\sigma_0}^{\sigma} u(\mathbf{x}) \delta u \, d\sigma , \qquad (2.16)$$

where the integral may be computed along the original unperturbed ray trajectory in the unperturbed reference medium.

Suppose a smooth interface is defined by $f_0(\mathbf{x}) = 0$. The perturbation of traveltime due to the interface perturbation, $\delta f(\mathbf{x})$, is given by the first term on the right hand side of equation (2.11). From the Legendre transformation equation (2.15), we have

$$\frac{\partial L}{\partial \dot{\mathbf{x}}} = \mathbf{p} . \qquad (2.17)$$

The perturbation of traveltime for a ray with fixed endpoints can then be expressed as

$$\delta T = [\mathbf{p} \cdot \delta \mathbf{x}]_{\sigma_0}^{\sigma} = [\mathbf{p} \cdot \delta \mathbf{x}]_{\sigma_0}^{\sigma_a^-} + [\mathbf{p} \cdot \delta \mathbf{x}]_{\sigma_a^+}^{\sigma} = [\mathbf{p} \cdot \delta \mathbf{x}]_{(\sigma_a^-)} - [\mathbf{p} \cdot \delta \mathbf{x}]_{(\sigma_a^+)} ,$$
$$(2.18)$$

where σ_a^- refers to the incident side of the interface and σ_a^+ refers to the reflected or transmitted side of the interface. This calculation of the effect on traveltime due to the change of boundary requires that the ray path be retraced through the layers. Approximating to first order and assuming that the effect on the traveltime is restricted to the effect of the extra distance travelled, we have

$$\delta T = [\mathbf{p}_0 - \hat{\mathbf{p}}_0] \cdot \delta \mathbf{x} , \qquad (2.19)$$

where $\mathbf{p}_0$ and $\hat{\mathbf{p}}_0$ are the slowness vectors along the unperturbed ray on the incident side and on the reflected/transmitted side of the interface, respectively. Developing the slowness vectors $\mathbf{p}_0$ and $\hat{\mathbf{p}}_0$ along the normal and the tangent planes to the interface, the difference between them is

$$\mathbf{p}_0 - \hat{\mathbf{p}}_0 = \frac{\langle \mathbf{p}_0 | \nabla f_0 \rangle \nabla f_0 - (\mathbf{p}_0 \times \nabla f_0) \times \nabla f_0}{\langle \nabla f_0 | \nabla f_0 \rangle}$$

$$- \frac{\langle \hat{\mathbf{p}}_0 | \nabla f_0 \rangle \nabla f_0 - (\hat{\mathbf{p}}_0 \times \nabla f_0) \times \nabla f_0}{\langle \nabla f_0 | \nabla f_0 \rangle}$$

$$= \frac{\langle \mathbf{p}_0 - \hat{\mathbf{p}}_0 | \nabla f_0 \rangle}{\langle \nabla f_0 | \nabla f_0 \rangle} \nabla f_0 \, , \tag{2.20}$$

which follows from the use of Snell's law

$$\mathbf{p}_0 \times \nabla f_0 = \hat{\mathbf{p}}_0 \times \nabla f_0 \, , \tag{2.21}$$

where $\langle \, | \, \rangle$ and $\times$ denote the inner product and the cross product respectively. Expanding the perturbed interface $f_0(\mathbf{x}) + \delta f(\mathbf{x}) = 0$ to first order, we get

$$\delta f + \langle \nabla f_0 | \delta \mathbf{x} \rangle = 0 \, . \tag{2.22}$$

Substituting equation (2.20) and equation (2.22) into equation (2.19), we then derive the variation of traveltime due to the interface perturbation,

$$\delta T = - \frac{\langle \mathbf{p}_0 - \hat{\mathbf{p}}_0 | \nabla f_0 \rangle}{\langle \nabla f_0 | \nabla f_0 \rangle} \, \delta f \, . \tag{2.23}$$

This formula is comparable with the one used by Bishop *et al.* (1985) and by Williamson (1990) for reflection inversion and has been used by Farra *et al.* (1989).

2.4. Propagator of paraxial rays and geometrical spreading

2.4.1 *Propagator of paraxial rays*

Assume a ray has been traced in a medium with slowness distribution, $u(\mathbf{x})$, using the ray-tracing algorithm described in section 2.2, and denote $\mathbf{y}_c(\sigma) = [\mathbf{x}_c(\sigma), \mathbf{p}_c(\sigma)]^T$, the canonical vector of that reference ray. Around this ray we can obtain paraxial rays by means of first-order perturbation theory. The analysis of this subsection follows that of previous authors (Thomson and Chapman, 1985; Farra and Madariaga, 1987; Virieux *et al.*, 1988; Farra *et al.*, 1989;

Virieux, 1991), but is summarized here in order to explain the modifications to
the theory introduced in this text.

For rays near the central ray,

$$\mathbf{y}(\sigma) = \mathbf{y}_\mathrm{c}(\sigma) + \delta\mathbf{y}(\sigma), \tag{2.24}$$

where $\delta\mathbf{y} = [\delta\mathbf{x}, \delta\mathbf{p}]^\mathrm{T}$ is the vector of perturbations to position and slowness
relative to the reference ray, the ray equation (2.13) can be linearized (Thomson
and Chapman, 1985; Nowack and Lutter, 1988) and rewritten as

$$\delta\dot{\mathbf{y}} = \mathbf{A}\,\delta\mathbf{y}\,, \tag{2.25}$$

where

$$\mathbf{A} = \begin{bmatrix} \nabla_\mathbf{x}\nabla_\mathbf{p}H & \nabla_\mathbf{p}\nabla_\mathbf{p}H \\ -\nabla_\mathbf{x}\nabla_\mathbf{x}H & -\nabla_\mathbf{p}\nabla_\mathbf{x}H \end{bmatrix}. \tag{2.26}$$

From the Hamiltonian (2.14), we have

$$\mathbf{A} = \begin{bmatrix} \mathbf{0} & \mathbf{I} \\ \mathbf{U} & \mathbf{0} \end{bmatrix}, \tag{2.27}$$

where $\mathbf{U}$ is a 3×3 symmetric matrix with elements given by

$$U_{ij} = \frac{\partial u}{\partial x_i}\frac{\partial u}{\partial x_j} + u\,\frac{\partial^2 u}{\partial x_i \partial x_j}\,. \tag{2.28}$$

The linear ray perturbation equations (2.26) can be solved using the standard
propagator techniques (Gilbert and Backus, 1966; Aki and Richards, 1980).
Any solution, $\delta\mathbf{y}(\sigma)$, can be written in terms of the propagator $\mathbf{\Pi}(\sigma, \sigma_0)$ and
the initial condition $\delta\mathbf{y}(\sigma_0)$ as

$$\delta\mathbf{y}(\sigma) = \mathbf{\Pi}(\sigma,\ \sigma_0)\,\delta\mathbf{y}(\sigma_0)\,, \tag{2.29}$$

where $\mathbf{\Pi}(\sigma, \sigma_0)$ is the 6×6 propagator matrix of the paraxial system, which
can be evaluated as an infinite series involving integrals of $\mathbf{A}$ along the ray path.
Truncating the series at the linear term, the propagator may be approximated
by the formula (Gilbert and Backus, 1966; Aki and Richards, 1980):

$$\mathbf{\Pi}(\sigma,\ \sigma_0) = \prod_{\sigma_0}^{\sigma} [\mathbf{I} + \mathbf{A}(\tau)\,d\tau]\,. \tag{2.30}$$

This is a discrete form of the product integration. For the properties of the
product integration, we may refer to Birkhoff (1937).

When the rays hit a discontinuity of zeroth- or first-order, we have to introduce appropriate boundary conditions for ray tracing (Červený, 1985; Chapman, 1985; Farra *et al.*, 1989). Assume a central ray $\mathbf{y}(\sigma)$ intersects the interface at sampling parameter σ_k, where the perturbation of the canonical vector of the paraxial ray in the incident medium is $\delta\mathbf{y}(\sigma_k)$. Let us denote variables in the reflected/transmitted ray with a caret above them. The continuity condition for perturbations of the canonical vectors $\delta\hat{\mathbf{y}}(\sigma_k)$ and $\delta\mathbf{y}(\sigma_k)$ across the interface may be expressed as

$$\delta\hat{\mathbf{y}}(\sigma_k) = \mathbf{\Sigma}(\sigma_k)\,\delta\mathbf{y}(\sigma_k)\,, \tag{2.31}$$

in terms of the transformation matrix $\mathbf{\Sigma}$.

Different versions of the transformation matrix have been given by Farra *et al.* (1989), Gajewski and Pšenčík (1990), Wang and Houseman (1995) and Farra and Le Bégat (1995). A generalized expression given by Farra and Le Bégat (1995) preserves the general properties of the propagator matrix (Thomson and Chapman, 1985). The generalized 6×6 transformation matrix can be written in a compact form as

$$\mathbf{\Sigma} = \begin{bmatrix} \mathbf{\Sigma}_{11} & \mathbf{\Sigma}_{12} \\ \mathbf{\Sigma}_{21} & \mathbf{\Sigma}_{22} \end{bmatrix}\,, \tag{2.32}$$

with submatrices defined explicitly as

$$\mathbf{\Sigma}_{11} = \mathbf{I} - \frac{|\nabla_{\mathbf{p}}H - \nabla_{\mathbf{p}}\hat{H}\rangle\langle\nabla f|}{\langle\nabla_{\mathbf{p}}H|\nabla f\rangle}\,, \tag{2.32a}$$

$$\mathbf{\Sigma}_{12} = \mathbf{0}\,, \tag{2.32b}$$

$$\mathbf{\Sigma}_{21} = \frac{|\nabla_{\mathbf{x}}H - \nabla_{\mathbf{x}}\hat{H}\rangle\langle\nabla f|}{\langle\nabla_{\mathbf{p}}H|\nabla f\rangle} + \frac{|\nabla f\rangle\langle\nabla_{\mathbf{x}}H - \nabla_{\mathbf{x}}\hat{H}|}{\langle\nabla_{\mathbf{p}}\hat{H}|\nabla f\rangle}$$

$$- \frac{(\langle\nabla_{\mathbf{x}}H|\nabla_{\mathbf{p}}\hat{H}\rangle - \langle\nabla_{\mathbf{x}}\hat{H}|\nabla_{\mathbf{p}}H\rangle)}{\langle\nabla_{\mathbf{p}}H|\nabla f\rangle\langle\nabla_{\mathbf{p}}\hat{H}|\nabla f\rangle}|\nabla f\rangle\langle\nabla f|$$

$$- \frac{\langle\mathbf{p}_0 - \hat{\mathbf{p}}_0|\nabla f\rangle}{\langle\nabla f|\nabla f\rangle}\left(\mathbf{I} - \frac{|\nabla f\rangle\langle\nabla_{\mathbf{p}}\hat{H}|}{\langle\nabla_{\mathbf{p}}\hat{H}|\nabla f\rangle}\right)\nabla\nabla f\left(\mathbf{I} - \frac{|\nabla_{\mathbf{p}}H\rangle\langle\nabla f|}{\langle\nabla_{\mathbf{p}}H|\nabla f\rangle}\right)$$

$$\tag{2.33c}$$

and

$$\mathbf{\Sigma}_{22} = \mathbf{I} + \frac{|\nabla f\rangle\langle\nabla_{\mathbf{p}}H - \nabla_{\mathbf{p}}\hat{H}|}{\langle\nabla_{\mathbf{p}}\hat{H}|\nabla f\rangle}\,, \tag{2.33d}$$

where $\hat{H}$ is the Hamiltonian in the reflected-transmitted medium, $\mathbf{I}$ is the 3×3 identity matrix and $\mathbf{0}$ is the 3×3 null matrix. The notation $|\mathbf{a}\rangle\langle\mathbf{b}|$ represents a matrix obtained by the tensor product of the vectors $\mathbf{a}$ and $\mathbf{b}$, $[|\mathbf{a}\rangle\langle\mathbf{b}|]_{ij} = a_i b_j$.

In the case of a ray crossing K interfaces, the propagator along the entire ray, taking transformation matrices at all K interfaces into account, is given by

$$\mathbf{\Pi}(\sigma, \sigma_0) = \mathbf{\Pi}(\sigma, \sigma_K) \prod_{k=K}^{1} \mathbf{\Sigma}(\sigma_k) \, \mathbf{\Pi}(\sigma_k, \sigma_{k-1}) \, . \tag{2.34}$$

Once $\mathbf{\Pi}(\sigma, \sigma_0)$ is calculated, the entire set of paraxial rays can be found by adjusting the initial conditions, which have to satisfy the condition,

$$\delta H = \nabla_{\mathbf{p}} H \cdot \delta \mathbf{p} + \nabla_{\mathbf{x}} H \cdot \delta \mathbf{x} = 0 \, , \tag{2.35}$$

derived from first order perturbation of the Hamiltonian $H = 0$ (Virieux, 1991).

2.4.2   *Ray geometrical spreading*

We have obtained the paraxial rays around a central ray. The ray geometrical spreading, $D(\sigma, \sigma_0)$, can be defined by

$$D(\sigma, \ \sigma_0) = \left\{ \det \left[\frac{\delta \mathbf{x}(\sigma)}{\delta \mathbf{x}(\sigma_0)} \right] \right\}^{1/2} . \tag{2.36}$$

Partitioning the propagator matrix as

$$\mathbf{\Pi}(\sigma, \sigma_0) = \begin{bmatrix} \mathbf{Q}_1(\sigma, \ \sigma_0) & \mathbf{Q}_2(\sigma, \ \sigma_0) \\ \mathbf{P}_1(\sigma, \ \sigma_0) & \mathbf{P}_2(\sigma, \ \sigma_0) \end{bmatrix} , \tag{2.37}$$

where $\mathbf{Q}(\sigma, \sigma_0)$ and $\mathbf{P}(\sigma, \sigma_0)$ are 3×3 matrices which act on $\mathbf{x}$ and $\mathbf{p}$ separately, the ray geometrical spreading can be written as

$$D(\sigma, \ \sigma_0) = \{ \det \left[\mathbf{Q}_1(\sigma, \ \sigma_0) + \mathbf{Q}_2(\sigma, \ \sigma_0) \mathbf{M}_0 \right] \}^{1/2} , \tag{2.38}$$

where $\mathbf{M}_0$ determines the initial shape of the ray beam,

$$\delta \mathbf{p}(\sigma_0) = \mathbf{M}_0 \delta \mathbf{x}(\sigma_0) \, . \tag{2.39}$$

For an initial point source in a constant slowness medium (Farra *et al.*, 1989),

$$\mathbf{M}_0 = \frac{u_0}{s_0} \left(\mathbf{I} - \frac{|\mathbf{p}_0\rangle\langle\mathbf{p}_0|}{\langle\mathbf{p}_0|\mathbf{p}_0\rangle} \right) , \tag{2.40}$$

where $u_0(\sigma_0)$ is the slowness at σ_0, and $s_0(\sigma_0)$ is the arc-length measured from $\sigma = 0$. In a non-dissipative system, the ray-amplitude in a seismogram is proportional to the inverse of the ray geometrical spreading function $D(\sigma, \sigma_0)$.

2.5. Ray perturbations due to model perturbations

In the iterative inversion procedure we need to calculate the Fréchet matrix, the Fréchet derivatives of the ray amplitudes with respect to the model parameters. The procedure can be linearized by assuming that the ray paths do not change during the inversion, but this approximation may not be valid for large amplitude slowness anomalies, in which case it is first necessary to calculate the perturbation to the ray trajectory. In this section, we use the first-order ray perturbation theory for determining the perturbed "two-point" ray in the perturbed medium.

Let u_0 and f_0 represent the reference slowness distribution and interface geometry, and Δu and Δf the slowness perturbation and interface perturbation. Note that Δ is used in this section to denote perturbations due to the structure, while lower case δ is used for paraxial perturbations (previous section) and variables with subscript "0" correspond to those in the unperturbed reference medium.

2.5.1 *"Shooting" perturbed ray due to slowness perturbation*

In the medium with perturbed slowness distribution

$$u(\mathbf{x}) = u_0(\mathbf{x}) + \Delta u(\mathbf{x}) ,$$

where the perturbation $\Delta u(\mathbf{x})$ is a smooth function that is assumed to have continuous second-order derivatives, calculation of the perturbed rays is similar to that of the paraxial rays in the reference medium. The perturbed ray with the canonical vector perturbation $\Delta \mathbf{y}(\sigma)$ may be found by the following linear system (Farra and Madariaga, 1987; Farra, 1990; Virieux and Farra, 1991; Farra, 1992):

$$\Delta \dot{\mathbf{y}} = \mathbf{A}_0 \Delta \mathbf{y} + \Delta \mathbf{B} , \qquad (2.41)$$

with a source $\Delta \mathbf{B}(\sigma)$ containing the Hamiltonian perturbations,

$$\Delta \mathbf{B} = \begin{bmatrix} \nabla_{\mathbf{p}} \Delta H \\ -\nabla_{\mathbf{x}} \Delta H \end{bmatrix} , \qquad (2.42)$$

where the perturbation of the Hamiltonian is given by

$$\Delta H = -u_0 \Delta u ,$$

if H is defined by equation (2.14). Then explicitly,

$$\Delta \mathbf{B} = \begin{bmatrix} 0 \\ u_0 \nabla_{\mathbf{x}} \Delta u + \Delta u \, \nabla_{\mathbf{x}} u_0 \end{bmatrix} . \qquad (2.43)$$

Equation (2.40) is similar to equation (2.26), except for the source term $\Delta \mathbf{B}$. The perturbation $\Delta \mathbf{y}(\sigma)$ of the central ray in the perturbed medium is then obtained by the standard propagator technique (Gilbert and Backus, 1966):

$$\Delta \mathbf{y}(\sigma) = \mathbf{\Pi}_0(\sigma, \sigma_0) \, \Delta \mathbf{y}(\sigma_0) + \int_{\sigma_0}^{\sigma} \mathbf{\Pi}_0(\sigma, \tau) \, \Delta \mathbf{B}(\tau) \, d\tau \,, \qquad (2.44)$$

where $\mathbf{\Pi}_0(\sigma, \sigma_0)$ is the 6×6 propagator matrix of the paraxial system equation (2.26) in the unperturbed medium. The integration term in equation (2.43) can be explicitly written as

$$\Delta \mathbf{x}_{\mathrm{B}}(\sigma) = \int_{\sigma_0}^{\sigma} (\sigma - \tau) \, (u_0 \, \nabla_{\mathbf{x}} \Delta u + \Delta u \, \nabla_{\mathbf{x}} u_0) \, d\tau \,,$$
$$\tag{2.45}$$
$$\Delta \mathbf{p}_{\mathrm{B}}(\sigma) = \int_{\sigma_0}^{\sigma} (u_0 \, \nabla_{\mathbf{x}} \Delta u + \Delta u \, \nabla_{\mathbf{x}} u_0) \, d\tau \,,$$

where $\Delta \mathbf{y}_{\mathrm{B}}(\sigma, \sigma_0) = [\, \Delta \mathbf{x}_{\mathrm{B}}, \Delta \mathbf{p}_{\mathrm{B}} \,]^{\mathrm{T}}$.

Upon reflection/transmission of the perturbed ray across an unperturbed interface, we may calculate the perturbation $\Delta \hat{\mathbf{y}}$ in the reflected or transmitted ray approximately, using equation (2.31),

$$\Delta \hat{\mathbf{y}}(\sigma_k) = \mathbf{\Sigma}_0(\sigma_k) \, \Delta \mathbf{y}(\sigma_k) \,, \qquad (2.46)$$

with the transformation matrix $\mathbf{\Sigma}_0(\sigma_k)$ calculated in the unperturbed medium.

2.5.2   *"Two-point" perturbed ray*

To trace the perturbed trajectory of the central ray with the same initial position vector as the unperturbed ray, we substitute

$$\Delta \mathbf{y}(\sigma_0) = \begin{bmatrix} 0 \\ \Delta \mathbf{p}(\sigma_0) \end{bmatrix} \qquad (2.47)$$

in equation (2.29), where $\Delta \mathbf{p} = (\Delta u / u)\mathbf{p}$ due to slowness perturbation at the source point. The "perturbed ray" will not, in general, hit the receiver R (see Figure 2.3). Assume that the perturbed ray hits the receiver level at R^*, near R. Denoting the unperturbed ray as ℓ_0 and the perturbed ray as ℓ^*, the perturbation of the position vector of the perturbed trajectory ℓ^* with respect to the unperturbed central ray ℓ_0, calculated by equations (2.43) and (2.45), is $\Delta \mathbf{x}^*(\sigma)$. The ray we want to obtain is a "two-point" perturbed ray, which is denoted by ℓ, connecting S and R. Referring to Figure 2.3, the "two-point" perturbed ray ℓ should be a paraxial ray of the perturbed central ray ℓ^*.

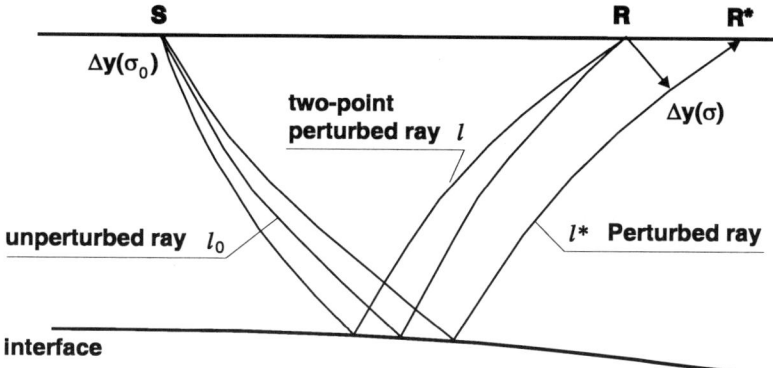

Figure 2.3. The geometry of ray perturbation. Suppose ℓ_0 is an unperturbed ray, ℓ^* and ℓ are, respectively, the "shooting" perturbed ray and the "two-point" perturbed ray, due to slowness perturbation.

The propagator matrix describing the paraxial ray propagation around the perturbed ray ℓ^* can be found from the solution of the linear system of equations (2.26) in the perturbed medium with $\mathbf{A}(\sigma) = \mathbf{A}_0(\sigma) + \Delta\mathbf{A}(\sigma)$. The perturbation of linear system, $\Delta\mathbf{A}$, can be written as

$$\Delta\mathbf{A}(\sigma) \approx \Delta\mathbf{A}_1(\sigma) + \Delta\mathbf{A}_2(\sigma) , \qquad (2.48)$$

in which $\Delta\mathbf{A}_1(\sigma)$ is a term due to ΔH, the perturbation of the Hamiltonian, and $\Delta\mathbf{A}_2(\sigma)$ is a term due to $\Delta\mathbf{y}$, the perturbation of the reference central ray (Farra and Madariaga 1987; Farra, 1989, 1990). They are given by

$$\Delta\mathbf{A}_1 = \begin{bmatrix} \nabla_{\mathbf{x}}\nabla_{\mathbf{p}}\Delta H & \nabla_{\mathbf{p}}\nabla_{\mathbf{p}}\Delta H \\ -\nabla_{\mathbf{x}}\nabla_{\mathbf{x}}\Delta H & -\nabla_{\mathbf{p}}\nabla_{\mathbf{x}}\Delta H \end{bmatrix} ,$$

$$\Delta\mathbf{A}_2 = [\Delta\mathbf{x}\,\nabla_{\mathbf{x}} + \Delta\mathbf{p}\,\nabla_{\mathbf{p}}]\,\mathbf{A}_0 . \qquad (2.49)$$

From equation (2.27), the equations (2.48) can be rewritten as

$$\Delta\mathbf{A}_1 = \begin{bmatrix} \mathbf{0} & \mathbf{0} \\ \Delta\mathbf{U} & \mathbf{0} \end{bmatrix} \quad \text{and} \quad \Delta\mathbf{A}_2 = \begin{bmatrix} \mathbf{0} & \mathbf{0} \\ \tilde{\mathbf{U}} & \mathbf{0} \end{bmatrix} ,$$

where the elements of $\Delta\mathbf{U}$ and $\tilde{\mathbf{U}}$ can be explicitly written as

$$[\Delta\mathbf{U}]_{ij} = u_0 \frac{\partial^2 \Delta u}{\partial x_i \partial x_j} + \Delta u \frac{\partial^2 u_0}{\partial x_i \partial x_j} + \frac{\partial u_0}{\partial x_i}\frac{\partial \Delta u}{\partial x_j} + \frac{\partial \Delta u}{\partial x_i}\frac{\partial u_0}{\partial x_j}$$

and

$$\left[\tilde{\mathbf{U}}\right]_{ij} = \Delta\mathbf{x} \cdot \nabla_{\mathbf{x}} U_{ij} \,,$$

respectively, where $\mathbf{U}$ is the 3×3 symmetric matrix defined by equation (2.28).

The solution of the linear system of equations (2.26) with $\mathbf{A}(\sigma) = \mathbf{A}_0(\sigma) + \Delta\mathbf{A}(\sigma)$ is given by equation (2.29) in which the perturbed propagator $\mathbf{\Pi}(\sigma, \sigma_0)$ can be obtained by the Born approximation (Aki and Richards, 1980),

$$\mathbf{\Pi}(\sigma, \sigma_0) = \mathbf{\Pi}_0(\sigma, \sigma_0) + \int_{\sigma_0}^{\sigma} \mathbf{\Pi}_0(\sigma, \tau) \,\Delta\mathbf{A}(\tau) \,\mathbf{\Pi}_0(\tau, \sigma_0) \, d\tau \,. \qquad (2.50)$$

Equation (2.49) determines the propagator of the perturbed ray ℓ^*. The position perturbation of ray ℓ with respect to ray ℓ^* is equal to $[-\Delta\mathbf{x}^*(\sigma)]$. Partitioning the propagator of equation (2.36), we write

$$[-\Delta\mathbf{x}^*(\sigma)] = [\mathbf{Q}_1(\sigma, \sigma_0) \quad \mathbf{Q}_2(\sigma, \sigma_0)] \begin{bmatrix} \Delta\mathbf{x}(\sigma_0) \\ \Delta\mathbf{p}(\sigma_0) \end{bmatrix} \,. \qquad (2.51)$$

To obtain the "two-point" perturbed ray, ℓ, we now reset the initial condition, $\Delta\mathbf{y}(\sigma_0) = [\Delta\mathbf{x}(\sigma_0), \Delta\mathbf{p}(\sigma_0)]^{\mathrm{T}}$ in equation (2.29), to

$$\begin{aligned} \Delta\mathbf{x}(\sigma_0) &= 0 \,, \\ \Delta\mathbf{p}(\sigma_0) &= -\mathbf{Q}_2^{-1}(\sigma, \sigma_0) \,\Delta\mathbf{x}^*(\sigma) \,, \end{aligned} \qquad (2.52)$$

where the 3×3 matrix $\mathbf{Q}_2$ is the submatrix of the propagator calculated from equation (2.49) for the perturbed central ray ℓ^*.

2.5.3   *Ray perturbation due to interface perturbation*

For the case of an interface perturbation,

$$f(\mathbf{x}) = f_0(\mathbf{x}) + \Delta f(\mathbf{x}) = 0 \,,$$

similar to the case of a slowness perturbation, a two-step strategy is presented here for the determination of the "two-point" perturbed ray. First, the perturbed trajectory with the same initial condition as the unperturbed ray is traced. Then, by adjusting the initial condition using equation (2.51), the "two-point" perturbed ray due to the perturbation of the interface is obtained.

Assume a reference ray intersects the unperturbed interface $f_0(\mathbf{x}) = 0$ at σ_a. The segment of the reference ray at $\sigma < \sigma_a$ is not perturbed. The smooth perturbation of interface Δf will cause perturbation of the reflection-transmission point. According to the ray-tracing system equation (2.13), i.e. the Hamilton formulation, the perturbations of the position vector and the slowness vector are, respectively,

$$\Delta \mathbf{x} = \Delta \sigma \, \nabla_{\mathbf{p}} H_0 \, ,$$
$$\Delta \mathbf{p} = -\Delta \sigma \, \nabla_{\mathbf{x}} H_0 \, ,$$

(2.53)

at the intersection point. Taking an inner product of the first equation in equation (2.52) with ∇f_0 and substituting it in equation (2.22), the expansion of the perturbed interface $f(\mathbf{x})$ to first order, we obtain the increment of the independent parameter,

$$\Delta \sigma = -\frac{\Delta f}{\langle \nabla f_0 | \nabla_{\mathbf{p}} H_0 \rangle} \, ,$$

(2.54)

from the intersection point $\mathbf{x}$ on interface $f_0(\mathbf{x})$, to the point $\mathbf{x} + \Delta \mathbf{x}$ on interface $f(\mathbf{x})$. Substituting equation (2.53) in equation (2.52), we then obtain

$$\Delta \mathbf{x} = -\frac{\Delta f}{\langle \nabla f_0 | \nabla_{\mathbf{p}} H_0 \rangle} \nabla_{\mathbf{p}} H_0 \, ,$$
$$\Delta \mathbf{p} = \frac{\Delta f}{\langle \nabla f_0 | \nabla_{\mathbf{p}} H_0 \rangle} \nabla_{\mathbf{x}} H_0 \, .$$

(2.55)

The perturbed slowness vector of the incident ray at the intersection point is $\mathbf{p} = \mathbf{p}_0 + \Delta \mathbf{p}$. The slowness vector, $\hat{\mathbf{p}}$, of the reflected-transmitted ray in the perturbed media satisfies Snell's law,

$$\hat{\mathbf{p}} \times \nabla f = \mathbf{p} \times \nabla f \, ,$$

(2.56)

where the gradient of the interface can be approximated, to first order, as

$$\nabla f = \nabla f_0 + (\nabla \nabla f_0) \, \Delta \mathbf{x} + \nabla \Delta f \, .$$

(2.57)

With the perturbed vectors, $\mathbf{p}$, $\hat{\mathbf{p}}$ and ∇f, we can now calculate the perturbed transformation matrix $\boldsymbol{\Sigma}(\sigma_a)$. The new condition at the perturbed point of reflection/transmission is then given by

$$\Delta \hat{\mathbf{y}}(\sigma_a) = \boldsymbol{\Sigma}(\sigma_a) \, \Delta \mathbf{y}(\sigma_a) \, .$$

(2.58)

The perturbed reference ray, ℓ^*, in actual fact the ray segment after the perturbed point due to the interface perturbation, can be traced by

$$\Delta \hat{\mathbf{y}}(\sigma) = \boldsymbol{\Pi}_0(\sigma, \, \sigma_a) \, \Delta \hat{\mathbf{y}}(\sigma_a),$$

(2.59)

using the unperturbed propagator $\boldsymbol{\Pi}_0(\sigma, \, \sigma_a)$. Again this perturbed ray ℓ^* will not hit the receiver, but must be adjusted at the source in order to construct the approximate two-point ray. In a 2-D case, using the initial conditions given by equation (2.51), the approximate "two-point" ray ℓ can be traced.

Once we have obtained the "two-point" perturbed ray trajectory ℓ, we can then evaluate the perturbed propagator matrix, the numerical integration along the ray path (equation 2.30), and the modified ray geometrical spreading $D(\sigma, \sigma_0)$ in equation (2.37).

2.6.   Ray amplitude

Using the concept of ray amplitude in the high-frequency approximation, it is assumed that the amplitude of a propagating pulse in a non-dissipative medium depends on the geometry of the ray tube and local reflection-transmission coefficients at interfaces. For tomographic inversion, the information content of the data lies in the dependence of the signal amplitude on the source-receiver offset, where the recorded amplitude is linearly proportional to the source amplitude, which is arbitrary.

Denoting the amplitude at σ_0 along the ray and close to the source point by $A(\sigma_0)$, we can express the recorded amplitude $A(\sigma)$ of a multiply reflected and transmitted ray as

$$A(\sigma) = A(\sigma_0) \left[\frac{v(\sigma_0)}{v(\sigma)} \right]^{1/2} \frac{C_Q}{D(\sigma, \sigma_0)} \prod_{k=0}^{K} C_k \,, \qquad (2.60)$$

where $v(\sigma)$ is the wave velocity, C_Q is the intrinsic inelastic attenuation factor, $D(\sigma, \sigma_0)$ is the ray geometrical spreading derived from the ray propagator describing the property of paraxial rays around a central ray (section 2.4), $\{C_k, k = 0, \cdots, K\}$ are the reflection and transmission coefficients at the interfaces, and K is the number of interface intersections of a particular ray path.

The complex displacement amplitude coefficients for reflection or transmission across an interface can be calculated by applying the Zöppritz amplitude equations, which take into account the P-SV interaction. In the next chapter, we derive the quadratic approximations to the Zöppritz equations, so that we can calculate analytically the Fréchet derivatives of the reflection amplitudes with respect to the elastic parameters when we invert for the interface geometry and the elastic parameters along a reflector.

In a homogeneous medium, the intrinsic attenuation due to the inelasticity of a medium appears to change exponentially with the travel distance r,

$$C_Q(\omega, r) = \exp\left[-\frac{\omega}{2} r \, u(\omega) \, Q^{-1}(\omega) \right] \,, \qquad (2.61)$$

where ω is the angular frequency, $u(\omega)$ is the phase slowness (the reciprocal of the phase velocity), and $Q(\omega)$ is called the quality factor and is usually frequency-dependent (Futterman, 1962; Aki and Richards, 1980). The wave energy is gradually absorbed by the medium, which is responsible for the eventual complete disappearance of the wave motion. In general, the losses of intensity due to geometrical spreading are more important than losses due to absorption for low frequencies at short distances from the source for ranges typical in exploration seismic. As the frequency and distance increase, however, absorption losses increase and eventually dominate. The increased absorption at higher frequencies results in change of waveshape with distance.

If the medium is inhomogeneous, the focusing effects of velocity (and density) variations should be included. Assuming that Q is large and that losses are small over one wavelength, equation (2.60) becomes

$$C_Q(\omega, \mathbf{x}) = \exp\left[-\frac{\omega}{2}\int_\gamma u(\omega, \mathbf{x})\, Q^{-1}(\omega, \mathbf{x})\, ds\right]\,, \qquad (2.62)$$

where the integral is performed along the ray path γ.

If we assume that Q is constant with respect to frequency (Kjartansson, 1979), i.e. $u(\omega, \mathbf{x}) = u(\mathbf{x})$, as exploration seismograms usually have low frequencies and are band limited, equation (2.61) can be rewritten as

$$C_Q(\mathbf{x}) = \exp\left[-\frac{\omega_0}{2}\int_\gamma u(\mathbf{x})\, Q^{-1}(\mathbf{x})\, ds\right]\,, \qquad (2.63)$$

where ω_0 represents the dominant frequency. Finally, we have

$$C_Q(\mathbf{x}) = \exp\left[-\int_\gamma \eta(\mathbf{x})\, ds\right]\,, \qquad (2.64)$$

where η is the absorption coefficient,

$$\eta(\mathbf{x}) = \frac{\omega_0}{2}\, u(\mathbf{x})\, Q^{-1}(\mathbf{x})\,. \qquad (2.65)$$

The structure of equation (2.63) is reminiscent of the traveltime problem. In an inversion problem using log amplitudes, we then can use the partial derivatives already calculated for the traveltime problem.

As far as Q is concerned, we have three options for the procedure of amplitude inversion:

1) If the pulse shapes are not significantly altered by attenuation, the simplified amplitudes (unaffected by frequency variation) could still be used, in principle, to provide constraints on interface and velocity variation, as well as the average attenuation, in the seismic inversion.

2) If the pulse shapes are altered significantly by attenuation, it might be necessary to carry out phase correction first. Inverse Q filtering for phase only
   is unconditionally stable and is easy to implement (Hargreaves and Calvert,
   1991).

3) If the Q value is estimated by physical measurement, such as downhole
   seismic, we may perform inverse Q filtering, compensating for both the
   phase and amplitude effects. We then perform amplitude inversion without
   the dependence of the earth Q effect.

In general, inclusion of amplitude compensation in inverse Q filtering may
cause numerical instability and generate undesirable artifacts in the seismic
traces. For stabilized inverse Q filtering that compensates for phase and amplitude effects simultaneously without boosting the noise, see Wang (2002,
2003).

Even if there is no intrinsic attenuation, the amplitude of reflected or transmitted pulses may be frequency-dependent for waves incident on thin layers
(e.g. MacDonald *et al.*, 1987), but we assume here the infinite-frequency amplitude obtained from the geometrical ray amplitude calculation (Červený and
Ravindra, 1971; Červený, 2001).

For the ray geometrical spreading, we may use either the explicit function
shown in Chapter 1 for a constant-velocity layer or the numerical algorithm for
heterogeneous media presented in this chapter, derived from the ray propagator
describing the property of paraxial rays around a central ray. The variation
of ray amplitude due to the model perturbation is calculated in terms of the
difference between the perturbed and unperturbed observations (logarithm of
amplitudes). The perturbed amplitude, in terms of the perturbation of the geometrical spreading and of the reflection coefficient, can be calculated along the
perturbed "two-point" ray, as described in the previous section.

Chapter 3

Amplitude coefficients and approximations

Abstract   For efficient inversion of reflection seismic amplitude data for elastic parameters, we derive the so-called pseudo-quadratic approximations to the Zöppritz
equations for calculating reflection and transmission coefficients as a function
of the ray parameter p. An immediate advantage of using these quadratic approximations in the tomographic inversion is that we can compute analytically,
but not necessarily numerically, the sensitivity matrix of Fréchet derivatives of
amplitudes with respect to elastic parameters. We also represent the amplitude
coefficients as a quadratic function of the elastic contrasts at an interface, and
compare them to the linear approximation used in conventional AVO analysis,
where we can invert for only two elastic parameters. We see that by using the
second-order approximation, the condition number of the Fréchet matrix for
three elastic parameters is improved significantly. Therefore, we can use these
quadratic approximations directly with amplitude information to estimate not
only two but three parameters: P-wave velocity contrast, S-wave velocity contrast, and the ratio of S-wave and P-wave velocities at an interface. We also see
that the approximation can provide instructive insight into pore-fluid content in
conventional AVO analysis.

3.1.   Introduction

Both tomographic amplitude inversion and conventional AVO analysis
depend on the Zöppritz equations (Knott, 1899; Zöppritz, 1919). In conventional AVO analysis (e.g. Ostrander, 1984; Rutherford and Williams, 1989;
Hilterman, 2001), various linear approximations of the Zöppritz equations with

respect to the elastic contrasts at an interface are used (Bortfeld, 1961; Chapman, 1976; Richards and Frasier, 1976; Aki and Richards, 1980; Shuey, 1985). In tomographic inversion, the exact Zöppritz equations are usually adopted to calculate reflection and transmission coefficients. To invert efficiently for elastic parameters, however, approximations to the Zöppritz equations with relatively higher accuracy are desirable in tomographic amplitude inversion. In this chapter we derive quadratic expressions for the *P-P* wave reflection and transmission coefficients, R_{PP} and T_{PP}, with respect to the relative contrast in elastic parameters. An immediate advantage of using these quadratic approximations, rather than the exact Zöppritz equations, in tomographic inversion is that the sensitivity matrix of Fréchet derivatives of amplitudes with respect to the elastic parameters can be computed analytically, but not necessarily numerically.

The quadratic approximations of amplitude coefficients with respect to the elastic contrasts at an interface are converted from the so-called pseudo-quadratic expressions with respect to the ray parameter p. We refer to the latter as the pseudo-formulae because the coefficients of the p^2 (and p^4) terms are defined as a function of vertical slownesses, which also depend on the ray parameter. Ursin and Dahl (1992) developed practical quadratic approximations of the Zöppritz equations as a function of the ray parameter. In their approximations, the coefficients of the Taylor series were computed directly from the medium parameters (not from the vertical slownesses we use here). However, an explicit expression of the fourth-order term as a function of the medium parameters was very cumbersome and therefore implemented only in a computer code. In this chapter, we provide an alternative to the approximations of Zöppritz equations, with explicit coefficients of both the p^2 and p^4 terms, which are in a fairly compact format.

We compare the quadratic expressions for the *P-P* wave reflection and transmission coefficients as a function of the elastic contrasts at the interface with the previously used linear formulae. Numerical analysis demonstrates that the quadratic approximations are more accurate than the previous linear approximate formulae. In using the quadratic approximation of the *P-P* wave reflection coefficient, we can potentially estimate the following three elastic parameters simultaneously: relative *P*-wave velocity contrast, $\Delta\alpha/\alpha$; relative *S*-wave velocity contrast, $\Delta\beta/\beta$; and the ratio of average *S*-wave to average *P*-wave velocities, β/α.

We also see that the approximation may provide an instructive insight into pore-fluid content in conventional AVO analysis.

3.2.   The Zöppritz equations

Let us consider a horizontal interface separating two half-space media in which the *P*-wave velocities are α_1 and α_2, the *SV*-wave velocities are β_1 and

β_2 and the densities are ρ_1 and ρ_2. For a plane wave propagating through the media, the ray parameter p is preserved in the primary and the mode-converted waves through Snell's law:

$$p = \frac{\sin\theta_1}{\alpha_1} = \frac{\sin\theta_2}{\alpha_2} = \frac{\sin\phi_1}{\beta_1} = \frac{\sin\phi_2}{\beta_2} , \tag{3.1}$$

where θ_1 and θ_2 are the *P-P* wave reflection and transmission angles and ϕ_1 and ϕ_2 are the *P-SV* wave reflection and transmission angles.

The exact formulae for *P-P* reflection and transmission coefficients can be expressed in terms of the ray parameter p as (Aki and Richards, 1980)

$$R_{PP}(p) = \frac{E + Fp^2 + Gp^4 - Dp^6}{A + Bp^2 + Cp^4 + Dp^6} , \tag{3.2}$$

and

$$T_{PP}(p) = \frac{H + Ip^2}{A + Bp^2 + Cp^4 + Dp^6} , \tag{3.3}$$

where

$A = (\rho_2 q_{\alpha_1} + \rho_1 q_{\alpha_2})(\rho_2 q_{\beta_1} + \rho_1 q_{\beta_2}) ,$

$B = -4\Delta\mu(\rho_2 q_{\alpha_1} q_{\beta_1} - \rho_1 q_{\alpha_2} q_{\beta_2}) + (\Delta\rho)^2 + 4(\Delta\mu)^2 q_{\alpha_1} q_{\alpha_2} q_{\beta_1} q_{\beta_2} ,$

$C = 4(\Delta\mu)^2 (q_{\alpha_1} q_{\beta_1} + q_{\alpha_2} q_{\beta_2}) - 4\Delta\mu\Delta\rho ,$

$D = 4(\Delta\mu)^2 ,$

$E = (\rho_2 q_{\alpha_1} - \rho_1 q_{\alpha_2})(\rho_2 q_{\beta_1} + \rho_1 q_{\beta_2}) ,$

$F = -4\Delta\mu(\rho_2 q_{\alpha_1} q_{\beta_1} + \rho_1 q_{\alpha_2} q_{\beta_2}) - (\Delta\rho)^2 + 4(\Delta\mu)^2 q_{\alpha_1} q_{\alpha_2} q_{\beta_1} q_{\beta_2} ,$

$G = 4(\Delta\mu)^2 (q_{\alpha_1} q_{\beta_1} - q_{\alpha_2} q_{\beta_2}) + 4\Delta\mu\Delta\rho ,$

$H = 2(\rho_2 q_{\beta_1} + \rho_1 q_{\beta_2})\rho_1 q_{\alpha_1}(\alpha_1/\alpha_2)$

and

$I = -4\Delta\mu(q_{\beta_1} - q_{\beta_2})\rho_1 q_{\alpha_1}(\alpha_1/\alpha_2) .$

These values are defined in terms of the *P*-wave and *SV*-wave vertical slownesses, q_{α_1}, q_{α_2}, q_{β_1} and q_{β_2}, the contrast in density $\Delta\rho$ and the contrast in shear moduli,

$$\Delta\mu = \rho_2\beta_2^2 - \rho_1\beta_1^2 . \tag{3.4}$$

3.3.   The pseudo-p^2 expressions

Using the Taylor expansion with respect to the ray parameter p, we can rewrite the amplitude coefficients equations (3.2) and (3.3) as

$$R_{PP}(p) = \frac{E}{A} + \left(\frac{F}{A} - \frac{BE}{A^2}\right)p^2 + \left(\frac{G}{A} - \frac{BF}{A^2} - \frac{CE}{A^2} + \frac{B^2E}{A^3}\right)p^4 + \cdots$$

(3.5)

and

$$T_{PP}(p) = \frac{H}{A} + \left(\frac{I}{A} - \frac{BH}{A^2}\right)p^2 - \left(\frac{BI}{A^2} + \frac{CH}{A^2} - \frac{B^2H}{A^3}\right)p^4 + \cdots.$$

(3.6)

Because the coefficients A, B, C, etc., defined below equation (3.3), also depend on p through the vertical slowness:

$$q_{\alpha_1} = \sqrt{1/\alpha_1^2 - p^2}\,, \qquad q_{\alpha_2} = \sqrt{1/\alpha_2^2 - p^2}\,,$$

and

$$q_{\beta_1} = \sqrt{1/\beta_1^2 - p^2}\,, \qquad q_{\beta_2} = \sqrt{1/\beta_2^2 - p^2}\,,$$

we refer to equations (3.5) and (3.6) as the pseudo-quartic approximations with respect to the ray parameter p.

Truncating the expressions at the p^2 term, we have the pseudo-quadratic expressions of the P-P wave reflection and transmission coefficients,

$$R_{PP}(p) \approx R_f - 2\frac{\Delta\mu}{\rho}p^2 + (1 - R_f)q_\alpha q_\beta \left(\frac{\Delta\mu}{\rho}\right)^2 p^2$$

(3.7)

and

$$T_{PP}(p) \approx T_f \frac{q_{\alpha_1}\alpha_1}{q_{\alpha_2}\alpha_2}\left[1 - q_\alpha q_\beta \left(\frac{\Delta\mu}{\rho}\right)^2 p^2\right],$$

(3.8)

with

$$R_f = \frac{\rho_2 q_{\alpha_1} - \rho_1 q_{\alpha_2}}{\rho_2 q_{\alpha_1} + \rho_1 q_{\alpha_2}}$$

(3.9)

and

$$T_f = 1 - R_f\,,$$

(3.10)

where q_{α_1} and q_{α_2} are the P-wave vertical slownesses, q_α is their average, q_β is the corresponding average of the SV-wave vertical slownesses, and ρ is the average of bulk densities.

In equation (3.9), R_f can be understood as the fluid-fluid reflection coefficient, which is the reflection coefficient between two media when the corresponding shear-wave velocities in both media are set to zero, and T_f in equation (3.10) is, correspondingly, the "fluid-fluid" transmission coefficient. In a similar formula for the *P-P* reflection coefficient (equation 3.7) derived by Mallick (1993), the term,

$$-R_f \, q_\alpha q_\beta \left(\frac{\Delta\mu}{\rho}\right)^2 p^2 \, ,$$

was omitted. In the following text, we refer to the *P-P* coefficients in equations (3.7) and (3.8) as the pseudo-p^2 expressions.

3.4. Quadratic expressions in terms of elastic contrasts

In this section, we convert the pseudo-p^2 approximations into quadratic expressions with respect to the elastic contrasts along the reflection interface. In conventional AVO analysis we often attempt to reveal the *contrasts* in elastic reflectivities and do not expect to determine the absolute values of the elastic parameters.

Following Snell's law, we have

$$\frac{\alpha_1}{\alpha_2} = \frac{\sin\theta_1}{\sin\theta_2} \, , \tag{3.11}$$

and using Taylor series expansions, accurate up to second order, we have

$$\frac{\alpha_1}{\alpha_2} \approx 1 - \frac{\Delta\alpha}{\alpha} + \frac{1}{2}\left(\frac{\Delta\alpha}{\alpha}\right)^2 \tag{3.12}$$

and

$$\frac{\sin\theta_1}{\sin\theta_2} \approx 1 - 2\frac{\tan\frac{\Delta\theta}{2}}{\tan\theta} + 2\left(\frac{\tan\frac{\Delta\theta}{2}}{\tan\theta}\right)^2 \, , \tag{3.13}$$

where

$$\Delta\theta = \theta_2 - \theta_1 \, , \quad \theta = \frac{1}{2}(\theta_1 + \theta_2) \, , \quad \Delta\alpha = \alpha_2 - \alpha_1 \, , \quad \alpha = \frac{1}{2}(\alpha_1 + \alpha_2) \, .$$

Equations (3.11), (3.12) and (3.13) leads to the quadratic equation,

$$\left(\frac{\tan\frac{\Delta\theta}{2}}{\tan\theta}\right)^2 - \frac{\tan\frac{\Delta\theta}{2}}{\tan\theta} + \frac{1}{2}\frac{\Delta\alpha}{\alpha} - \frac{1}{4}\left(\frac{\Delta\alpha}{\alpha}\right)^2 \approx 0 \, . \tag{3.14}$$

Solving this equation, we obtain the following formula:

$$\tan\frac{\Delta\theta}{2} \approx \frac{1}{2}\frac{\Delta\alpha}{\alpha}\tan\theta \,. \tag{3.15}$$

Using equation (3.15), we can rewrite the fluid-fluid reflection coefficient, equation (3.9), in terms of the relative contrasts of density and P-wave velocity as

$$R_f \approx \frac{1}{2}\left(\frac{\Delta\rho}{\rho} + \sec^2\theta\frac{\Delta\alpha}{\alpha}\right) + \mathcal{O}(\chi^3) \,, \tag{3.16}$$

where χ represents the relative contrast of an elastic parameter at the interface. Note that both equations (3.15) and (3.16) are accurate up to the second order terms, which are zero valued, in the relative contrasts.

In addition, for nonevanescent waves, p is small and then the vertical slownesses, q_{α_1} and q_{α_2}, are real. The following expression is approximated also up to the p^2 term,

$$\frac{q_{\alpha_1}\alpha_1}{q_{\alpha_2}\alpha_2} \approx \frac{1 - \frac{1}{2}\alpha_1^2 p^2 - \frac{1}{8}\alpha_1^4 p^4}{1 - \frac{1}{2}\alpha_2^2 p^2 - \frac{1}{8}\alpha_2^4 p^4} \approx 1 + \alpha\,\Delta\alpha\,p^2 + \mathcal{O}(p^4) \,. \tag{3.17}$$

The contrast in shear modulus can also be read as

$$\Delta\mu = \beta^2\Delta\rho + 2\rho\beta\Delta\beta + \frac{1}{4}\Delta\rho(\Delta\beta)^2 \,, \tag{3.18}$$

where β and $\Delta\beta$ are the S-wave velocity average and difference, respectively.

We now substitute equations (3.15)–(3.18) into equations (3.7) and (3.8), and obtain respectively the P-P reflection and transmission coefficients,

$$R_{PP} \approx \left[\frac{1}{2} - 2\left(\frac{\beta}{\alpha}\right)^2\sin^2\theta\right]\frac{\Delta\rho}{\rho} + \frac{1}{2}\sec^2\theta\frac{\Delta\alpha}{\alpha} - 4\left(\frac{\beta}{\alpha}\right)^2\sin^2\theta\frac{\Delta\beta}{\beta}$$

$$+ \left(\frac{\beta}{\alpha}\right)^3\cos\theta\sin^2\theta\left(\frac{\Delta\rho}{\rho} + 2\frac{\Delta\beta}{\beta}\right)^2 \tag{3.19}$$

and

$$T_{PP} \approx 1 - \frac{1}{2}\frac{\Delta\rho}{\rho} + \frac{1}{2}(\tan^2\theta - 1)\frac{\Delta\alpha}{\alpha} - \left(\frac{\beta}{\alpha}\right)^3\cos\theta\sin^2\theta\left(\frac{\Delta\rho}{\rho} + 2\frac{\Delta\beta}{\beta}\right)^2 \,, \tag{3.20}$$

where both equations are quadratic approximations with respect to the relative contrasts in three elastic parameters: the bulk density ρ, the P-wave velocity α and the S-wave velocity β.

Table 3.1. Model parameters for the calculation of reflection and transmission coefficients

Material	density ρ (g/cm^3)	P-wave velocity α (m/s)	S-wave velocity β (m/s)	Poisson's ratio σ
Sand	2.65	3780	2360	0.18
Limestone	2.75	3845	2220	0.25
Shale	2.25	3600	1585	0.38
Anhydrite	2.95	6095	3770	0.19

3.5. Accuracy of the quadratic approximations

Let us now compare the accuracy of the approximations (3.19) and (3.20) with those of previously published linear approximations. If we ignore the terms containing $(\Delta\rho/\rho + 2\Delta\beta/\beta)^2$ in equations (3.19) and (3.20), we obtain the following linear expressions,

$$R_{PP}(\theta) \approx \left[\frac{1}{2} - 2\left(\frac{\beta}{\alpha}\right)^2 \sin^2\theta\right]\frac{\Delta\rho}{\rho} + \frac{1}{2}\sec^2\theta\frac{\Delta\alpha}{\alpha} - 4\left(\frac{\beta}{\alpha}\right)^2 \sin^2\theta\frac{\Delta\beta}{\beta}$$

(3.21)

and

$$T_{PP}(\theta) \approx 1 - \frac{1}{2}\frac{\Delta\rho}{\rho} + \frac{1}{2}\left(\tan^2\theta - 1\right)\frac{\Delta\alpha}{\alpha},$$

(3.22)

which are linear in each of the three elastic contrast terms. These expressions are equivalent to the approximations previous published by Bortfeld (1961), Richards and Frasier (1976) and Aki and Richards (1980). Shuey (1985) then made a further modification to equation (3.21) by replacing β and $\Delta\beta$ with σ and $\Delta\sigma$, where σ is Poisson's ratio, given by

$$\sigma = \frac{1/2 - (\beta/\alpha)^2}{1 - (\beta/\alpha)^2}.$$

(3.23)

Let us examine the difference between alternative approximations. Table 3.1 lists the elastic parameters for a group of sedimentary minerals. Figures

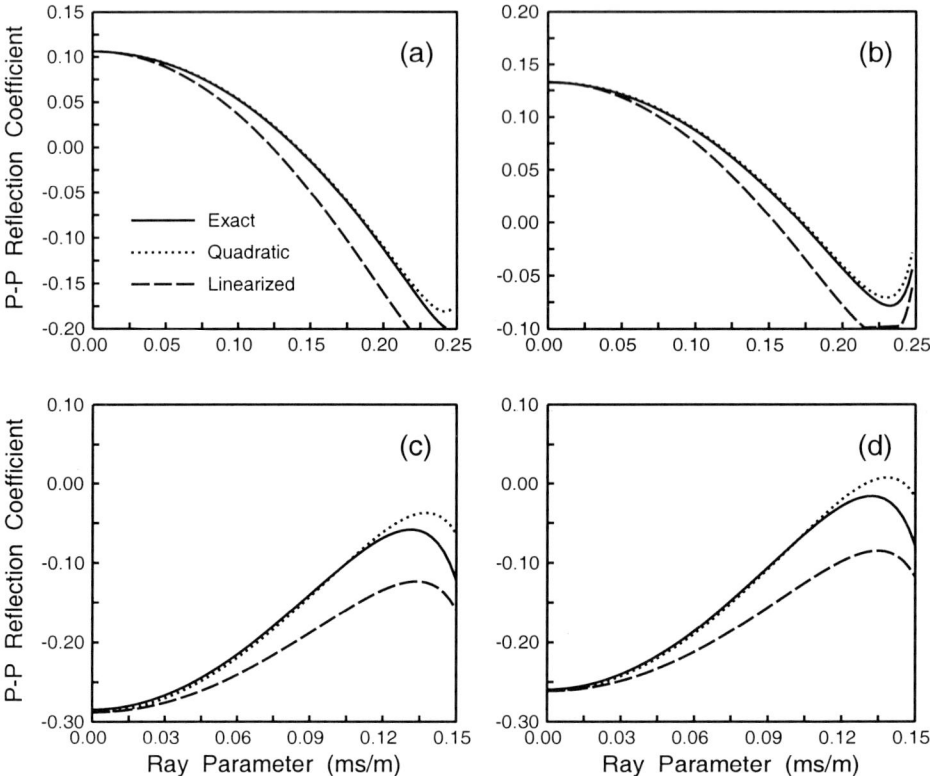

Figure 3.1.   The *P-P* wave reflection coefficients. Solid lines represent the results calculated from the exact Zöppritz equation, dotted and dashed lines are those from the quadratic and linearized approximations, respectively. Reflection coefficients corresponding to interface models of shale/sand (a), shale/limestone or dolomite (b), anhydrite/sand (c), and anhydrite/limestone or dolomite (d) are shown.

3.1 and 3.2 show the *P-P* reflection and transmission coefficients for four interface examples between these sedimentary layers. The four interface examples represent shale/sand, shale/limestone (or dolomite), anhydrite/sand and anhydrite/limestone (or dolomite) interfaces. In the interface examples, some reasonable "worst cases", such as shale/limestone (or dolomite) and evaporite (anhydrite)/sandstone, are included, to demonstrate the accuracy of the approximations.

The *P-P* reflection and transmission coefficients shown in Figures 3.1 and 3.2 are obtained from the exact Zöppritz equations (equations 3.1 and 3.2), the quadratic approximations (equations 3.19 and 3.20) and the linear approximations (equations 3.21 and 3.22). As expected, the quadratic approximations in general have better accuracy than linear ones. For the shale/sand

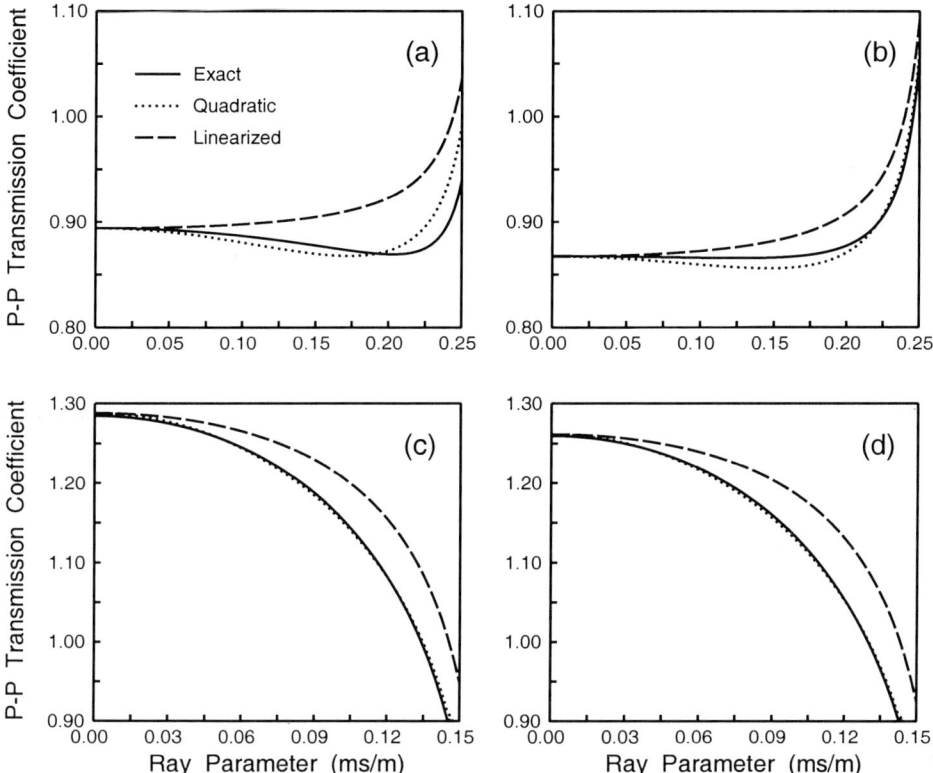

Figure 3.2. The P-P wave transmission coefficients. Solid lines represent the results calculated from the exact Zöppritz equation, dotted and dashed lines are those obtained from the quadratic and the linearized approximations, respectively. Transmission coefficients corresponding to interface models of shale/sand (a), shale/limestone or dolomite (b), anhydrite/sand (c), and anhydrite/limestone or dolomite (d) are shown.

and shale/limestone (or dolomite) cases, the quadratic approximations provide very good results. For the "worst case" with an evaporite (Figures 3.1c and 3.1d), the quadratic expression for the reflection coefficient achieves good accuracy for small and intermediate p values. When the incidence angle approaches the critical angle, there is a considerable difference between the reflection coefficients obtained from the exact calculation and the quadratic approximation.

From Figure 3.1 we see that ignoring the term containing $(\Delta\rho/\rho + 2\Delta\beta/\beta)^2$ in equation (3.19) causes a significant error, and reflection amplitudes become more negative in these examples. From Figure 3.2 we see that ignoring this term also causes the absolute amplitude of the transmission coefficient to be overestimated. For a multi-layered structure, the cumulative error in the amplitude estimate will be significant. The errors in linear approximations arise from assuming that the contrasts of all elastic parameters are small relative to their averages.

3.6.   Amplitude coefficients represented as a function of three elastic parameters

When using the linearized approximation of the reflection coefficient R_{PP} (equation 3.21) in the inversion, it is difficult to estimate more than two elastic parameters (Stolt and Weglein, 1985; de Nicolao *et al.*, 1993; Ursin and Tjåland, 1993). In contrast, Ursin and Tjåland (1996) showed that, by using the exact Zöppritz equations, up to three parameters could be estimated from pre-critical *P-P* reflection coefficients. To examine the capacity of the quadratic approximation for the estimation of three parameters, which are essential in the description of the elastic properties of a reflector, let us first modify equations (3.19) and (3.20) to write the coefficients as a function of three elastic parameters.

As we know, a simple systematic relationship exists between the *P*-wave velocity and the bulk density of many sedimentary rocks (Gardner *et al.*, 1974). As a result, reflection and transmission coefficients can be estimated satisfactorily from velocity information alone. As shown by Gardner *et al.* (1974), the density ρ in sedimentary rocks may often be considered to be proportional to the fourth root of the velocity α, i.e.

$$\rho \approx 0.31\alpha^{1/4} , \tag{3.24}$$

where the units of α and ρ are m/s and g/cm^3, respectively. Equation (3.24) is representative of a large number of laboratory and field observations of different brine-saturated rock types (excluding evaporites). This empirical relationship leads to

$$\frac{\Delta\rho}{\rho} = r\frac{\Delta\alpha}{\alpha} , \qquad \text{where} \qquad r \approx 1/4 . \tag{3.25}$$

Substituting equation (3.25) in equations (3.19) and (3.20), we obtain the following expressions

$$R_{PP} \approx \frac{1}{2}\left[r + 1 + \tan^2\theta - \left(\frac{\beta}{\alpha}\right)^2\sin^2\theta\right]\frac{\Delta\alpha}{\alpha} - 4\left(\frac{\beta}{\alpha}\right)^2\sin^2\theta\frac{\Delta\beta}{\beta}$$

$$+ \left(\frac{\beta}{\alpha}\right)^3\cos\theta\sin^2\theta\left(r\frac{\Delta\alpha}{\alpha} + 2\frac{\Delta\beta}{\beta}\right)^2 \tag{3.26}$$

and

$$T_{PP} \approx 1 - \frac{1}{2}(r + 1 - \tan^2\theta)\frac{\Delta\alpha}{\alpha} - \left(\frac{\beta}{\alpha}\right)^3 \cos\theta \sin^2\theta \left(r\frac{\Delta\alpha}{\alpha} + 2\frac{\Delta\beta}{\beta}\right)^2 .$$

(3.27)

We now have the formulae for the reflection and transmission coefficients expressed in terms of three elastic parameters: the contrasts $\Delta\alpha/\alpha$ and $\Delta\beta/\beta$, and the ratio of S-to-P wave average velocities, β/α.

Note that, although departures from the systematic relationship between velocity and density, given by equation (3.24), exist and may in some cases be significant, the reflection and transmission coefficients in general are not sensitive to the perturbation of the bulk density. In addition, in the application of equations (3.26) and (3.27), the coefficient $r \approx 1/4$ is a good approximation used for the variation of the bulk density. For example, using the lithologic relationships given by Castagna (1993), we have

Sand:	$\rho \approx 0.273\alpha^{0.261}$,	$\frac{\Delta\rho}{\rho} \approx 0.261\frac{\Delta\alpha}{\alpha}$;
Shale:	$\rho \approx 0.279\alpha^{0.265}$,	$\frac{\Delta\rho}{\rho} \approx 0.265\frac{\Delta\alpha}{\alpha}$;
Limestone:	$\rho \approx 0.317\alpha^{0.225}$,	$\frac{\Delta\rho}{\rho} \approx 0.225\frac{\Delta\alpha}{\alpha}$;
Dolomite:	$\rho \approx 0.302\alpha^{0.243}$,	$\frac{\Delta\rho}{\rho} \approx 0.243\frac{\Delta\alpha}{\alpha}$;
Anhydrite:	$\rho \approx 0.725\alpha^{0.160}$,	$\frac{\Delta\rho}{\rho} \approx 0.160\frac{\Delta\alpha}{\alpha}$.

In the last column, i.e. the relationship between the contrast in density and the contrast in velocity, all the coefficients (except that of the anhydrite) vary from 0.225 to 0.261. Therefore, the errors obtained using the coefficient $r \approx 0.250$ arc under 2.5%.

3.7.   Three elastic parameters from amplitude inversion

We now consider the capacity of the quadratic approximation for the estimation of three elastic parameters.

Let us consider a linearized inverse problem, using reflection amplitude data, represented as

$$\delta\mathbf{d} = \mathbf{F}\delta\mathbf{m} ,$$

(3.28)

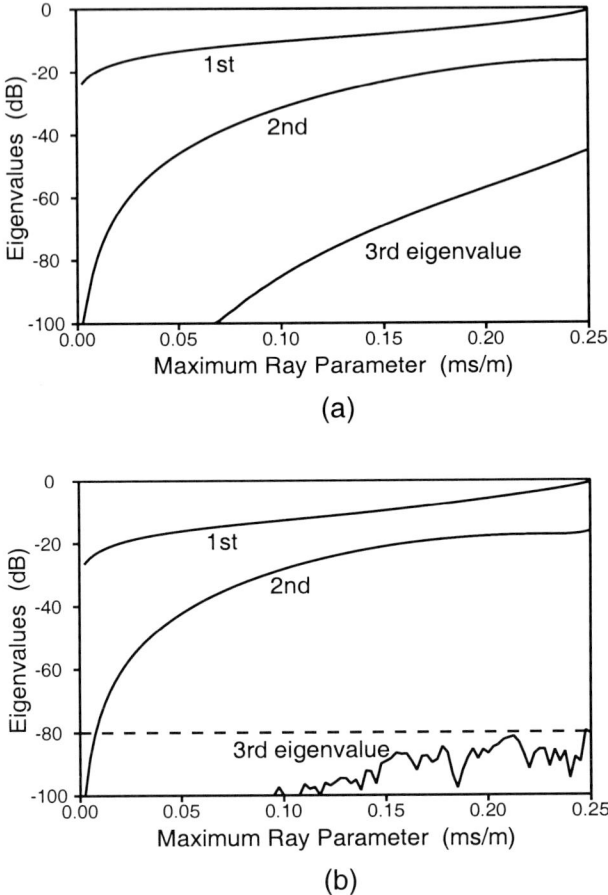

Figure 3.3. Eigenvalues of the Fréchet matrix versus the maximum ray parameter p_{max}: (a) and (b) correspond to the inverse problems using, respectively, the quadratic expression and the linearized approximation of the P-P reflection coefficient. The Fréchet matrix is evaluated based on the shale/sand interface model.

where $\delta \mathbf{d}$ is the data residual vector, $\mathbf{F}$ is the sensitivity matrix of Fréchet derivatives of the model responses with respect to the model parameters, and $\delta \mathbf{m}$ is a small model perturbation to ensure that the linear map between the model and the data perturbations is valid. In the numerical analysis here, we consider the data space ranging from vertical incidence to a maximum ray parameter p_{max} which is assumed to be smaller than that at the critical incidence angle, and we sample the data uniformly over p; this assumption simplifies the computation because the data do not depend on the interface depth in a specified model. Let us evaluate the Fréchet matrix around the solution point of the shale/sand interface model given in Table 3.1.

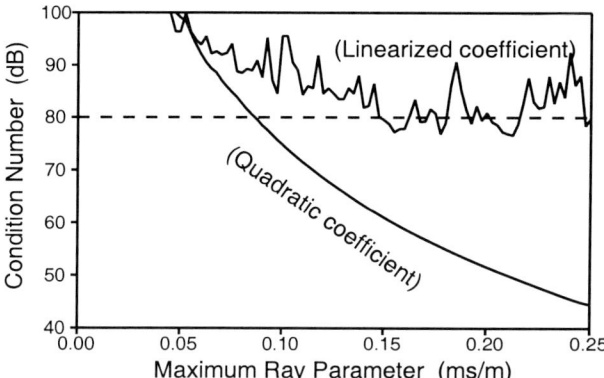

Figure 3.4. The condition numbers of the Fréchet matrix in the inverse problem using, respectively, the quadratic expression and the linearized approximation of the P-P reflection coefficient.

As we know, in the inversion with a linearized approximation to the reflection coefficient, difficulty arises when we estimate more than two elastic parameters, due to ill conditioning of **F**. The matrix **F** for the three-parameter inverse problem has a very large condition number, where the condition number of a matrix can be defined as the ratio of the maximum eigenvalue to the minimum eigenvalue.

Figure 3.3 shows the eigenvalues of the Fréchet matrix versus the maximum ray parameter, p_{max}, considered. Figure 3.3a depicts the result using the quadratic approximation of the P-P reflection coefficient and Figure 3.3b shows the result of the linearized R_{PP} approximation. With the linearized approximation, the third eigenvalue is too small, close to or less than the machine epsilon. In Figure 3.3b the dashed line represents an example truncating value applied in the numerical calculation.

Figure 3.4 displays the condition numbers for the cases of the quadratic formula and the linear approximation. For all p_{max}, the condition number for the quadratic case is smaller than the condition number for the linear case, and the condition number for the linear case is larger than 80 dB (dashed line). In the numerical computation, if the reciprocal condition number, `reconum`, of a matrix is small enough so that the logical expression,

$$\{1.0 + \texttt{reconum} == 1.0\}\,,$$

is true, then the matrix is regarded as singular to working precision.

Most previously published work (e.g. de Nicolao *et al.*, 1993; Ursin and Tjåland, 1993) on linearized inversions of reflection coefficients only solve for two elastic parameters and often assume that the S- to P-wave velocity ratio is known *a priori*. Such a constraint could cause the inverse problem to be inaccurate and biased. From Figure 3.4 we see that, when p_{max} increases, the condition number for the linear case does not change, but there is a linear decrease in the condition number for the quadratic case. Therefore, when data include reflections with moderate and large offsets, the quadratic equation (3.26) could be used in the amplitude inversion, in principle, to estimate not only two (in conventional AVO) but three key parameters, provided an appropriate inverse algorithm is adopted.

3.8.   Implication for fluid substitution modelling

In this section, we see that the new approximation also has implications for pore-fluid prediction with fluid substitution modelling.

Bortfeld (1961) derived several formulae for calculating reflection and transmission coefficients as a function of incident angle. He claimed that his approximations of Zöppritz's equations were accurate to within a few degrees of the critical angle. His formula for the *P-P* reflection coefficient is

$$R_{PP} \approx \underbrace{\frac{1}{2}\ln\left(\frac{\alpha_2\rho_2\cos\theta_1}{\alpha_1\rho_1\cos\theta_2}\right)}_{\text{fluid-fluid}} + \underbrace{\left(\frac{\sin\theta_1}{\alpha_1}\right)^2 (\beta_1^2 - \beta_2^2)\left(2 + \frac{\ln(\rho_2/\rho_1)}{\ln(\beta_2/\beta_1)}\right)}_{\text{rigidity}} .$$

(3.29)

An important aspect of this equation is the insight that it gives the interpreter into predicting how amplitude varies with offset as a function of rock properties. The first term in Bortfeld's *P-P* wave reflection equation is the fluid-fluid reflection coefficient equation. The second term has been called the rigidity term because of its dependence on the *S*-wave velocity, and thus on the shear-rigidity modulus. Readers may refer to Hilterman (2001) for the examples of pore-fluid prediction by Bortfeld's equation.

Although Bortfeld's introduction of the fluid-fluid and rigidity terms provided rock-property insight into how the AVO curves change during pore-fluid substitution, this occurred only after the AVO curves were computer generated. The fluid-fluid and rigidity terms themselves are still rather difficult to evaluate mentally.

However, the pseudo-p^2 expression we have derived in this chapter is quite instructive. Let us take the first two terms of equation (3.7), i.e.

$$R_{PP}(\theta) \approx \underbrace{\frac{1}{2}\left(\frac{\Delta\rho}{\rho} + \sec^2\theta\frac{\Delta\alpha}{\alpha}\right)}_{\text{fluid-fluid}} - \underbrace{\frac{2\Delta\mu}{\rho\alpha^2}\sin^2\theta}_{\text{rigidity}} ,$$ (3.30)

where the fluid-fluid reflectivity term is approximated using equation (3.16). In expression (3.30), the fluid-fluid term indicates that only the *P*-wave velocity can change the way the amplitude increases or decreases with offset or the incident angle θ in fluid-substitution models, because the density contribution is uniform. In the rigidity term, $\Delta\mu$ is the same for the hydrocarbon- and water-saturated states. The term $\sin^2\theta/\alpha^2$ is the square of the ray parameter p, and this basically remains the same for hydrocarbon- or water-saturated states. The only change that the rigidity terms offers during fluid substitution is through the small change in the average density ρ.

The simplification of equation (3.30), compared to Bortfeld's equation (3.29), is obtained by introducing the elastic constant μ, rather than using velocity and density variables exclusively. In the conventional linear approximation equations (Aki and Richards, 1980), one of the assumptions is that

$$\frac{\Delta\rho}{\rho} \ll 1\,, \quad \frac{\Delta\alpha}{\alpha} \ll 1 \quad \text{and} \quad \frac{\delta\beta}{\beta} \ll 1\,.$$

That is, when all the contrasts in the elastic parameters are small, the linear approximation agrees well with the exact Zöppritz equations. Such a match fails when modelling a single interface with large contrasts in elastic parameters. However, the derivation in this chapter clearly indicates that it is the $\Delta\mu/\rho$ term that determines whether the linear and exact reflection coefficients will match, as we have seen that the linear expression (3.30) is obtained from the quadratic expression (3.7) by dropping the term containing $(\Delta\mu/\rho)^2$. In essence, the linear approximation corresponds well with the exact Zöppritz equations even for large contrasts in *P*- and *S*-wave velocities and density, as long as the change in shear rigidity is small, i.e.

$$\frac{\Delta\mu}{\rho} \ll 1 \quad \text{or} \quad \Delta\mu \approx 0\,,$$

which occurs in shallow clastic young sediments. This can be easily demonstrated numerically.

Chapter 4

Amplitude inversion for interface geometry

Abstract   The amplitude of a reflected seismic wave is determined partly by the reflection coefficients and partly by the curvature of the reflector. The latter causes the spherical divergence of the seismic rays to be modified (focused or defocused) at the reflection point. In this chapter, we see that the information contained in amplitude versus offset data suffices to constrain accurately the geometry of an arbitrary 2-D reflector separating constant velocity layers. We use a model parameterization in which the interface is described as a discrete Fourier series with fixed upper and lower bounds on the wavenumber. The most effective inversion method is a subspace gradient algorithm, in which we allocate the model parameters to separate subspaces firstly on the basis of different physical dimensionality. We also find that, based on the sensitivity analysis, declaring separate subspaces for those parameters defining short, intermediate and long wavelength components of the interface geometry is effective in accelerating convergence and obtaining a more accurate solution.

4.1.   Introduction

Our ultimate aim is to investigate the use of simplified amplitude data in order to improve on the results of traveltime inversion. This is an intermediate step between traveltime inversion and waveform inversion. As a practical application, it uses more information than traveltime inversion to model subsurface structure, and we expect that it will provide better velocity resolution than is possible with traveltime data alone, without excessive consumption of computational time. In the present chapter, we assume that amplitudes are more sensitive to the geometry of interval reflection surfaces than they are to continuously varying velocity anomalies, and explore models containing variable geometry reflectors separating constant velocity layers. To illustrate the uses of amplitude data we initially exclude all traveltime information from the inversion. We aim to show that the amplitude inversion method has potential as a practical, flexible and robust technique, separately or in conjunction with traveltime data.

Several gradient-based inversion algorithms such as steepest descent, conjugate gradient and subspace gradient methods can be used in amplitude inversion. All of these algorithms can rapidly converge for the parameters that determine the velocity contrast at the interface, since the amplitude data is significantly influenced by the velocity contrasts (with the largest singular value of the Fréchet matrix). However, for the convergence of parameters that determine the geometry of the interface, the subspace gradient method (Kennett *et al.*, 1988) is found to be the most efficient.

The subspace method is ideally suited to problems in which the model space includes parameters of different dimensionality (e.g. velocity and depth). Kennett *et al.* (1988) applied the subspace method to a non-linear traveltime reflection problem involving *P*-wave velocity and reflector depth parameters and also to a linearized inversion of earthquake arrival times for *P*- and *S*-wave velocities and hypocentral location parameters. Further examples of the application of the subspace method were provided by Williamson (1990) with reference to the reflection problem and by Sambridge (1990) with reference to the joint hypocentre/velocity problem. Sambridge *et al.* (1991) also proposed using the subspace method in waveform inversion.

In this chapter, we explore different strategies in the subspace gradient method of amplitude inversion for the determination of interface geometry, where the interface between layers is approximated by a discrete Fourier series with fixed upper and lower bounds on wavenumber. We find that declaring separate subspaces for those parameters defining short, intermediate and long wavelength components of the interface geometry is very effective in accelerating convergence and obtaining a more accurate solution.

4.2.   Parameterization and forward modelling

We consider a 2-D stratified velocity structure consisting of variable-thickness homogeneous isotropic layers and we specify the interface $z(x)$ between two layers by a truncated Fourier series:

$$z(x) = d + \sum_{i=1}^{M} a_i \sin(2\pi k_i x + \varphi_i) , \qquad (4.1)$$

where d is the horizontal depth, a_i, k_i and φ_i are the amplitude, wavenumber and phase of the ith basis function, and M is the number of harmonic terms. We assume that each interface crosses the model from left to right without crossing another interface and without zero- and first-order discontinuities. Within each layer, the P-wave velocity α is constant. The S-wave velocity, $\beta(\alpha)$, and the density, $\rho(\alpha)$, can be evaluated by predetermined relations within each layer.

When the radius of curvature of the interface configuration is not too small, the intersections of rays and interfaces are determined by Fermat's principle. We can obtain a set of coupled equations:

$$\frac{\partial t^i}{\partial x_j^i} = 0 , \qquad \text{for} \quad \begin{cases} i = 1, \cdots, M, \\ j = 1, \cdots, K, \end{cases} \qquad (4.2)$$

where t^i is the traveltime of the ith ray, x_j^i is the x-coordinate of intersection of the ith ray and the jth interface. We can use a modified Newton method to solve this problem (Broyden, 1969).

However, sometimes the centre of curvature of the reflecting interface is found inside the layer, between interfaces. In this case we can generally use the shooting method to obtain the multiple ray paths to one receiver reflected from different points on the same interface. The shooting method is much slower than the modified Newton method for the multiple layer problem. When computing the Fréchet matrix used in the inversion procedure we need to trace ray paths for a set of perturbed models. We use the ray paths of the current model estimate, and then perturb the take-off angle to obtain the ray paths for the perturbed model.

If the ray crosses or is reflected by interfaces across which the velocity is discontinuous, the divergence of the rays is modified, depending on the velocity contrast, the angle of incidence and the local curvature of the interface. Curvature of the interface may have a large effect because of consequent focusing or defocusing of the beam. In Chapter 1 we have given an analytical expression for a modified ray geometrical spreading function $L(\ell)$, which accounts for these effects. Although the structure is considered two-dimensional, geometrical spreading of the rays in 3-D is assumed.

We assume that the principal radii of curvature of the wavefront along the interfaces are continuous in the neighbourhood of points of reflection or refraction on any interfaces, and that the radii of curvature of the interfaces are large in comparison with the acoustic wavelength so that diffraction effects can be neglected. Attenuation and inhomogeneous scattering can also be neglected.

With these assumptions, we represent the recorded amplitude in terms of the geometrical spreading function $L(\ell)$ along the ray path, where ℓ is the distance from the source, and the reflection or transmission coefficients C_i at the interfaces. That is,

$$A(\ell) = A_0 \frac{\ell_0}{L(\ell)} \prod_{i=0}^{K} C_i \,, \tag{4.3}$$

where A_0 is the ray amplitude at distance ℓ_0 from the source, and K is the number of interface intersections of a particular ray path. In a uniform medium and in the absence of attenuation, $K = 0$, $C_0 = 1$ and $L(\ell) = \ell$. The vertical component of amplitude recorded by a receiver which lies inside the medium under the Earth's surface (neglecting the free surface effect) is

$$A_{\mathrm{v}} = A\cos(\varphi_{\mathrm{I}}) \,, \tag{4.4}$$

where φ_{I} is the incident angle at the receiver. In the inversion examples below we minimize the misfit of $\log_{10} A_{\mathrm{v}}$ values.

To describe the quality of an inversion result in this chapter, let us define two measures of how well the current interface estimate z matches the synthetic model z_0:

$$\Delta z_{\mathrm{rms}} = \left[\frac{1}{X} \int_X |z(x) - z_0(x)|^2 \, dx \right]^{1/2} \,, \tag{4.5}$$

where X is the distance of ray coverage, and

$$\Delta z_{\mathrm{max}} = \max_X |z(x) - z_0(x)| \,. \tag{4.6}$$

4.3.   Subspace gradient inversion method

4.3.1   *Gradient-based inversion techniques*

Most inverse problems may be stated in terms of an optimization problem (*cf.* Tarantola, 1987b). Usually we define an objective function by the misfit function $J(\mathbf{m})$ based on the residual of the forward calculation prediction, $\mathbf{d} = f(\mathbf{m})$, relative to the observed data $\mathbf{d}_{\mathrm{obs}} = \log_{10} A_{\mathrm{obs}}$. That is,

$$J(\mathbf{m}) = [f(\mathbf{m}) - \mathbf{d}_{\mathrm{obs}}]^{\mathrm{T}} \mathbf{C}_{\mathrm{D}}^{-1} [f(\mathbf{m}) - \mathbf{d}_{\mathrm{obs}}] \,, \tag{4.7}$$

where the matrix $\mathbf{C}_{\mathrm{D}}$ is the data covariance matrix and the vector $\mathbf{m}$ is a set of unknown parameters that describe the properties of the underground medium. In the case here, the data are the discrete amplitude samples of the reflection seismic response. The aim of a geophysical inversion is to deduce an optimum model $\mathbf{m}$ which minimizes the misfit function. If $J(\mathbf{m})$ is a smooth function of the model parameters, we can make a locally quadratic approximation about some current model $\mathbf{m}$ by truncating the Taylor series for J:

$$J(\mathbf{m} + \delta\mathbf{m}) \approx J(\mathbf{m}) + \hat{\mathbf{g}}^{\mathrm{T}} \delta\mathbf{m} + \frac{1}{2} \delta\mathbf{m}^{\mathrm{T}} \mathbf{H} \delta\mathbf{m} \tag{4.8}$$

in terms of the gradient vector $\hat{\mathbf{g}}$ and the Hessian matrix $\mathbf{H}$.

The gradient of the misfit function at a given point $\mathbf{m}$ is defined by

$$\hat{\mathbf{g}} = \nabla_{\mathbf{m}} J(\mathbf{m}) = \mathbf{F}^{\mathrm{T}} \mathbf{C}_{\mathrm{D}}^{-1} [f(\mathbf{m}) - \mathbf{d}_{\mathrm{obs}}] , \qquad (4.9)$$

where

$$\mathbf{F} = \nabla_{\mathbf{m}} f(\mathbf{m}) \qquad (4.10)$$

is the matrix of the Fréchet derivatives of the wave amplitudes $f(\mathbf{m})$ with respect to the model parameters. Element (i, j) of matrix $\mathbf{F}$ represents the first-order perturbation of the ith sample of seismic amplitude data $\mathbf{d}$ due to a small perturbation of the jth parameter of model $\mathbf{m}$.

The Hessian matrix $\mathbf{H}$ of the misfit function can be calculated by

$$\mathbf{H} = \nabla_{\mathbf{m}} \nabla_{\mathbf{m}} J(\mathbf{m})$$

$$= \mathbf{F}^{\mathrm{T}} \mathbf{C}_{\mathrm{D}}^{-1} \mathbf{F} + \nabla_{\mathbf{m}} \mathbf{F}^{\mathrm{T}} \mathbf{C}_{\mathrm{D}}^{-1} [f(\mathbf{m}) - \mathbf{d}_{\mathrm{obs}}] . \qquad (4.11)$$

Since $\nabla_{\mathbf{m}} \mathbf{F} = \nabla_{\mathbf{m}} \nabla_{\mathbf{m}} f(\mathbf{m})$ appears with the data misfit, its significance should diminish as minimization proceeds and it is often neglected at the outset (Kennett *et al.*, 1988).

Introducing $\mathbf{C}_{\mathrm{M}}$ a model covariance matrix with unit (*model parameter*)^2, we get the steepest ascent vector $\boldsymbol{\Gamma}$ in the model space in terms of the gradient vector $\hat{\mathbf{g}}$:

$$\boldsymbol{\Gamma} = \mathbf{C}_{\mathrm{M}} \hat{\mathbf{g}} . \qquad (4.12)$$

The steepest descent (SD) method updates the model parameters $\mathbf{m}$ along the steepest descent direction $(-\boldsymbol{\Gamma})$. For approximately linear inverse problems, we would like to apply the conjugate gradient (CG) algorithm (Nolet *et al.*, 1986; Scales, 1987; Kormendi and Dietrich, 1991), which has been found to speed convergence at practically no extra computational cost. However, the CG method is often not as effective as the SD method for strongly non-linear problems. Both SD and CG methods ignore the differences between different parameter types. Where the model depends on parameters of different dimensionality (e.g. velocity parameters and depth parameters), applying a single step length to all parameters can result in very slow convergence.

A very effective approach known as the subspace gradient method (Kennett *et al.*, 1988; Sambridge *et al.*, 1991) is developed by restricting the local minimization of the quadratic approximation to the misfit functional J to a relatively small q-dimensional subspace of model parameters. We also attempt to apply this method in the following amplitude inversion problem. For completeness, we summarize briefly the subspace gradient approach.

4.3.2   *Subspace method*

The subspace method is analogous to the steepest descent method in choosing the step length that minimizes J in the steepest descent direction. In the subspace method, however, the steepest ascent vector $\mathbf{\Gamma}$ is partitioned into several independent subvectors and the optimal step length is chosen for each of them. Following Kennett *et al.* (1988), we introduce q basis vectors $\{\mathbf{a}^{(j)}\}$ and a projection matrix $\mathbf{A}$ composed of the components of these vectors:

$$\mathbf{A}_{ij} = a_i^{(j)}, \qquad \text{for} \quad \begin{cases} i = 1, \cdots, N, \\ j = 1, \cdots, q, \end{cases} \tag{4.13}$$

where N is the length of the basis vectors (equal to the length of the model parameter vector $\mathbf{m}$) and q is the number of subspace directions. We then construct the following perturbation to the current model in the space spanned by the $\{\mathbf{a}^{(j)}\}$:

$$\delta\mathbf{m} = \sum_{j=1}^{q} \alpha_j \mathbf{a}^{(j)} \equiv \mathbf{A}\vec{\alpha}. \tag{4.14}$$

The coefficients α_j are determined by minimizing J for this class of perturbation. That is, by setting

$$\frac{\partial J}{\partial \alpha_j} = 0, \qquad \text{for} \quad j = 1, \cdots, q,$$

we obtain the coefficient vector,

$$\vec{\alpha} = -(\mathbf{A}^{\mathsf{T}}\mathbf{H}\mathbf{A})^{-1}\mathbf{A}^{\mathsf{T}}\hat{\mathbf{g}}, \tag{4.15}$$

assuming the inverse of $(\mathbf{A}^{\mathsf{T}}\mathbf{H}\mathbf{A})$ exists. Compared to the full matrix inversion an immediate advantage of this approach is that only the $q \times q$ matrix $(\mathbf{A}^{\mathsf{T}}\mathbf{H}\mathbf{A})$ needs to be inverted. This small $q \times q$ projected Hessian matrix is generally well conditioned with judicious choices for the basis vectors $\{\mathbf{a}^{(j)}\}$, which will normally be related to the steepest ascent vector $\mathbf{\Gamma}$.

Assume the model parameters can be classified as several different parameter types, say $\mathbf{m}_I$, for $I = A, B, \cdots$. Concentrating on one class of model parameter at a time, the gradient component is

$$\hat{\mathbf{g}}_I = \nabla_{\mathbf{m}_I} J(\mathbf{m}), \qquad \text{for} \quad I = A, B, \cdots. \tag{4.16}$$

We construct the corresponding steepest ascent vector in full model space, i.e.

$$\Gamma_A = \mathbf{C}_{\mathrm{M}} \begin{bmatrix} \hat{\mathbf{g}}_A \\ 0 \\ \cdot \\ \cdot \\ \cdot \\ 0 \end{bmatrix} , \quad \Gamma_B = \mathbf{C}_{\mathrm{M}} \begin{bmatrix} 0 \\ \hat{\mathbf{g}}_B \\ 0 \\ \cdot \\ \cdot \\ 0 \end{bmatrix} , \quad \cdots . \tag{4.17}$$

The Γ_I are projection vectors of the gradient components $\hat{\mathbf{g}}_I$, for each parameter type. The basis vectors $\{\mathbf{a}^{(j)}\}$ are built up using these vectors $\hat{\mathbf{g}}_I$:

$$\mathbf{a}^{(1)} = \|\Gamma_A\|^{-1}\Gamma_A , \quad \mathbf{a}^{(2)} = \|\Gamma_B\|^{-1}\Gamma_B , \quad \cdots . \tag{4.18}$$

In the case $q = 1$ the subspace gradient method is equivalent to the steepest descent method. In the case $q = N$ (the number of model parameters), the subspace gradient method is equivalent to a Newton iteration.

4.4. A simple example of reflection amplitude inversion

4.4.1 Model definition

As a first example of reflection amplitude inversion, let us consider a simple model (earth model A) with an interface consisting of a simple harmonic function $z(x)$ separating two constant velocity layers, defined by six arbitrarily chosen parameters as specified in Table 4.1. The dataset consists, in this example, of the set of relative amplitudes recorded at the receiver locations illustrated in Figure 4.1. In the singular value analysis described below we also use, for comparison, the dataset consisting of the set of traveltimes for the same ray paths.

Traveltime inversion using reflected rays can, in principle, determine five of the model parameters: the mean depth of the interface (d), the amplitude, wavenumber and phase of the interface (a, k, φ), and the formation velocity above the interface (α_1). Traveltime data do not constrain α_2. Reflection amplitude data are, however, sensitive to the acoustic impedance contrast at the interface and therefore depend on α_1 and α_2. In the amplitude inversion procedure for this synthetic model we can attempt to invert for the six free parameters defined above, or we can arbitrarily fix one velocity value and invert for the velocity of the other layer. We consider both cases (referred to as strategies A and B below) and see which one can handle the inversion problem better.

Table 4.1.   Earth model A in which an interface is defined as a simple harmonic function separating two constant velocity layers.

interface	$z = d + a\sin(2\pi kx + \varphi)$
	where
	$d = 2000$ m
	$a = 50$
	$k = 0.2222$ km^{-1}
	$\varphi = -\pi/2$
velocity above interface	$\alpha_1 = 2500$ m/s
velocity below interface	$\alpha_2 = 2800$ m/s

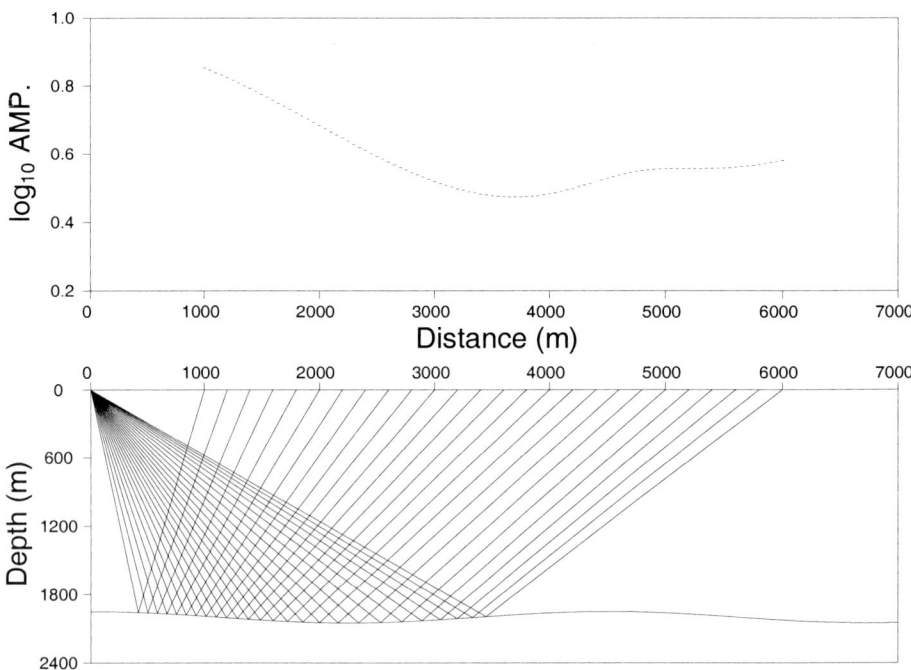

Figure 4.1.   Synthetic amplitudes and ray-path geometry for earth model A.   The absolute amplitude scale is arbitrary.

4.4.2   Singular value analysis

A useful measure of the sensitivity of amplitudes and traveltimes to velocity and reflector structure is afforded by means of singular value (SV) analysis. Following Jupp and Vozoff (1975), the parameters in a matrix inversion can be classified as important, unimportant or irrelevant, based on the order of magnitude of their SVs. "Irrelevant parameters", corresponding to zero SVs, have no influence on the predicted data values at any of the observation points. "Important parameters" corresponding to larger SVs strongly influence the model prediction. The "unimportant parameters" with smaller SVs can undergo very large changes without significant change in the predicted data values. The latter property can cause instability in an iterative inversion method and may induce model artefacts. For a non-linear inversion problem, SV analysis depends on the current estimate of the solution (which is updated at every iteration). The SVs given below are for the matrix of Fréchet derivatives evaluated at the synthetic model and do not necessarily give an accurate guide to the overall convergence rate from an arbitrary initial estimate.

For earth model A, the sequence of SVs of the matrix of Fréchet derivatives (FDs) of traveltimes with respect to the parameters, ordered in decreasing size and normalized relative to the maximum SV, is

$$\{\alpha_1, \, d, \, \varphi, \, a, \, k\} = \{1, 0.409, 0.257, 0.459 \times 10^{-1}, 0.171 \times 10^{-1}\} \, .$$
(4.19)

In contrast, the sequence of SVs of the matrix of FDs of ray amplitudes (using logarithms) with respect to the six model parameters of strategy A is

$$\{\alpha_1(\text{or } \alpha_2), \, k, \, \varphi, \, a, \, d, \, \alpha_2(\text{or}) \, \alpha_1)\}$$
$$= \{1, 0.722, 0.405, 0.317 \times 10^{-1}, 0.201 \times 10^{-1}, \, 0.603 \times 10^{-3}\} \, ,$$
(4.20)

while the sequence of SVs of the matrix of FDs of ray amplitudes with respect to the five model parameters of strategy B is

$$\{\alpha_1(\text{or } \alpha_2), \, k, \, \varphi, \, a, \, d\} = \{1, 0.969, 0.556, 0.435 \times 10^{-1}, 0.275 \times 10^{-1}\} \, .$$
(4.21)

Firstly, the comparison of SVs of the matrices of FDs of traveltimes and amplitudes shows that, in both cases, the velocity is the most "important parameter" in terms of its influence on both traveltime and ray amplitude data. From equation (4.20) we see, however, that inverting simultaneously for both velocities (strategy A) causes the matrix to be relatively ill conditioned. The matrix is presumably ill conditioned because the amplitude is sensitive to some linear combination of α_2 and α_1 but is relatively insensitive to the absolute values of α_1 and α_2. If we eliminate one velocity parameter from the inversion (strategy *B*), the condition number is improved significantly (equation 4.21). In the examples below we arbitrarily fix α_1 and invert for α_2. The choice is arbitrary in a mathematical sense, but might be justified by assuming that travel-time data would otherwise constrain α_1.

Secondly, the interface wavenumber k is the next most "important parameter" influencing amplitude data, because the geometrical focusing and defocusing caused by interface shape has a strong influence on the amplitude data. The average horizontal depth d is, on the other hand, a relatively "unimportant parameter" for reflection amplitudes. In dealing with traveltime data, however, the parameter d is the second most "important parameter" and k is a relatively "unimportant parameter". Thus, reflection amplitudes and traveltimes do indeed contain some independent information, being sensitive to different features of the model. In the following sections, we explore further the characteristics of inversions based only on amplitude data.

4.4.3   Amplitude inversion test

We now invert the synthetic amplitude data from earth model *A* (Figure 4.1, Table 4.1), using the misfit function defined by equation (4.7) (with $\mathbf{C_D}$ = constant × unit matrix). Note the focusing and defocusing of the seismic wave energy in Figure 4.1 by the systematic curvature of the interface, which modifies the otherwise monotonic decrease in amplitude with distance.

For the inversion of the synthetic data of earth model *A* with only five unknown parameters, we use the Newton method (equivalent to the subspace gradient inversion formulation of the previous section but using a separate subspace for each parameter). An initial guess for the solution is specified in Table 4.2. The result of this inversion after 50 iterations is shown by solid lines in Figure 4.2 and it accurately matches (overlies) the synthetic model, shown by dotted lines in the figure, for amplitude, reflection coefficients, traveltimes and interface geometry (Δz_{rms} = 0.11 m, Δz_{max} = 0.23 m).

Table 4.2 shows the convergence of model parameters. Inversion with different initial estimates can also converge steadily, although the convergence rates are dependent on the initial estimates. This simple example demonstrates that accurate amplitude data can be used to provide accurate interface geometry inversions—even without traveltime data.

In the amplitude inversion described above, the Fréchet derivative matrix is computed by a finite-difference method at each iteration. A practical constraint on the initial estimate is that a, the amplitude of the interface, should be non-zero in order to calculate the Fréchet derivatives with respect to k and φ. In calculating the steepest ascent vector, $\mathbf{C_M}$ is set to be the unit matrix.

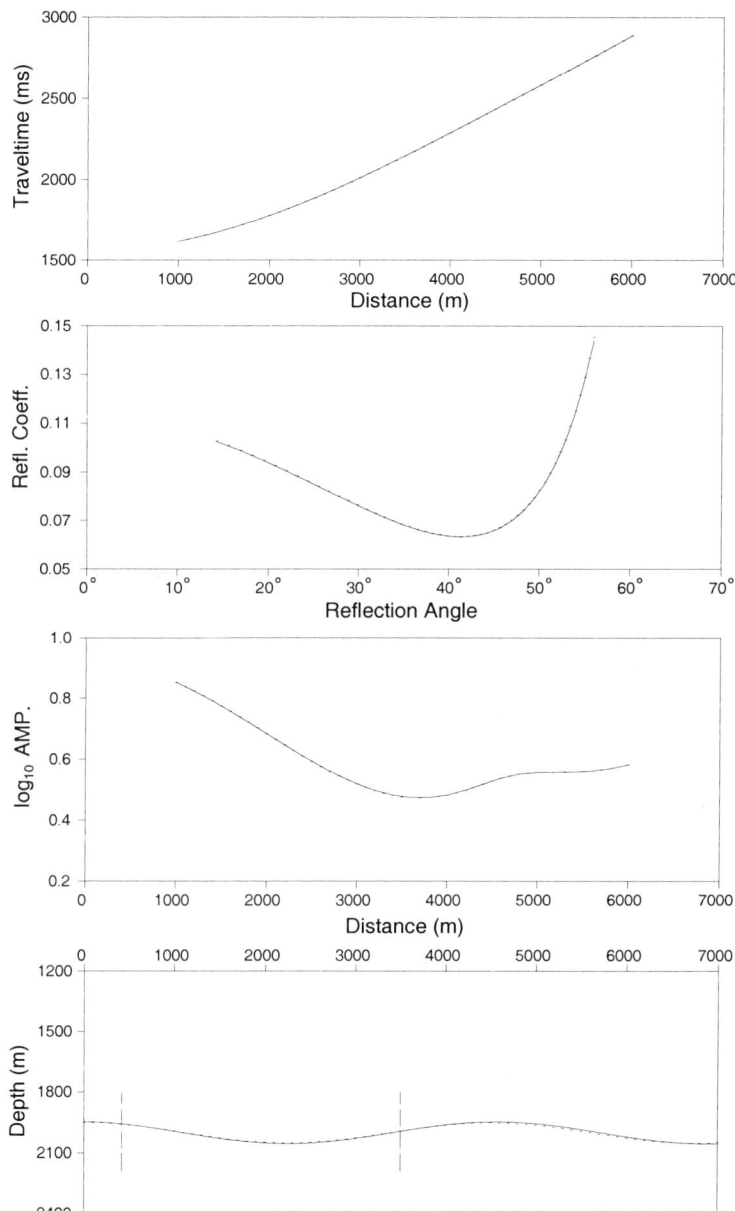

Figure 4.2. Traveltimes, reflection coefficients, amplitudes at receivers and the interface geometry for earth model *A*: comparison of the synthetic model (dotted lines) and a converged model (solid lines) after 50 iterations by seismic amplitude inversion. The dotted lines here plot on top of the solid lines and are hardly visible. Vertical marks on the reflector define the limit of ray coverage.

Table 4.2.   Convergence of model parameters for amplitude inversion of earth model A, using a 5-D Newton algorithm.

Model parameters	depth d (m)	amplitude a (m)	wavenumber k (km^{-1})	phase φ	v contrast α_2/α_1	Δz_{rms} (m)	Δz_{max} (m)
True model	2000.00	50.0	0.2222	-1.57	1.12		
Initial	1800.00	10.0	0.10	0.0	1.04	211.97	240.13
Iter. 5	2003.26	40.54	0.2419	-1.87	1.12	4.83	7.06
Iter. 10	2000.32	43.24	0.2359	-1.77	1.12	5.12	6.45
Iter. 30	1999.98	48.85	0.2243	-1.60	1.12	0.95	1.17
Iter. 50	1999.84	50.30	0.2216	-1.56	1.12	0.11	0.23

Table 4.3.   Convergence of model parameters for amplitude inversion of earth model A from two inversion strategies: (A) to invert for both α_1 and α_2, and (B) to invert for α_2, with α_1 fixed in error.

Model parameters	depth d (m)	amplitude a (m)	wavenumber k (km^{-1})	phase φ	velocity α_1	velocity α_2	Δz_{rms} (m)
True model	2000.0	50.0	0.2222	-1.57	2500.0	2800.0	
strategy A							
Initial	1800.0	10.0	0.10	0.0	2600.0	2700.0	212.0
Iter. 5	1999.6	32.9	0.2621	-2.13	2515.6	2819.9	13.6
Iter. 10	2000.0	36.6	0.2510	-1.97	2515.8	2819.6	10.4
Iter. 30	2000.5	42.8	0.2361	-1.77	2515.8	2818.7	5.2
Iter. 50	2000.5	46.8	0.2281	-1.65	2516.0	2818.4	2.1
strategy B							
Initial	1800.0	10.0	0.10	0.0	2600.0	2700.0	212.0
Iter. 5	1993.8	43.4	0.2365	-1.79	-	2910.7	11.5
Iter. 10	2000.7	44.4	0.2329	-1.72	-	2912.8	3.8
Iter. 30	2000.0	48.8	0.2244	-1.60	-	2912.1	1.0
Iter. 50	1999.7	50.7	0.2210	-1.55	-	2911.9	0.3

4.4.4   *Constraints on absolute velocity*

In the amplitude inversion above, we have fixed α_1 at the correct value and inverted for α_2, the velocity of the lower layer. Let us now consider two possible inversion strategies when the correct value of α_1 is not known. Firstly, we invert for both α_1 and α_2, with six free parameters (strategy A); secondly, we apply strategy B with five free parameters as above and α_1, set in error. Convergence of the misfit for these two strategies is shown in Table 4.3 and Figure 4.3. Figure

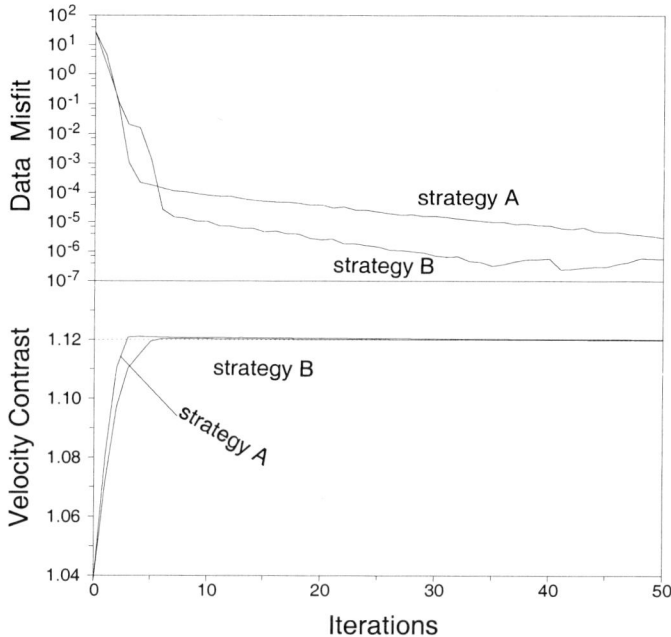

Figure 4.3. Comparison of convergence rates of amplitude inversions using strategies A and B. Strategy A inverts for both unknown velocity parameters and strategy B inverts for α_2 with α_1 fixed arbitrarily in error. The upper part of the diagram shows the misfit function and the lower part shows the velocity contrast (α_2/α_1). The thin and thick lines correspond to strategies A and B, respectively, and the dotted lines is the synthetic model.

4.3 shows the misfit F and the convergence of the velocity contrast in relation to the actual synthetic model value. We see that in both strategies the velocity contrast (α_2/α_1) of 1.12 is approximately obtained. However, the geometrical parameters of the interface (a, k, φ) are obtained more accurately ($\Delta z_{rms} = 1.0$ m rather than 5.2 m after 30 iterations) when one of the velocities is fixed, even though both velocities are systematically in error. Moreover the inversion with strategy B is faster and more stable than with strategy A and for some initial estimates we used, the convergence of strategy A failed completely.

As a final experiment in this section, we try a further inversion using strategy A with the initial estimate defined by the converged solution of strategy B after 50 iterations (Table 4.3). Given accurate interface geometry values, we wish to see whether the systematic error in velocity values could then be corrected using strategy A. However, we see no further convergence from the experiment (not shown). In summary, we conclude that strategy B provides superior results in inversions for interface geometry if we acknowledge the possibility of systematic error in interval velocities caused by the inaccurate choice of α_1. An accurate estimate for the velocity contrast (α_2/α_1) is obtained even though both velocities may be systematically high or low. The absolute value of α_1 could, of course, be strongly constrained by the addition of traveltime data, but we concentrate here on the amplitude inversion problem.

We see in the next section that amplitude inversion can produce good results for a generalized 2-D interface geometry with many parameters, and we show how the subspace method can be applied to produce more efficient inversions.

4.5.   Inversion for an interface represented as a sum of harmonic functions

4.5.1   General interface representation

Ideally we require that an interface of any specified configuration can be resolved by amplitude inversion. One way to parameterize an arbitrary interface uses discrete Fourier series. We have seen above that we can invert satisfactorily if the interface is defined by a single harmonic term of unknown wavelength. We now consider the general case of an interface defined by an arbitrary discrete Fourier series,

$$z = d + \sum_{i=1}^{M} a_i \sin(2\pi i \Delta k x + \varphi_i) \,. \tag{4.22}$$

The series consists of M terms, each with horizontal wavenumber equal to an integral multiple of the fundamental wavenumber Δk. The "$i = 1$" term corresponds to wavelength $1/\Delta k$. The series is truncated at a value of M which provides adequate resolution of the horizontal structure of the surface. In the inversion, the parameters Δk and M are assumed known *a priori*. We discuss below a strategy for the choice of these parameters in the case of an arbitrary unknown interface geometry. In the limiting case $\Delta k \to 0$ and $M \to \infty$, any interface geometry can be represented, but practical considerations compel us to minimize the number of unknown parameters a_i and φ_i.

Because the number of unknown parameters ($2M + 2$, including the mean depth d and the velocity α_2) may be large, we now use an inversion formulation based on the subspace gradient method as described above. We illustrate the method using several examples and consider different ways to allocate the unknown parameters in a subspace of limited dimensionality.

4.5.2   General interface inversion

In earth model *B* (Table 4.4), we define the interface using a sum of three harmonic terms for which synthetic amplitudes and ray geometries are shown

Table 4.4. Earth model B in which the interface is given, in general, by a sum of harmonic terms.

interface	$$z = d + \sum_{i=1}^{3} a_i \sin(2\pi k_i x + \varphi_i)$$

with

$d = 2000$ m

$a_1 = 100$ m	$k_1 = 0.1$ km^{-1}	$\varphi_1 = 0$
$a_2 = 50$ m	$k_2 = 0.2$ km^{-1}	$\varphi_2 = -\pi/2$
$a_3 = 10$ m	$k_3 = 0.4$ km^{-1}	$\varphi_3 = 0$

velocity above interface $\alpha_1 = 2500$ m/s

velocity below interface $\alpha_2 = 2800$ m/s

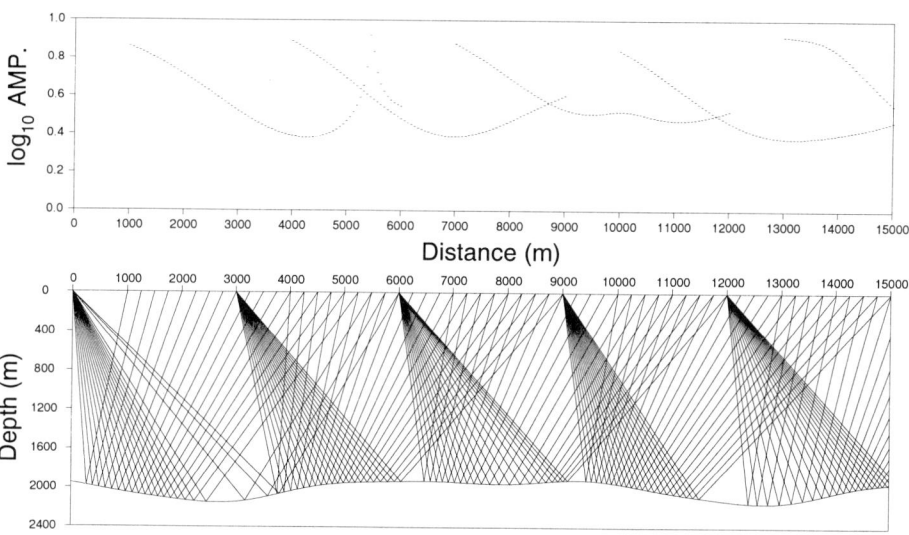

Figure 4.4. Synthetic amplitudes and ray geometries from the forward calculation of earth model B.

in Figure 4.4. The synthetic data consists of 105 observations from five source points with 21 receivers per source. The source-source and receiver-receiver spacings are 3000 m and 250 m, respectively. The ray coverage of the interface is roughly from offset 250 m to 15000 m. In the interface inversion demonstrated below, we use the subspace gradient method with a residual function defined by equation (4.7) as in the previous section. This example of intermediate difficulty

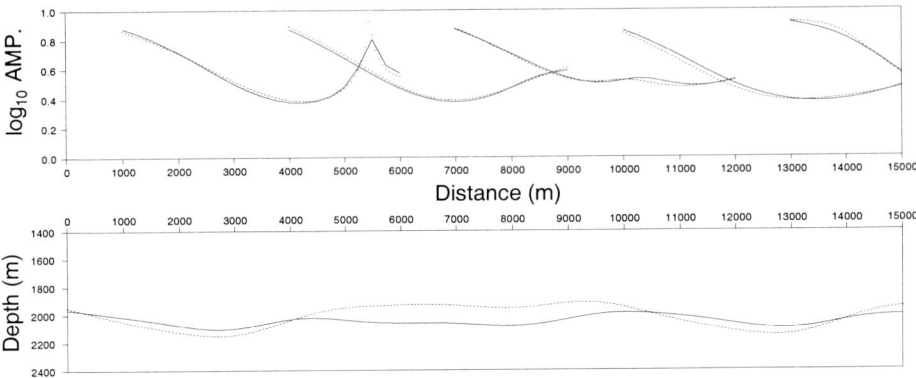

Figure 4.5. Inversion of model B after 50 iterations (solid lines), using only amplitude data and the 4-D subspace gradient approach, compared with the synthetic model (dotted lines).

will provide some insight into how best to allocate the model parameters in a subspace of limited dimensionality.

In the first attempt to invert for earth model B, we assume $\Delta k = 0.05$ km^{-1} (half the smallest non-zero wavenumber in model B) and $M = 10$. The range of wavenumbers in the parameterization includes the three terms which describe earth model B implying that an exact solution is possible. We declare a 4-D subspace based on different parameter types (from the previous section, we follow the strategy of assuming a value for α_1 and inverting for α_2):

$$d; \{a_i\}; \{\varphi_i\}; \alpha_2, \qquad \text{for} \quad i = 1, \cdots, M.$$

Figure 4.5 (solid lines) shows the result of this inversion of the synthetic amplitude data from earth model B. The comparison with the synthetic model (dotted lines) in Figure 4.5 shows that this procedure is partially effective: the amplitude curves, as well as the interface geometry are only approximately determined. The actual parameter values after 50 iterations are shown in Table 4.5 ("4-D inversion"), and there is clearly a significant error present in those wavenumbers for which the synthetic amplitude a_i is zero.

Examining the upward concave part of the interface at about 3000 m in Figure 4.5, we see that this part of the interface has been approximately reconstructed, and focusing of the reflection energy (note the shape of the amplitude curve at 5500 m) has appeared. Examination of the interface geometry shows, however, that significant errors remain. The a_i values of the basis functions with lowest wavenumber (longer wavelength) are not accurately determined, as shown in Table 4.5. The velocity value α_2 is obtained accurately after only a few iterations, as is the mean depth d, presumably because each of these parameters has its own subspace.

Table 4.5. Amplitude inversion of synthetic earth model B (50 iterations) using 4-D and 5-D subspace gradient methods with the interface modelled as a fixed wavenumber Fourier series ($\Delta k = 0.05$ km^{-1} and $M = 10$). The correct value of $\alpha_1 = 2500$ m/s is assumed known.

Model parameters	depth d	$(i = 1)$ $k_1 = 0.05$ a_1, φ_1	$(i = 2)$ $k_2 = 0.1$ a_2, φ_2	$(i = 3)$ $k_3 = 0.15$ a_3, φ_3	$(i = 4)$ $k_4 = 0.2$ a_4, φ_4	$(i = 5)$ $k_5 = 0.25$ a_5, φ_5
True model	2000.0	-	100.0, 0.0	-	50.0, $-\pi/2$	-
Initial	2200.0	1.0, 0.0	1.0, 0.0	1.0, 0.0	1.0, 0.0	1.0, 0.0
4-D inv.	2019.8	1.17, 0.066	4.94, 0.426	16.11, 0.974	27.81, -1.910	-5.15, 0.918
5-D inv.	2007.6	31.11, 1.613	106.59, 0.015	4.85, 1.151	44.49, -1.601	3.95, -1.035

Model parameters	$(i = 6)$ $k_6 = 0.3$ a_6, φ_6	$(i = 7)$ $k_7 = 0.35$ a_7, φ_7	$(i = 8)$ $k_8 = 0.4$ a_8, φ_8	$(i = 9)$ $k_9 = 0.45$ a_9, φ_9	$(i = 10)$ $k_{10} = 0.5$ a_{10}, φ_{10}	velocity contrast α_2/α_1
True model	-	-	10.0, 0.0	-	-	1.1200
Initial	1.0, 0.0	1.0, 0.0	1.0, 0.0	1.0, 0.0	1.0, 0.0	1.0400
4-D inv.	11.98, -2.969	-3.22, 0.156	7.915, -0.002	2.10, -0.635	1.79, -0.851	1.1217
5-D inv.	1.76, 1.339	0.09, 0.239	9.66, 0.034	-0.92, -0.271	0.18, 1.210	1.1196

From the data listed in Table 4.5 we can see that convergence of the amplitude and phase values of harmonic terms with higher wavenumbers (shorter wavelength) is better than those with lower wavenumber (longer wavelength) in this 4-D subspace parameterization. This observation is confirmed by singular value decomposition. Figure 4.6a shows the singular values (SV) of the Fréchet derivatives (FD) of ray amplitudes with respect to a_i (for $i = 0, \cdots, 10$ with $a_0 = d$). We can see that the misfit function J is more sensitive to those parameters defining the large wavenumber basis functions, i.e. $\{a_{10}, a_9, \cdots, a_6\}$ are relatively "important parameters" compared to $\{a_2, a_1, d\}$. In contrast, the SVs of the FDs with respect to φ_i show no such comparable trend as a function of wavenumber (Figure 4.6b).

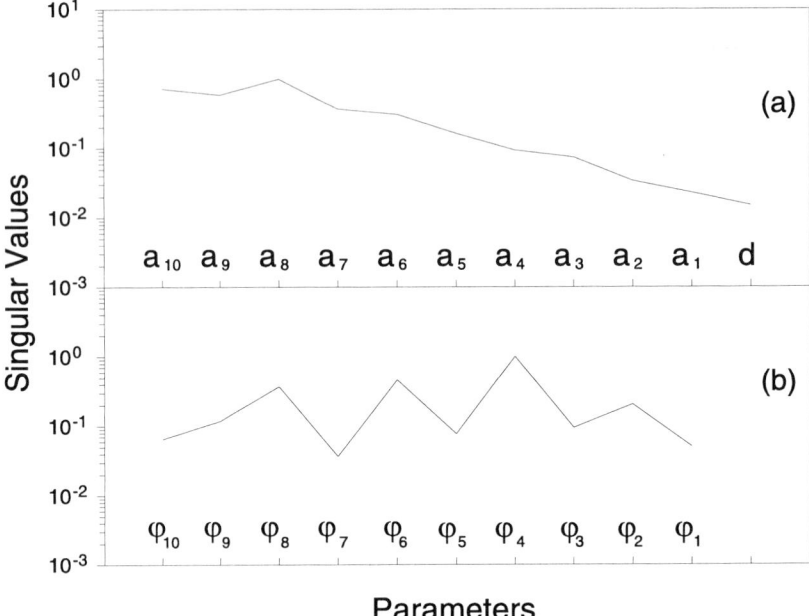

Figure 4.6.   Singular values of the Fréchet derivatives of ray amplitudes with respect to amplitude a_i and phase parameters φ_i for earth model B. For $i = 1, \cdots, 10$, the wavenumber is $k_i = 0.05i$ (km^{-1}). d is mean depth of the interface.

These considerations lead to the design of a new subspace partitioning (5-D), which recognizes the dependence of convergence properties on wavenumber, i.e.

$$\{d, \ a_1, \ a_2\}; \ \{a_3, \ a_4, \ a_5\}; \ \{a_6, \ \cdots \ a_M\}; \ \{\varphi_i\}; \ \alpha_2 \ .$$

After 50 iterations with this 5-D parameterization (earth model B, same Δk, M and initial solution estimate as in the preceding 4-D inversion), we see much improved convergence, as shown in Figure 4.7: the derived amplitude curves and interface geometry (solid lines) closely match the synthetic model (dotted lines). Table 4.5 shows the accuracy of parameter values obtained with this 5-D subspace inversion, compared with those from the previous 4-D inversion.

Although Table 4.5 shows that there remains considerable error in the estimated amplitude values a_i of some basis functions, particularly those with lower wavenumbers (e.g. $k = 0.05$, 0.10), Figure 4.7 shows a satisfactory inversion result. Figure 4.8 shows the progress of convergence of the interface geometry in this inversion, where the dotted lines show the "true" interface configuration of the synthetic model (earth model B) and the solid lines are current estimates of the solution. Although for the initial estimate $\Delta z_{\mathrm{rms}} = 197.7$ m, after four

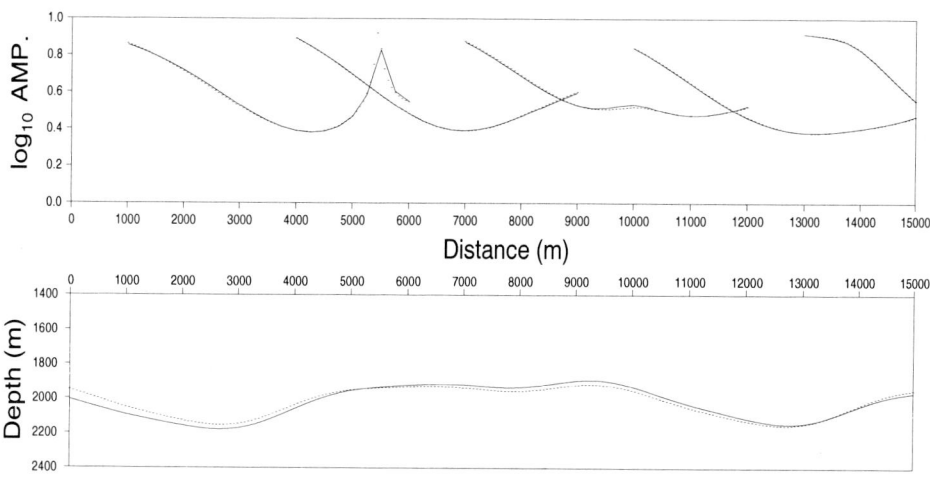

Figure 4.7. Interface inversion after 50 iterations (solid lines), using only amplitude data from the synthetic model B (dotted lines) and the 5-D subspace gradient approach.

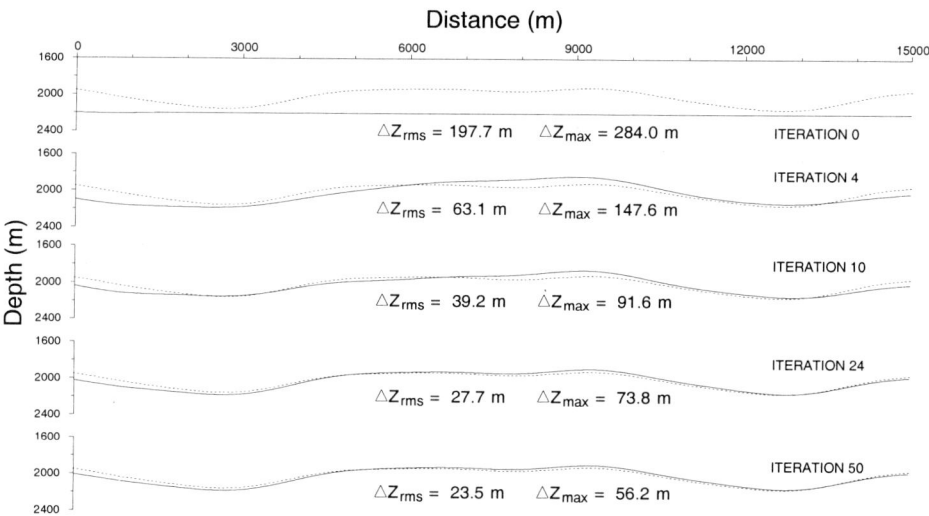

Figure 4.8. Convergence of the interface inversion using the 5-D subspace gradient method. The current estimate is shown as a solid line and the synthetic model as a dotted line.

iterations $\Delta z_{rms} = 63.1$ m, and finally after 50 iterations $\Delta z_{rms} = 23.5$ m and the largest errors are evidently near the limits of ray coverage.

Figure 4.9 shows the dramatic improvement of convergence from 4-D to 5-D subspace parameter allocations. We can see that the dimensionality of

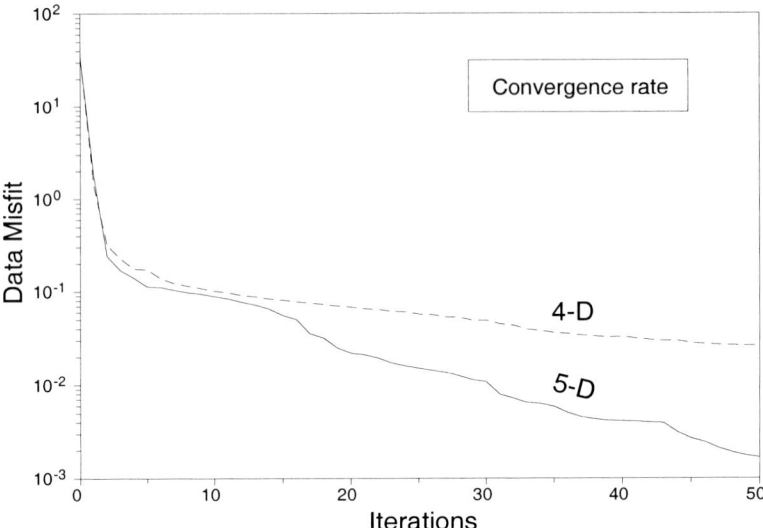

Figure 4.9.   Comparison of convergence rates between 4-D and 5-D subspace gradient inversion of earth model B.

the subspace used for the inversion is an important consideration. However, another major consideration is the inversion parameterization using a set of fixed wavenumber components of the interface and the choice of Δk and M in that parameterization. We consider this problem in section 4.7 after describing in the next section the stability of the above amplitude inversion example with respect to both poor initial estimate and data noise.

4.6.   Stability of the amplitude inversion

To test the dependence on the initial estimate, we repeat the above 5-D subspace inversion with several different initial estimates. In each case a flat reflector at a depth between 1200 and 2400 m is used (note $d = 2000$ m in model B). Figure 4.10 shows the interface configurations of the synthetic model and solution estimates obtained after 50 iterations using the 5-D subspace gradient method with different initial estimates. The figure shows that if the initial estimate of interface depth is shallower than that of the true model, the inversion converges faster than in the case where the initial estimate is deeper than the actual interface. Inversions with initial interface depth estimates of 1100 m and 2500 m, get stuck in local minima and are not satisfactory (not shown). However, satisfactory convergence can be obtained over a relatively wide range of initial interface depth estimates (1200-2400 m).

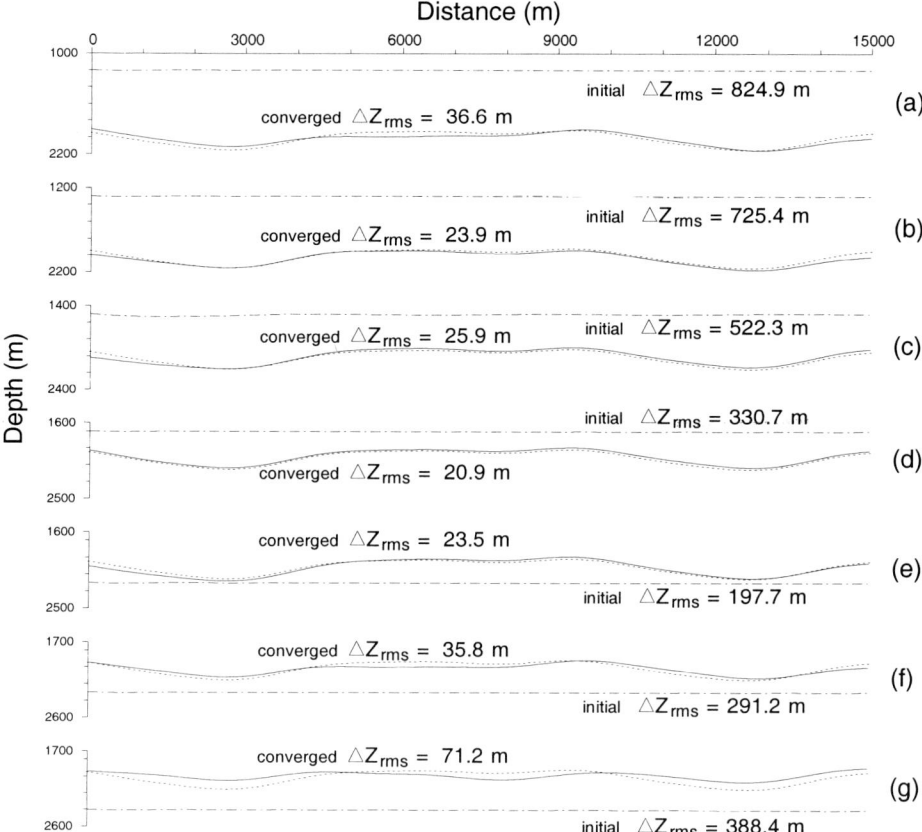

Figure 4.10. Inversion solutions (after 50 iterations) of the interface configuration with different initial estimates, using the 5-D subspace gradient method. The initial estimates of interface depth are (a) 1200, (b) 1300, (c) 1400, (d) 1700, (e) 2200, (f) 2300 and (g) 2400 m, respectively, shown as dashed lines. The solid lines and dotted lines correspond to estimated solutions and the synthetic model, respectively.

The stability of the amplitude inversion procedure in the presence of data noise can also be tested. Let us repeat the 5-D subspace inversion of earth model *B* data as described in the previous section, with synthetic noise added to the synthetic data. Figure 4.11 shows the synthetic data, random noise signals with white spectrum, and the synthetic data with noise added. If we assume the magnitude range of synthetic amplitudes is between $10^{0.3}$ to $10^{0.9}$ (2 to 8) arbitrary units, the amplitude range of the data noise is ± 0.6 in the same units.

Even with added noise the 5-D subspace inversion of model *B* quickly converges to the velocity α_2. The interface configuration is also obtained approxi-

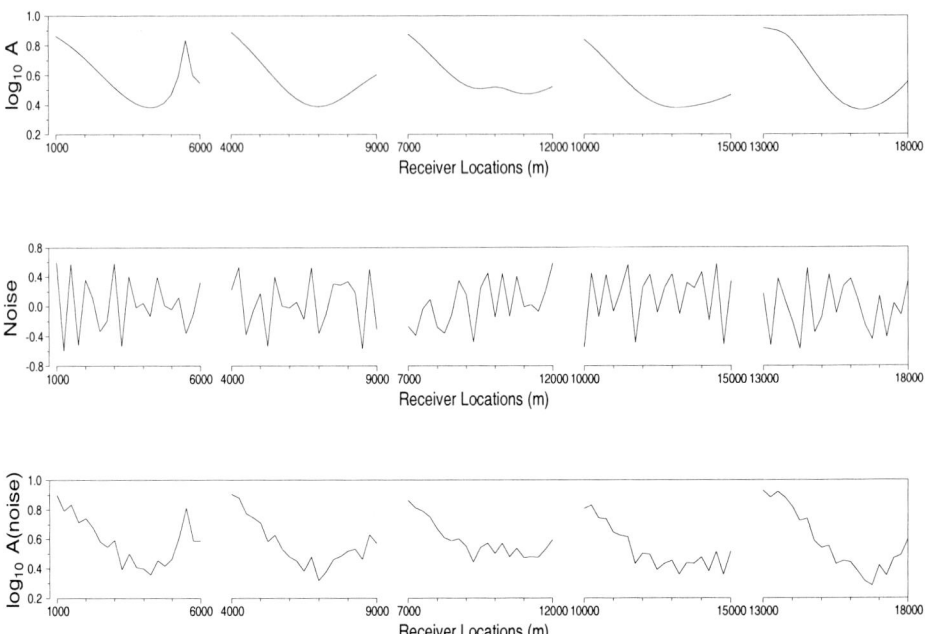

Figure 4.11.   Examples of synthetic amplitudes from earth model B ($\log_{10}$ scale), random noise signals (linear scale) and the synthetic amplitudes with those noise signals added and used as input data ($\log_{10}$ scale) for the amplitude inversion. Absolute units are arbitrary as the problem is linear in source amplitude. Each receiver record was corrupted by a unique uncorrelated noise record.

mately, as shown in Figure 4.12. The "rms" difference of actual and estimated interfaces after iteration 50, Δz_{rms}, is 37.9 m, almost 50% greater than Δz_{rms} for the same inversion without data noise. This test shows that amplitude inversion with the interface modelled as a fixed wavenumber Fourier series is stable even in the presence of significant data noise.

4.7.   Strategy for the choice of Δk and M

4.7.1   The choice of Δk and M

The interface defined by earth model B has a power spectral density defined by 3 terms which can be exactly represented in the inversion parameterization used above. In general the unknown interface will have, however, a continuous spectrum which must be approximated by our inversion parameterization, and therefore the quality of any estimated solution is constrained by the choice of

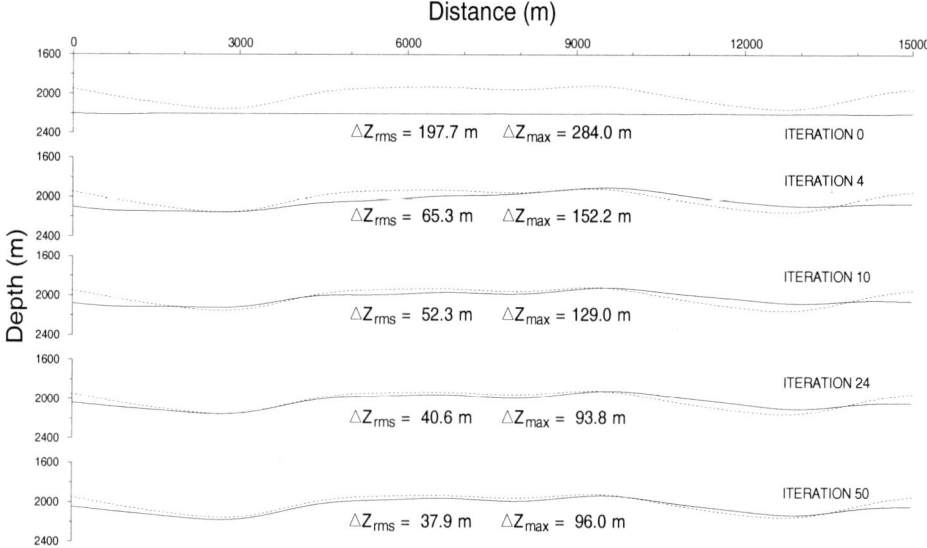

Figure 4.12. Convergence of the interface of model B in the presence of data noise using 5-D subspace gradient inversion. The input of the inversion is the synthetic data with added noise. Dotted and solid lines correspond to the synthetic model and the current estimate respectively (compare Figure 4.8).

Δk and M. We now propose the following strategy to ensure that appropriate values of Δk and M are used in the inversion.

Step 1: preliminary estimate of interface geometry. We need firstly a rough estimate of the power spectral density of the interface geometry in order to set Δk, which determines the longest wavelength that can be represented in the solution. The interface geometry could be approximately established using zero-offset traveltime data. Alternatively, we can use amplitude data in conjunction with the simplest ray geometrical spreading model.

If the amplitude of the received signal is assumed to be inversely proportional to the ray path length for a particular shot, then the following approximate relationship between amplitude A_i and receiver offset y can be obtained for shot i:

$$\frac{d}{dy}(\log_{10} A_i) = -\frac{y}{y^2 + 4h_i^2} \log_{10} e , \qquad (4.23)$$

where h_i is the depth to the interface at shot point i and we neglect slope and curvature of the interface. From the slope of the observed $d(\log_{10} A_i)/dy$ at the near offset, equation (4.23) can be inverted to give an estimate of h_i near each shot i. Joining up the set of h_i estimates for the different shots using straight

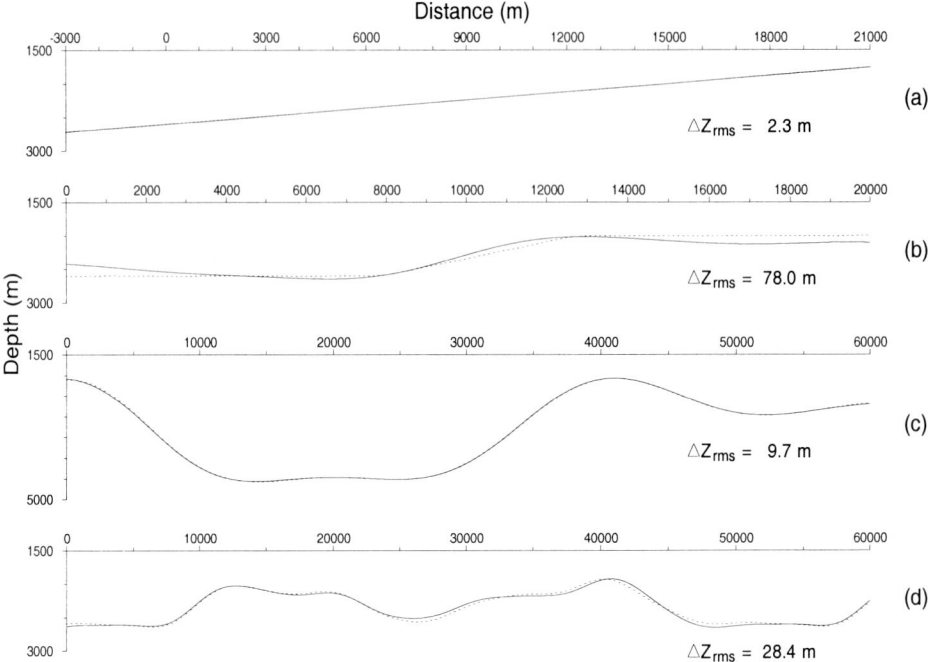

Figure 4.13.   Four examples of amplitude inversion using 5-D subspace gradient inversion. The solution of inversion (after 50 iterations) are shown as solid lines, compared with the synthetic models shown as dotted lines. In each case $M = 10$ and Δk was chosen using the method described in the text.

line segments, we obtain a preliminary estimate of the interface geometry which may be used as an initial estimate in the subsequent amplitude inversion.

The use of measured $d(\log_{10} A_i)/dy$ values avoids the problem of variation in shot energy and shot to ground coupling and allows best-fit estimates to be used in the presence of receiver noise.

Step 2: estimation of Δk. The power spectral density of the interface can be estimated using the maximum entropy method applied to the above preliminary estimate. Δk should be set small enough to resolve any wavenumber component for which the estimated power is a significant component of the solution. In the examples below (Figure 4.13), and in most physically relevant examples, the power spectral density of the preliminary estimate is characterized by a number of local maxima (Figure 4.14, solid lines). In the examples below we choose to set Δk equal to half the wavenumber at which the first local maximum ($k > 0$) in the power spectral density occurs.

Step 3: inversion with $M = 10$. The preliminary estimate of interface geometry gives no indication of the required short wavelength resolution. Using M

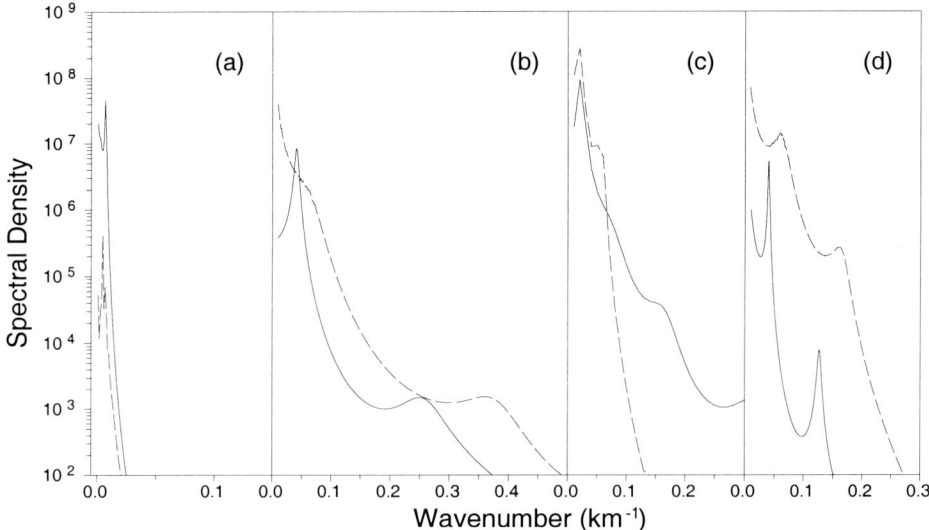

Figure 4.14. Comparison of power spectral densities of interface geometries of preliminary estimates (solid lines) and synthetic models (dashed lines) of the four structures illustrated in Figure 4.13 (dotted lines). These spectral densities were computed using the maximum entropy method with 20 poles.

= 10 permits a full order of magnitude variation of the represented wavelengths and is proposed here as the basis for the first inversion cycle.

Step 4: double M and repeat inversion. If the $M = 10$ inversion has not adequately resolved the short wavelength structure of the reflector, inversion with $M = 20$ should show an improved data misfit. In principle, step 4 could be repeated iteratively until there ceases to be any improvement in the data misfit.

4.7.2   Examples

Let us now test the above strategy on the four examples shown in Figure 4.13. These synthetic examples may be described as (a) a constant gradient ramp, (b) a monocline, (c) and (d) hand-drawn syncline/anticline structures, and each has a continuous power spectrum (Figure 4.14, dashed lines). In these examples the inversion strategy appears to be quite satisfactory using only the first three steps above. In Figure 4.13, the solutions after 50 iterations (solid lines) are compared to the synthetic models (dotted lines). In each case $M = 10$ and the Δk (km^{-1}) are, respectively, (a) 0.005, (b) 0.02, (c) 0.01 and (d) 0.02, as derived from the preliminary estimate spectra shown in Figure 4.14.

In the two examples with the worst misfit (Figures 4.13b and 4.13d), we also apply step 4 of the inversion procedure, using $M = 20$. In these cases the convergence parameters after 50 iterations compare as follows:

(b) $M = 10$: $\Delta z_{rms} = 78.0$ m, $F = 8.592 \times 10^{-2}$;

      $M = 20$: $\Delta z_{rms} = 91.5$ m, $F = 5.314 \times 10^{-2}$;

(d) $M = 10$: $\Delta z_{rms} = 28.4$ m, $F = 2.577 \times 10^{-3}$;

      $M = 20$: $\Delta z_{rms} = 23.6$ m, $F = 1.787 \times 10^{-3}$.

In both of these cases, the $M = 20$ inversion gives a relatively small improvement in data misfit, but Δz_{rms} is worse for the case of (b). A significantly better result can be found when this example is repeated with $\Delta k = 0.04$ km^{-1} and $M = 10$ ($\Delta z_{rms} = 50.4$ m, $F = 5.366 \times 10^{-2}$). The optimum strategy of choosing Δk and M could thus be improved with further experimentation. The power spectral densities of synthetic models shown in Figure 4.14 (dashed lines) indicate why the inversions for (a), (c) and (d) are quite satisfactory using only $M = 10$.

For the above inversions with $M = 20$, the question of subspace parameter allocation again arises. Without extensive testing, we may follow the logic of our previous result and chose to partition the 42 parameters into seven different subspaces, based on wavenumber groupings as

$$\{d, a_1, a_2\}; \ \{a_3, a_4, a_5\}; \ \{a_6, ..., a_{10}\}; \ \{a_{11}, \cdots, a_{15}\}; \ \{a_{16}, \cdots, a_{20}\};$$
$$\{\varphi_i\}; \ \alpha_2 .$$

The $M = 20$ inversions require approximately twice the computation time of the $M = 10$ inversions, because we calculate the Fréchet derivative matrix **F** at each iteration.

4.8.   Discussion

In this chapter we have demonstrated that amplitude data can be used to constrain effectively the subsurface geometry of a 2-D reflector separating two constant velocity layers. Singular-value decomposition shows that amplitude data contain information that is complementary to traveltime data. Where possible, both datasets should be used, but in this chapter, we investigate the effectiveness of amplitude-only inversion.

The amplitude calculation used to generate synthetic data is based on ray theory, with the model parameterized as a 2-D homogeneously isotropic layered velocity structure with zero attenuation. These simplifying assumptions are appropriate here where the aim is to demonstrate the use of amplitude inversion, and to design a suitable model parameterization and subspace parameter allocation. Each of these simplifying assumptions can, in principle, be lifted, and should be the subject of future development.

Amplitude data provide a strong constraint on the velocity contrast of the interface (α_2/α_1) but only weakly constrain the absolute values of α_1 and α_2. The best strategy for dealing with this ambiguity is to assume an *a priori* value for α_1 and invert for unknown α_2 and interface geometry parameters. The result of this procedure may be a systematic error in α_1 and α_2, but the value of α_2/α_1 and interface geometry parameters can be accurately determined. Including traveltime data also removes the ambiguity.

We have seen that a parameterization of the general unknown interface using discrete Fourier series can be effectively used in the interface inversion, provided the range of wavenumbers $\{ k_i = i \Delta k, \text{ for } i = 1, \cdots, M \}$ is adequate to represent the interface geometry. The set of unknown model parameters in the resulting inversion consists then of the Fourier amplitude and phase coefficients $\{ a_i, \varphi_i,$ for $i = 1, \cdots, M \}$, the mean depth d and the unknown velocity α_2. Satisfactory results have been obtained for a series of synthetic examples with smoothly varying, geologically relevant, interface geometries.

For the inversions we have used a subspace gradient inversion method which is based on a local quadratic approximation of a misfit function between the calculated and observed $\log_{10}$(amplitude) data. There is, however, considerable flexibility (even ambiguity) concerning the partitioning of the model vector into subspaces. An initial conclusion is that parameters of different dimensionality and different order of magnitude of SVs of the Fréchet derivative are best put in different subspaces. Tests have shown that a poor choice of subspace partitioning can cause the inversion to converge very slowly or to get stuck in a local minimum. The most effective partitioning of the vector of unknown model parameters separates the amplitude variables into separate subspaces on the basis of short, intermediate and long wavelength Fourier components. Significant errors in individual interface description parameters appear to partially cancel each other out so as to give satisfactory inversion of the interface as a whole, as determined by the maximum and rms differences between inversion results and the synthetic model.

The inversion method is relatively stable in the presence of data noise and for a range of initial estimates. Inclusion of traveltime data will produce better constrained results, since amplitude and traveltime data are sensitive to different features of a model and both inversions are complementary. We see the combined use of traveltime and ray amplitude data as offering a cost-effective improvement on current traveltime inversion methods for reflection seismic data, without resorting to the computationally expensive strategy of waveform inversion.

Chapter 5

Amplitude inversion for velocity variation

Abstract   In this chapter we focus on the amplitude inversion for 2-D velocity variations. By comparing amplitude perturbations arising from slowness perturbations along the whole ray path with those arising from the slowness perturbation close to the interface, we see that the data residuals have most effect on velocity anomalies near the interface in an inversion of reflection seismic amplitude data. We demonstrate the efficacy of amplitude inversion for velocity variation, using the subspace inversion method, with a 2-D Fourier series parameterization of the slowness distribution, and find that the inversion efficiency lies in a judicious partitioning of model parameters into subspaces. A stable strategy for the parameter partitioning is to separate parameters on the basis of the magnitude of rms values of the Fréchet derivatives of ray-amplitudes with respect to the model parameters. We see from numerical examples that the amplitudes of reflected signals are sensitive to the location of the velocity anomalies, and that amplitude data contain information that can constrain unknown velocity variation.

5.1.   Introduction

In the previous chapter, we explored the use of amplitude data to invert models containing variable geometry reflectors separating constant velocity layers. In this chapter, we investigate the tomographic inversion of amplitude data for 2-D continuously varying velocity with a known reflector geometry. We use a model parameterization in which the subsurface velocity distribution consists of layers within which the velocity varies continuously, separated by surfaces across which the velocity changes discontinuously. In the examples, we invert for an unknown velocity distribution in a single layer, bounded below by a horizontal reflection surface (planar). Within a layer the velocity distribution is parameterized using a 2-D Fourier series.

A numerical comparison shows that the amplitude of a reflected wave is more sensitive to the slowness perturbation in the vicinity of the reflection point than to a comparable perturbation at any other point on the ray path. Therefore, even though some quite good results have been obtained from inversion of the amplitude data of direct seismic arrivals (e.g. Nowack and Lutter, 1988), it is difficult to reconstruct interval velocity variation from the amplitudes of reflected arrivals because of this property. Singular value analysis indicates that amplitude data are most sensitive to the short-wavelength Fourier components of the velocity model.

We then present examples of inversion for velocity variation of synthetic reflection seismic amplitude data, using the subspace inversion method. The subspace method is ideally suited to the problem in which the model space includes parameters of different dimensionality. As we saw in the previous chapter, however, even when all parameters have the same physical dimension, appropriate partitioning of different parameter components into separate subspaces may be effective in accelerating convergence and obtaining an accurate solution. In the amplitude inversions presented in this chapter, we partition the parameters into different subspaces on the basis of the different sensitivities of the ray-amplitude values with respect to the model parameters. We demonstrate the application of reflection seismic amplitude inversion for velocity variation using synthetic datasets obtained from a range of simple models with 2-D velocity variation.

5.2.   Amplitude dependence on slowness perturbation

5.2.1   *Parameterization of slowness variation*

For ray-amplitude calculations the model slowness distribution must vary smoothly within a layer. In a medium with smoothly varying slowness, any ray is composed of an arc with continuously varying curvature. Traveltime and its derivatives with respect to the model parameters can be calculated, and if the ray-tube around the reference ray smoothly diverges, calculation of the ray-amplitude is also stable.

For the ray-tracing calculations used to generate the synthetic data, we represent a variable slowness field by a set of discrete slowness values on a regular 2-D grid. We then determine the slowness within the cells defined by the grid points using bicubic interpolation, so that the values of the function and the specified derivatives change continuously as the interpolating point crosses from one grid cell to another. Bicubic interpolation requires us to specify at each grid point not just the slowness function $u(\mathbf{x})$, where $\mathbf{x}$ is the Cartesian coordinate, but also the gradient $\partial u/\partial x_i$ and the cross derivative $\partial^2 u/\partial x_i \partial x_j$. We can determine these derivatives at each grid point centred finite difference of the values of u on the grid.

While ray tracing requires a densely sampled slowness distribution, the inversion procedure must be designed to obtain solutions for long wavelength slowness variation, and avoid the computational instability that can occur where short length scale variations in slowness are permitted. Therefore, in the inversions described below, we use the following truncated 2-D Fourier series representation of the slowness field within a given layer rather than the cell representation frequently used in traveltime tomography:

$$u = \sum_{i=0}^{N} \sum_{j=0}^{N} [a_{ij}\cos(i\Delta k\pi x)\cos(j\Delta k\pi z) + b_{ij}\cos(i\Delta k\pi x)\cos(j\Delta k\pi z)$$

$$+ c_{ij}\cos(i\Delta k\pi x)\cos(j\Delta k\pi z) + d_{ij}\cos(i\Delta k\pi x)\cos(j\Delta k\pi z)] ,$$

$$(5.1)$$

where a, b, c and d with subscripts i and j are amplitude coefficients of the (i, j)th harmonic term and N is the number of harmonic terms in each dimension. We define the wavenumbers in the series as integer multiples of the fundamental wavenumber Δk. For a given velocity field, we also can use equation (5.1) to construct the regular mesh of constant slowness values used in ray tracing.

At an interface between layers, the assumption of a smooth interface (i.e. the existence of its partial derivatives of the first and second order) is necessary in order to calculate the transformation of the ray and paraxial rays. Energy partitioning due to reflection or transmission at the interfaces and the effects of focusing and defocusing of the ray tube at interfaces are the crucial factors in the determination of ray-amplitude in a reflection seismogram. In Chapter 2 we derived the expression for the complete transformation of the ray-tube on a curved interface, but for the inversion examples here, we assume that the reflection surface is horizontal and planar, in order to separate the influence of reflector curvature and internal velocity variation.

5.2.2 Linearized approximation of amplitude in a simple example

Before undertaking amplitude inversion for velocity variation, we first investigate the dependence of amplitude on the perturbations to the slowness field, so as to gain some insight into how the inversion procedure is affected by the reflection configuration. In our tests of amplitude inversion, we use $\log_{10}$ of the vertical component of the displacement amplitude at the surface recorder. The perturbation of the ray-amplitude (using logarithms) due to the model perturbation may be expressed as

$$\Delta(\log_{10}A) = \log_{10}\left(\frac{D_0}{D}\right) + \log_{10}\left(\frac{C}{C_0}\right) , \qquad (5.2)$$

in terms of the ray geometrical spreading function D and the product of reflection and transmission coefficients C, where variables with subscript "0" correspond to those in the unperturbed reference medium.

For the simple case of a model with constant slowness distribution (the initial estimate used in the following inversions), the following linearized approximation of $\log_{10}(D_0/D)$ can be obtained (Wang and Houseman, 1995):

$$\log_{10}\left(\frac{D_0}{D}\right) \approx \frac{\log_{10}e}{(\sigma - \sigma_0)}\int_{\sigma_0}^{\sigma}\frac{\Delta u(\tau)}{u_0(\tau) + \Delta u(\tau)}d\tau , \qquad (5.3)$$

where Δu is the slowness perturbation along the ray path from σ_0 to σ. We now consider a two-layer structure with a smoothly varying interface on which the ray is reflected towards the receiver on the earth's surface, and assume the slowness difference between the media above and below the interface to be $u_1 - u_2$. A first-order estimate of $\log_{10}(C/C_0)$ can be obtained as

$$\log_{10}\left(\frac{C}{C_0}\right) = \log_{10}e\left[\frac{\Delta u_k}{(u_1 - u_2)}\right] , \qquad (5.4)$$

where Δu_k is the slowness perturbation near the reflection point in the incident medium.

Note the special significance of velocity anomalies adjacent to the reflection point (u_k). We now compare the magnitudes of $\log_{10}(D_0/D)$ and $\log_{10}(C/C_0)$. If we assume a constant slowness perturbation along the whole ray path across a reference model with an average slowness $\bar{u}_0$, the relative influence on the ray-amplitude perturbation of the two factors, given by equations (5.3) and (5.4), can be estimated as

$$\frac{\Delta E_1}{\Delta E_2} \equiv \frac{\log_{10}(D_0/D)}{\log_{10}(C/C_0)} = \frac{u_1 - u_2}{\bar{u}_0} . \qquad (5.5)$$

If $\bar{u}_0 = 300$ ms/km and $u_1 - u_2 = 3$ ms/km, then $\Delta E_1/\Delta E_2 = 1/100$.

An important conclusion can be drawn from equation (5.5): that the perturbation of ray-amplitude depends more significantly on the slowness perturbations near the reflection point. Therefore, in an inversion of reflection seismic amplitude data, the data residuals will have most effect on velocity anomalies near the reflecting interface. If we use the common model parameterization of dividing a velocity structure into rectangular cells, we can anticipate that a straightforward inversion algorithm would cause slowness anomalies within the layer to appear concentrated in the velocity cells adjacent to the reflector.

5.2.3   Singular value analysis

Using the model parameterization of slowness in terms of 2-D Fourier series, we now try to show which Fourier components of the model can be better resolved by the amplitude inversion. In a linearized iterative inversion procedure, the inversion problem is characterized by the Fréchet matrix. Singular value analysis of the Fréchet matrix is a useful measure of the sensitivity of the model response to model parameters. In Chapter 4, we used it for the analysis of the reflection seismic amplitude inversion problem for interface geometry. The singular value analysis of the Fréchet matrix is also informative in the case of amplitude inversion for slowness variation.

In the singular value analysis, we set $N = 5$ in the slowness distribution (equation 5.1), but the accompanying Figure 5.1 shows only the 36 slowness parameters $\{a_{ij}, \text{ for } i,j = 0, 1, \cdots, 5\}$ (the cosine-cosine coefficients). The Fréchet derivatives are evaluated in the solution space (with a constant background slowness). The eigenvectors and associated singular values (SVs) of the Fréchet matrix of the ray-amplitudes with respect to these 36 slowness parameters a_{ij} are shown diagrammatically in Figure 5.1, in order of decreasing magnitude from left to right. The SVs, shown in the upper graph, are normalized relative to the maximum SV. The corresponding eigenvectors, shown diagrammatically below the graph, span the parameter space. In each eigenvector square, the wavenumbers of the Fourier components increase from left to right (horizontal wavenumber), and from top to bottom (vertical component). Darker tone shows that a particular component of an eigenvector is greater in magnitude.

From Figure 5.1 we can see that the ray-amplitude data are most sensitive (i.e. the singular values are greatest) for those Fourier components with shorter wavelengths in both x- and z-directions. Components with longer wavelength in both x- and z-directions have only small SVs and it is therefore more difficult to invert for them. Because many of the eigenvectors are sensitive to a broad range of wavenumbers in the z-direction, the ability of an inversion procedure to resolve the influence of different vertical wavenumber components in the model is relatively weak.

5.3.   Inversion algorithm

In the following sections, giving examples of amplitude inversion for velocity variation, we assume that the actual reflection interface is known *a priori*. The

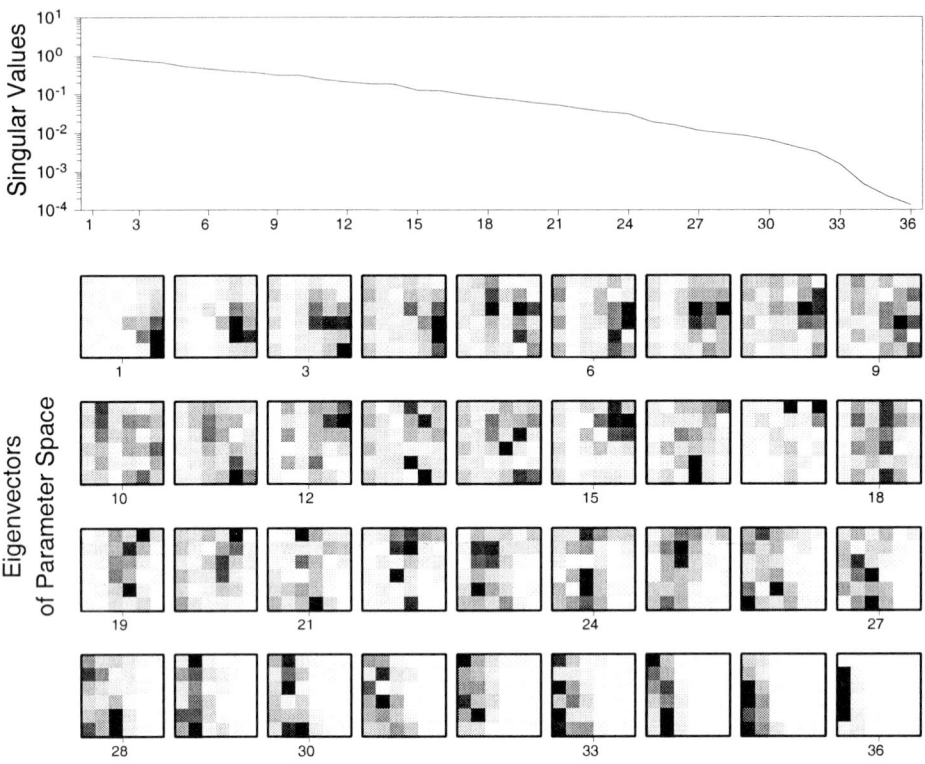

Figure 5.1. Singular value analysis of the Fréchet matrix of ray amplitudes with respect to model parameters for the slowness distribution defined by equation (5.1) with $N = 5$. Only the 36 cosine-cosine coefficients a_{ij} are represented. See text for interpretation.

free parameters to be determined in the inversion are the coefficients of the basis functions (equation 5.1) which are referred to as the "model" $\mathbf{m}$. The dimension of the model space (allowing for the components which are zero when $i = 0$ or $j = 0$) is

$$M = 1 + 4N(N + 1) = 1 + 8 \sum_{n=1}^{N} n . \tag{5.6}$$

The inverse problem is reduced to finding a vector, $\mathbf{m} \in R^M$, which adequately reproduces the observations, $\mathbf{d}_{\text{obs}}$.

The subspace method (Kennett *et al.*, 1988) is used again in the following examples to invert for the velocity variation. The inversion is implemented iteratively. At each iteration, the model update may be expressed in terms of the gradient vector of data misfit and the Hessian matrix as

$$\delta\mathbf{m} = -\mathbf{A}(\mathbf{A}^T\mathbf{H}\mathbf{A})^{-1}\mathbf{A}^T\hat{\mathbf{g}} , \qquad (5.7)$$

where the "−" sign indicates that the model will be updated along the steepest descent directions.

The success or failure of a subspace approach hinges upon a judicious selection of the spanning vectors for the activated subspace. For the general case of M model parameters described in equation (5.1), we allocate them systematically into N_{sub} subspaces:

$$N_{sub} = 5 + \sum_{n=3}^{N} n . \qquad (5.8)$$

When $N = 2$, $N_{sub} = 5$. The subspaces contain respectively (and arbitrarily)

$$3, 4, 5, 6, 7, 8, \cdots, \text{and } 8$$

model parameters. The basis vectors $\{\mathbf{a}^{(j)}\}$ are constructed in terms of components of the steepest ascent of data misfit corresponding to those parameters.

Although the number of subspaces and the number of parameters allocated in each subspace are factors influencing the efficiency of the inversion (e.g. Oldenburg *et al.*, 1993), the major factor will be how to group the parameters within each subspace. In restricting the dimension of the model space for each iteration, it may occur that vectors which are important in finding the global minimum of the desired objective function are not available and convergence to the solution is slowed or prevented. In the examples presented below, following Chapter 4, we partition model parameters into separate subspaces on the basis of sensitivity of the amplitude data to variation of the model parameters as defined by the matrix of Fréchet derivatives. In the linearized iterative inversion, as we know, the data residual influences the update of the model parameters at a rate that depends on the sensitivity of the data to those parameters. Thus it is desirable to partition the model parameters into several subspaces based on the magnitude of their influence on the output data.

It is difficult to measure the sensitivities for each model parameter individually, as seen in Figure 5.1. An empirical quantity indicating the sensitivity of the amplitude with respect to a model parameter is the rms value of the Fréchet derivatives for the complete set of observations. In the iterative inversion procedure we recalculate the matrix of the Fréchet derivatives of ray-amplitudes with respect to the model parameters at each iteration. We also regroup the model parameters for every subspace at each iteration, depending on the changes to the Fréchet derivatives calculated at each iteration. Using the reflection seismic amplitude data from several synthetic models, we now demonstrate the effectiveness of this strategy in the inversion of amplitude data for velocity variation.

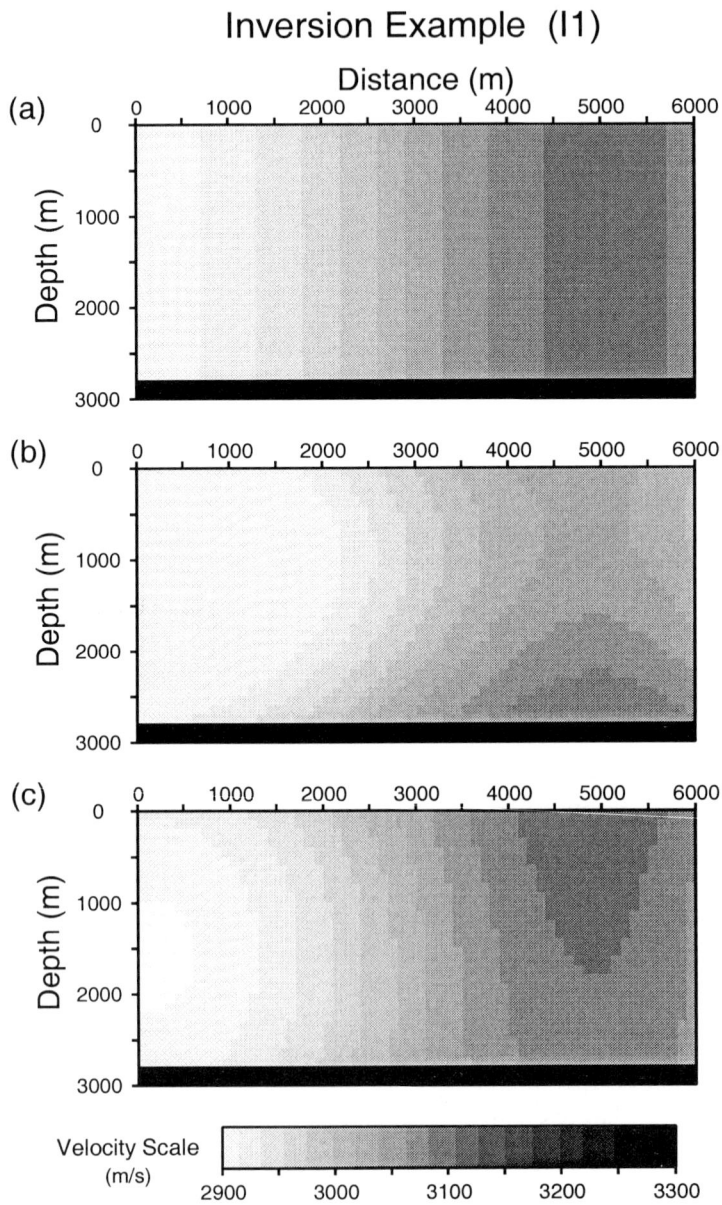

Figure 5.2. Inversion example I1: (a) the synthetic model with 1-D velocity variation in a horizontal direction; (b) the inversion solutions after five iterations and (c) after 50 iterations.

5.4.   Inversion example of 1-D slowness distribution

We now demonstrate the application of amplitude inversion using synthetic amplitude data, firstly obtained from models with a variable slowness distribution given by harmonic functions overlying a reflector, and then obtained from an arbitrary model with localized velocity anomalies which include Fourier components at all wavenumbers.

All data sets of synthetic reflection amplitudes are generated with the reflection configuration illustrated in Figure 2.2 and defined as follows: 40 receivers spaced at intervals of 50 m, for each of 6 shots at the free surface, with a horizontal shot-separation of 1000 m. The ray coverage at the reflector is roughly between 0 and 6000 m on the horizontal coordinate. A total of 240 synthetic "observed data" is used in the amplitude inversion.

We first consider two examples of synthetic models with 1-D velocity variation and investigate the inversion's ability to recover slowness variation in the horizontal and vertical directions. Figure 5.2a shows one model (referred to as inversion model I1) which has a slowness variation in the horizontal direction defined by

$$u = a_0 + b_0\cos(\pi k_0^x x) , \tag{5.9}$$

where $a_0 = 333.3$ ms/km, $b_0 = 10$ ms/km and $k_0^x = 0.2$ km^{-1}. The velocity below the interface, which is simply represented as a flat plane at 2800 m depth, is 3300 m/s. Although this model is 1-D, we adopt a 2-D model parameterization defined by the Fourier series of equation (5.1) and ask whether an inversion of the synthetic data can recover the original velocity distribution if the interface geometry is assumed known. We set the wavenumber increment $\Delta k = 0.1$ km^{-1} and $N = 2$ in equation (5.1). Equation (5.6) shows that there are 25 parameters to be estimated in this inversion.

The inversion procedure requires an initial estimate in which we set the velocity (α_1) of the top layer constant. It is obtained from a simple traveltime estimate, $\alpha_1 = \ell/t$, where t is the traveltime and ℓ is the length of the ray path. For a receiver at near offset h with a horizontal planar reflector, we have $\ell = (h^2 + 4d^2)^{1/2}$, where d is the depth to the interface. The velocity below 2800 m is also assumed known. For the inversion of the synthetic amplitude data from model I1, we consider an initial model with constant velocity $\alpha_1 = 2980$ m/s (and $\alpha_2 = 3300$ m/s).

The result of the inversion after 5 and 50 iterations is shown in Figures 5.2b and 5.2c. We can see that the inversion procedure first recovers the slowness variation near the interface. This observation is consistent with the conclusion drawn from equation (5.5) above, i.e. in an inversion of reflection seismic amplitude data, the data residuals have most effect on velocity anomalies near the interface.

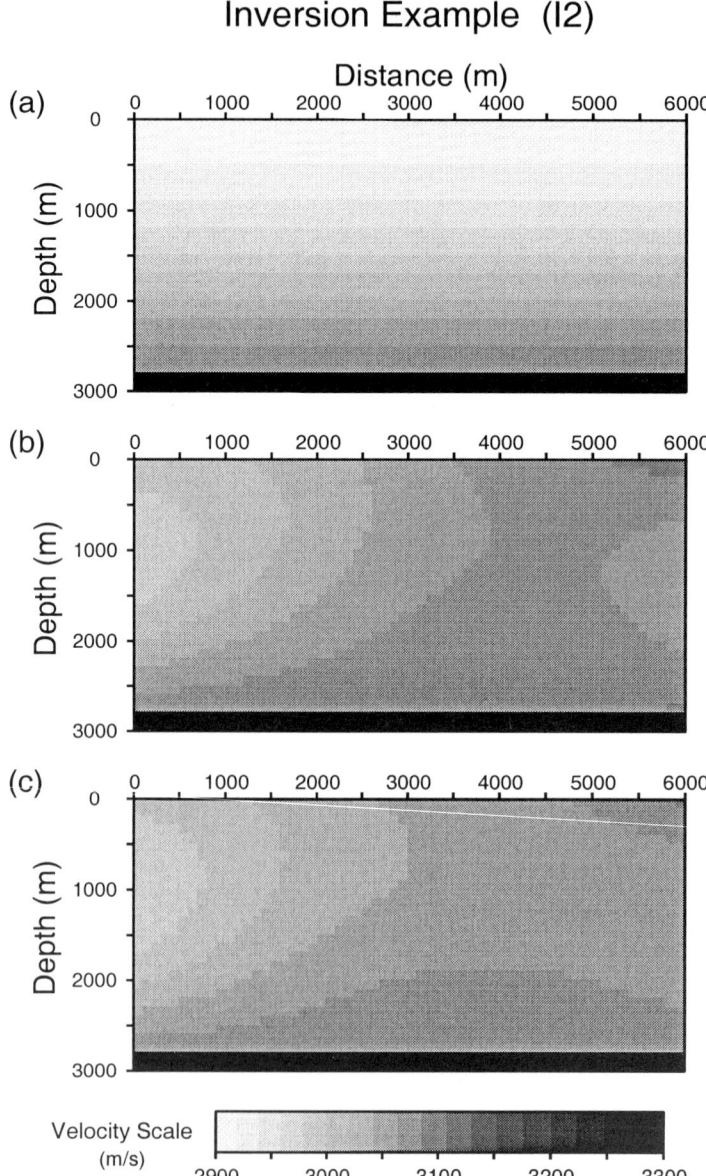

Figure 5.3. Inversion example I2: (a) the synthetic model with 1-D velocity variation in the vertical direction; (b) the inversion solutions after five iterations and (c) after 50 iterations.

Figure 5.3 shows a second example (model I2), this time with 1-D velocity variation in the vertical direction, given by

$$u = a_0 + b_0 \cos(\pi k_0^z z) , \qquad (5.10)$$

where

$$a_0 = 333.3 \text{ ms/km}, \quad b_0 = 10 \text{ ms/km} \quad \text{and} \quad k_0^z = 0.3 \text{ km}^{-1}.$$

In the inversion of model I2, we use the model parameterization of equation (5.1) with $\Delta k = 0.15$ km^{-1} and $N = 2$, but the same initial model estimate as used in the inversion of model I1. Figure 5.3 shows a comparison of the synthetic model (a), with inversion solutions after five iterations (b) and after 50 iterations (c). This time there is no variation along the interface but the inversion first recovers the slowness contrast across the interface.

Comparing Figures 5.2 and 5.3 we would conclude that amplitude inversion with the reflection configuration is better able to reconstruct velocity variation in the horizontal direction than variation in the vertical direction. However, the following examples show that such a conclusion is over simplistic.

5.5. Constraining higher wavenumber components

Two other synthetic examples of 1-D velocity variation with higher wavenumber are shown in Figures 5.4 and 5.5. One model (I3) has a slowness variation in the horizontal direction defined by equation (5.13) and the other (I4) has slowness varying in the vertical direction defined by equation (5.14) but with $a_0 = 333.3$ ms/km, $b_0 = 6.67$ ms/km and $k_0^x = k_0^z = 0.7$ km^{-1}.

In the inversions of models I3 and I4, we use the model parameterization of equation (5.1) with $\Delta k = 0.35$ km^{-1} and $N = 2$. In Figures 5.4 and 5.5 we compare (a) the synthetic models, (b) the inversion solution after five iterations and (c) the inversion solution after 50 iterations for these two examples. We see that both inversions have recovered the velocity variations with Fourier components at high wavenumber, while model I4 is better than model I3. In the inversion example I3, the velocity variation in the horizontal direction is recovered but an artificial signal with variation in the vertical direction is also introduced.

Ideally all Fourier components with any higher wavenumber can be constrained in the inversion if those components are included in the model parameterization. However, we would not expect to constrain the higher wavenumbers in the inversion without an increased spatial density of data sampling. One constraint on the maximum wavenumber of the slowness distribution that can be adequately resolved should derive from the receiver spacing.

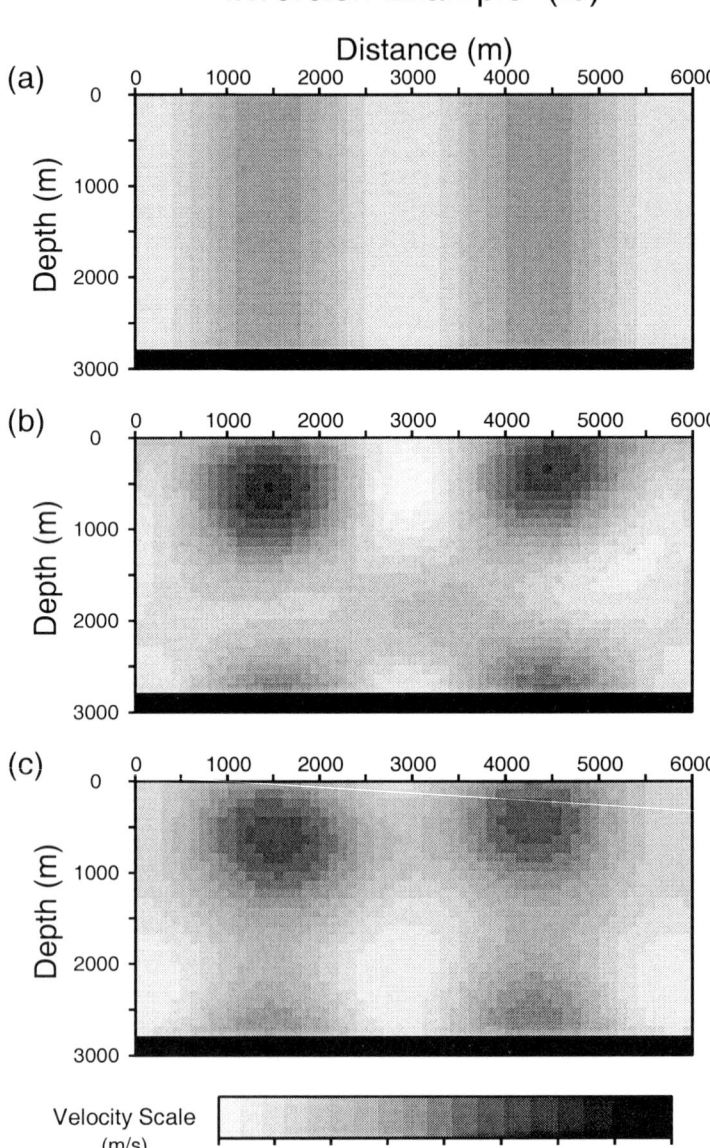

Figure 5.4. Inversion example I3: (a) the synthetic model with 1-D velocity variation and a high wavenumber component (0.7 km^{-1}) in a horizontal direction; (b) the inversion solutions after five iterations and (c) after 50 iterations.

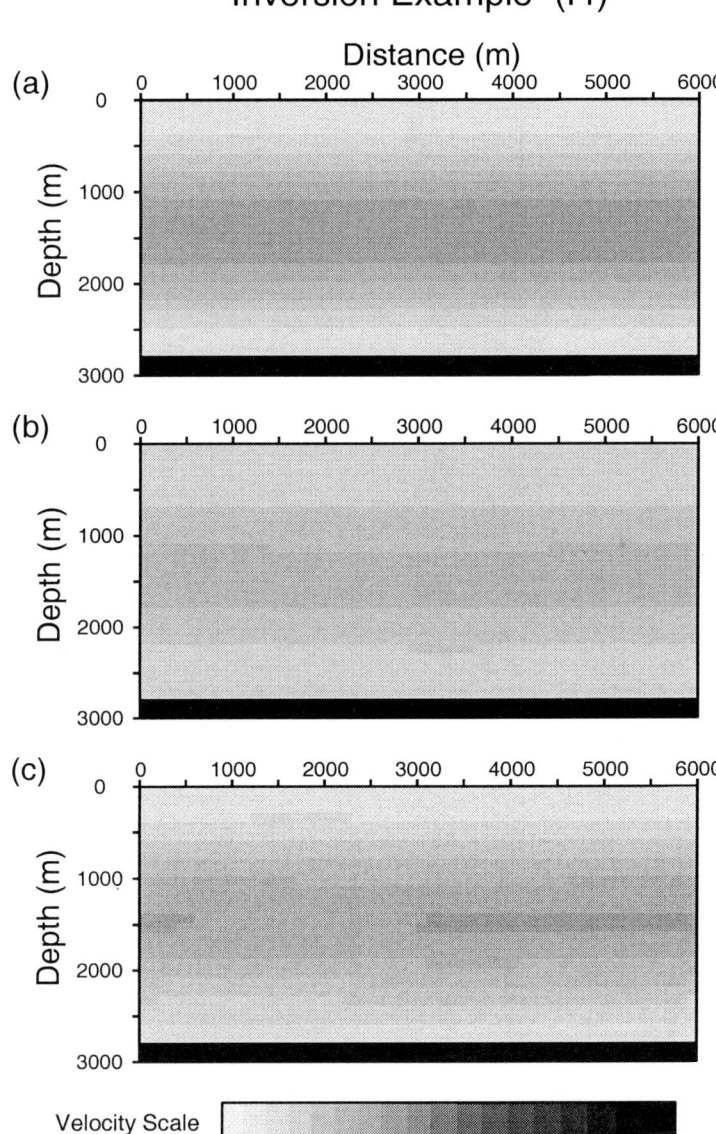

Figure 5.5. Inversion example I4: (a) the synthetic model with 1-D velocity variation and a high wavenumber component (0.7 km^{-1}) in the vertical direction; (b) the inversion solutions after five iterations and (c) after 50 iterations.

Inversion
Example
(I5)

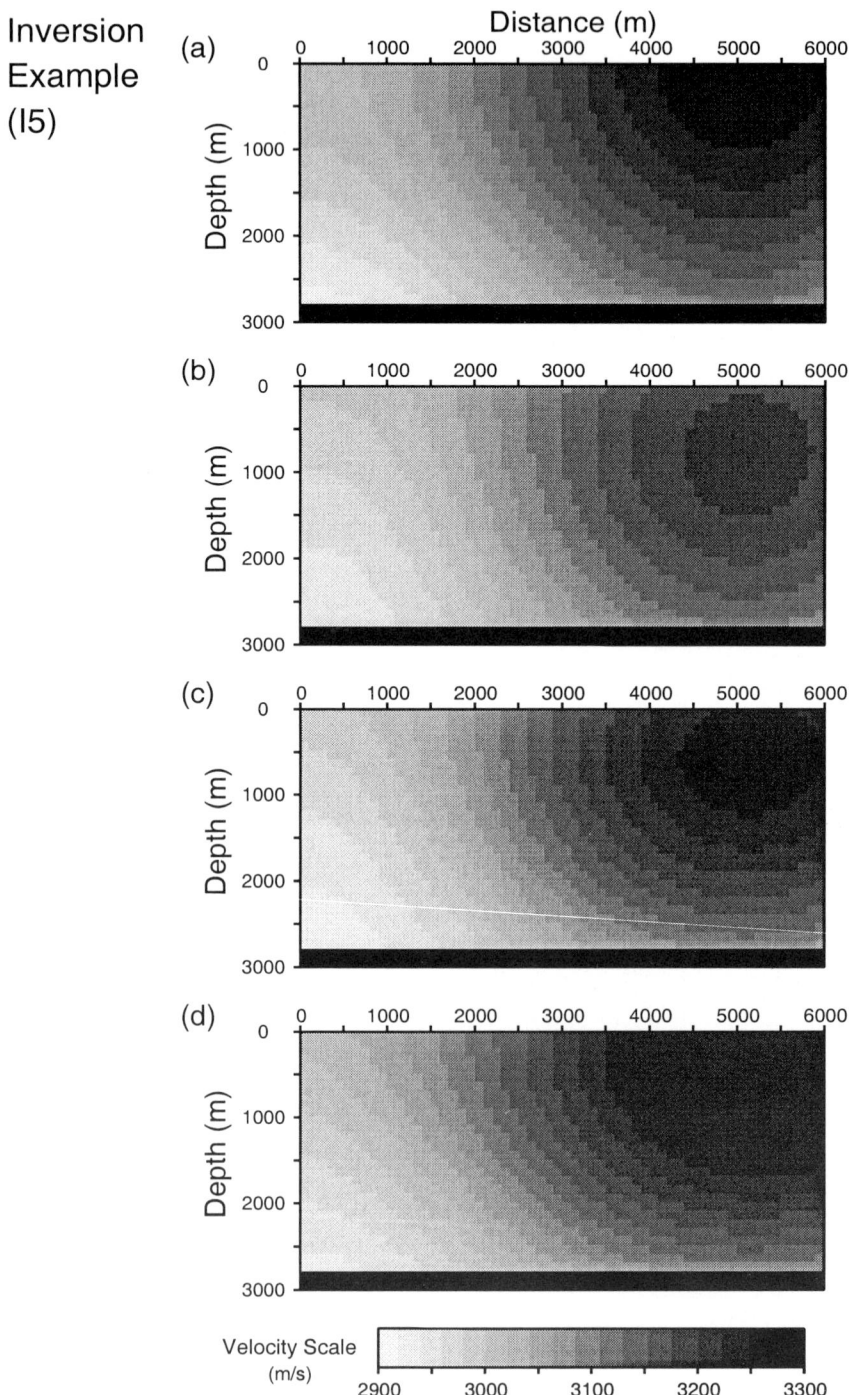

Figure 5.6. Inversion example I5: (a) the synthetic model with a 2-D variable slowness distribution given by harmonic function (equation 5.15); (b) the inversion solutions after five iterations,
(c) after 20 iterations and (d) after 50 iterations.

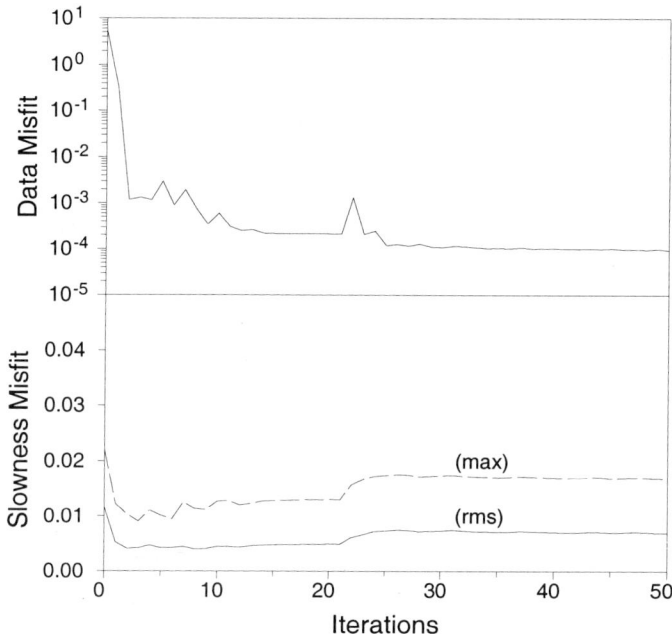

Figure 5.7. Convergence rate of amplitude inversion I5, shown by data misfit defined by equation (5.7) and the rms and (absolute) maximum differences between the synthetic model and the current estimate in the inversion.

5.6. Robustness of the inversion in the presence of model error or data noise

We now consider an example (I5) with a 2-D variable slowness distribution given by harmonic functions, i.e.

$$u = a_0 + b_0\cos(\pi k_0 x) + c_0\cos(\pi k_0 z) + d_0\cos(\pi k_0 x)\cos(\pi k_0 z) \,,$$
(5.11)

with a_0, b_0, c_0 and d_0 equal to 333.3, 10, -10 and 2 (ms/km), respectively, and k_0 = 0.2 km^{-1} (Figure 5.6a). In this inversion we use the model parameterization of equation (5.1) with $N = 3$ and $\Delta k = 0.1$ km^{-1}. From equation (5.6) there are 49 parameters to be estimated. The solutions of the inversion after 5, 20 and 50 iterations are shown in Figures 5.6b–d. Compared with the synthetic model we can see that the inversion recovers the main features of velocity variation within the layer. Figure 5.7 shows the convergence rate of the inversion by means of the data misfit and slowness differences (rms and max). We see that the inversion is stable with this subspace partitioning strategy. Unfortunately, this example shows that although our final estimate has the minimum data misfit, the model misfit has slightly deteriorated since the estimate obtained after 20 iterations.

Inversion
with reflector
depth in error

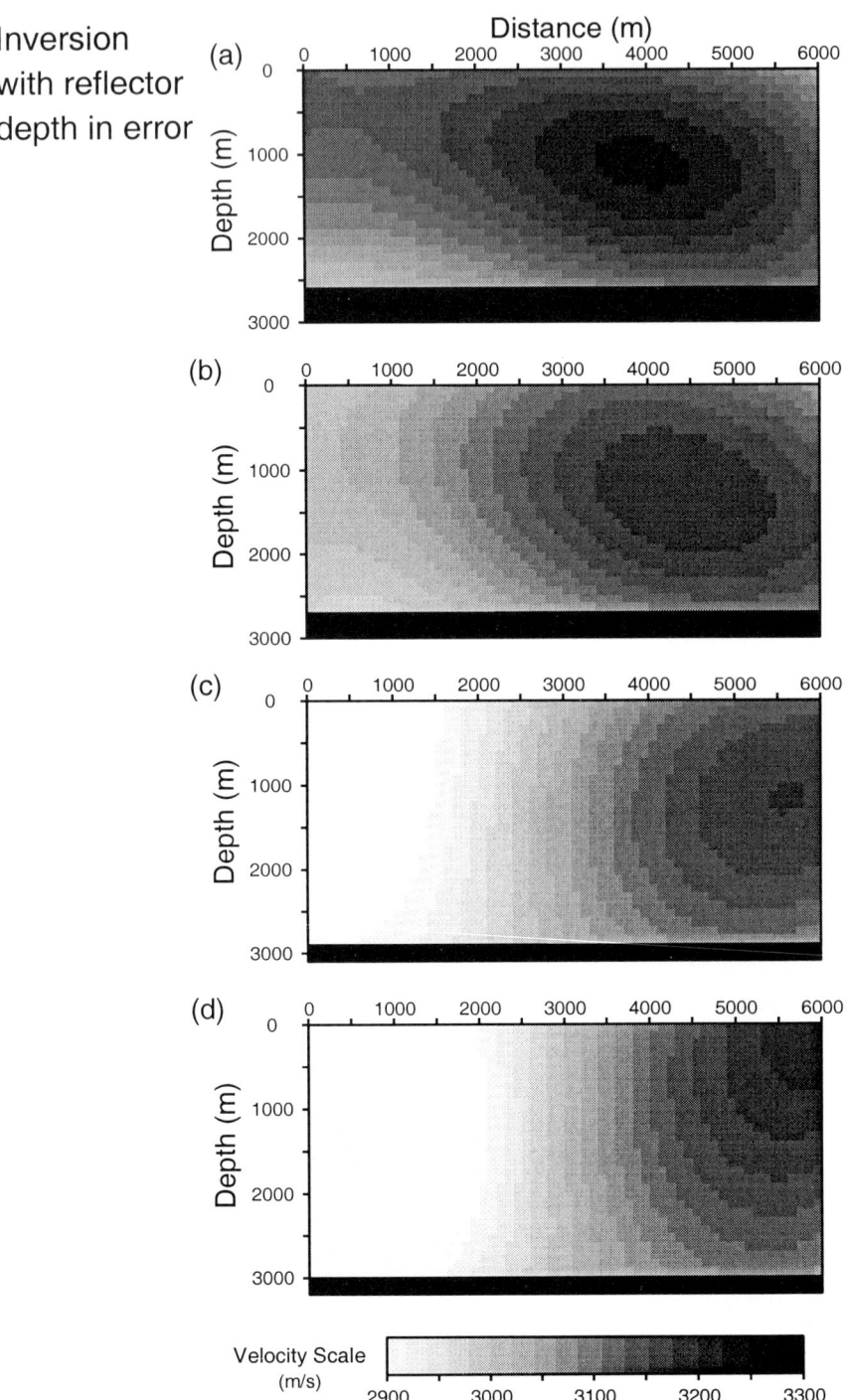

Figure 5.8. Inversion tests (synthetic model I5) with the reflector depth set in error, where (a), (b), (c) and (d) are the inversion results (after 20 iterations) with the reflector depth at 2600, 2700, 2900 and 3000 m, respectively.

In the amplitude inversions for the velocity variation above we have assumed that the depth of the reflector is known *a priori*. We now test the stability of the inversion if the mean depth of the interface (actually at 2800 m) is given in error. Figure 5.8 shows the inversion solutions, after 20 iterations, for assumed reflector depths at (a) 2600 m, (b) 2700 m, (c) 2900 m and (d) 3000 m, respectively. Comparing Figure 5.8 with Figure 5.6a, the synthetic model, and Figure 5.6c, the inversion estimate after 20 iterations with the correct reflector depth, we see that this error introduces a major distortion to the inverted velocity variation. When the reflector is too shallow, the horizontal variation of the velocity field is artificially suppressed and when it is too deep we see that the horizontal variation of the field is artificially enhanced. Clearly it is important for the velocity inversion that reflector depth (and geometry) are accurately constrained. While the amplitude signal also contains information about reflector depth (Chapter 4), the possibility of simultaneous inversion of amplitude data for both interval velocity and reflector geometry has not yet been explored.

To test the robustness of the method in the presence of data noise, we simply repeat the inversion example I5, with the synthetic amplitude data modified by the addition of 1%, 3% and 5% white noise. The S/N ratio here is defined as

$$\nu = \left(\sum_i n_i^2 \bigg/ \sum_i s_i^2 \right)^{1/2}, \tag{5.12}$$

where n_i is the noise amplitude and s_i is the signal amplitude (no logarithms). The range of synthetic amplitudes s_i is between $10^{0.20}$ and $10^{0.75}$ (1.5–5.5) arbitrary units, the amplitude ranges of 1%, 3% or 5% noise n_i are ±0.08, ±0.2 or ±0.4 in the same units.

Figure 5.9 shows the inversion results after 20 iterations. The amplitude inversion of the data with 1% noise added (Figure 5.9a) converges very well. In the inversions of data with 3% and 5% noise added, although both inversions have recovered approximately the velocity distribution in the area close to the reflector, the vertical velocity variation is poorly represented in these inversion solutions. The inversion of the data with 3% noise (Figure 5.9b) has determined the general shape of the model feature but overestimated the amplitude of variation. The inversion with 5% noise (Figure 5.9c) appears to have suppressed the vertical velocity variation.

5.7. Inversion of arbitrary smooth velocity anomalies

In the above examples (I1–I5) the actual solution could be represented exactly using the model parameterization. We now consider two more complex cases in which the model velocity distribution cannot be represented exactly using

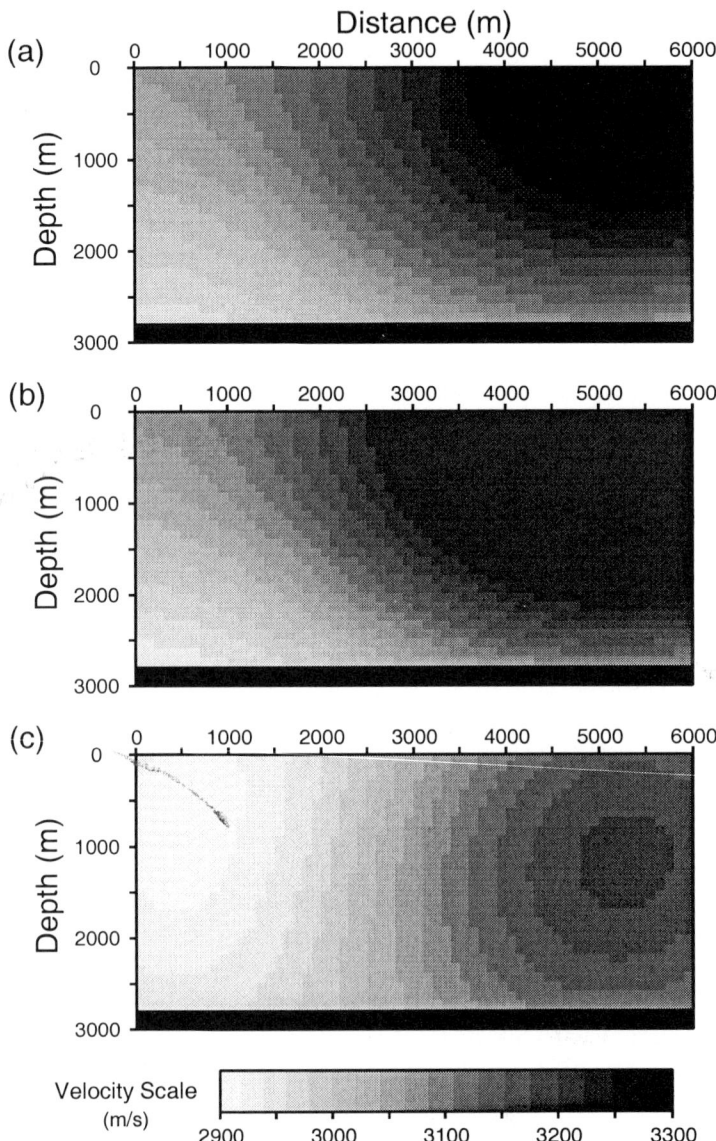

Inversion Models
(from data with 1%, 3% & 5% white noise)

Figure 5.9. A test of the inversion method in the presence of data noise, where (a) (b) and (c) are the inversion results (after 20 iterations) of the synthetic amplitude data from example I5 with 1, 3, and 5 per cent white noise added, respectively.

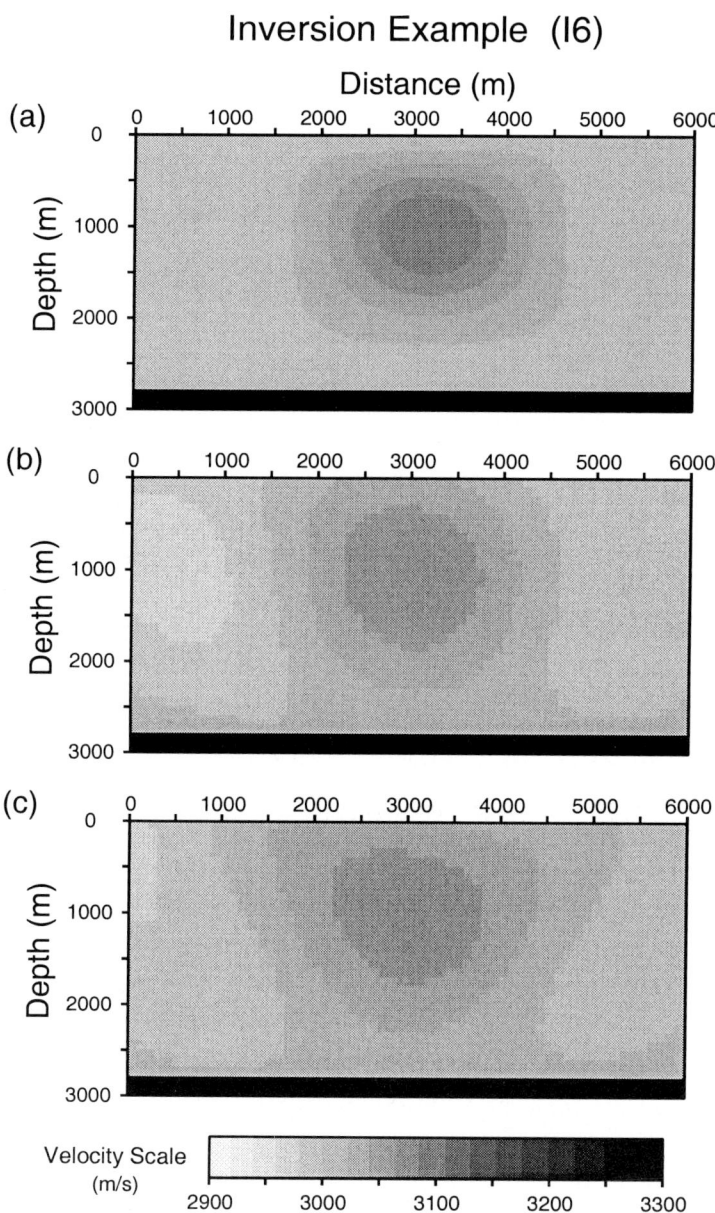

Figure 5.10. Inversion example I6: (a) the synthetic model with an arbitrary localized velocity anomaly over constant background velocity distribution; (b) the inversion solutions after five iterations and (c) after 20 iterations.

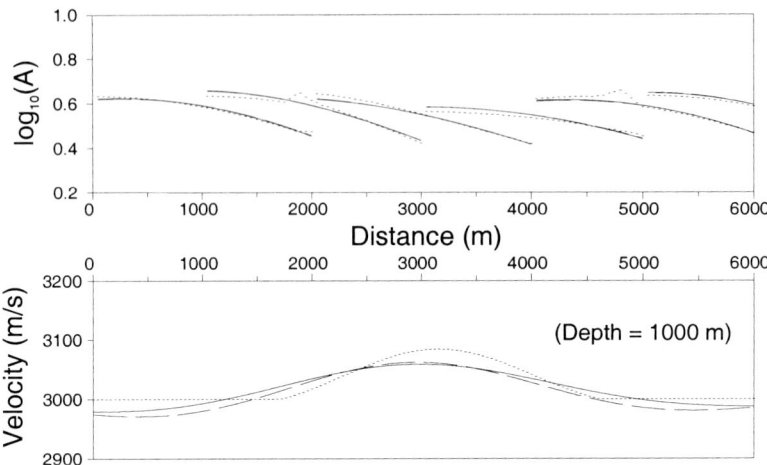

Figure 5.11. Bottom: comparison of velocity profiles for the synthetic model (dotted) and model estimates after five (dashed) and 20 (solid) iterations. Top: comparison of the corresponding synthetic (dotted) and estimated amplitude curves (dashed and solid).

the model Fourier series because it includes components at all wavenumbers, including high wavenumbers that are not represented in the model parameterization. Model I6 (Figure 5.10a) is an arbitrary model with a localized velocity anomaly. Model I6 has relatively stronger velocity gradients than the preceding models (see profile in the bottom of Figure 5.11 and note the rapid change in gradient at horizontal distance 1700 m and 4600 m), implying Fourier components at high wavenumbers.

In the inversion of amplitude data from synthetic model I6, we set $\Delta k = 0.25$ km^{-1} and $N = 2$ in the Fourier series parameterization. After only five iterations the velocity anomaly appears approximately in the solution. Comparing Figures 5.10b and 5.10c which show the inversion solutions after 5 and 20 iterations, with the synthetic model of Figure 5.10a, we see that the main anomaly is reasonably well reconstructed. Figure 5.11 compares velocity profiles (at depth 1000 m) and amplitude curves from the current estimate (after 5 and 20 iterations) with those from the synthetic model. The data misfit is a minimum after four iterations. If the inversion iterations are continued beyond that minimum, the data misfit remains at the same level. As the comparison of amplitude curves in Figure 5.11 shows, there is little further improvement in the model or in the fit to the data between 5 and 20 iterations. The data misfit remains relatively large because the Fourier series parameterization cannot represent the higher wavenumbers present in the original model. However, for most purposes, the inversion illustrated in Figure 5.10 would be considered a successful representation of the original model.

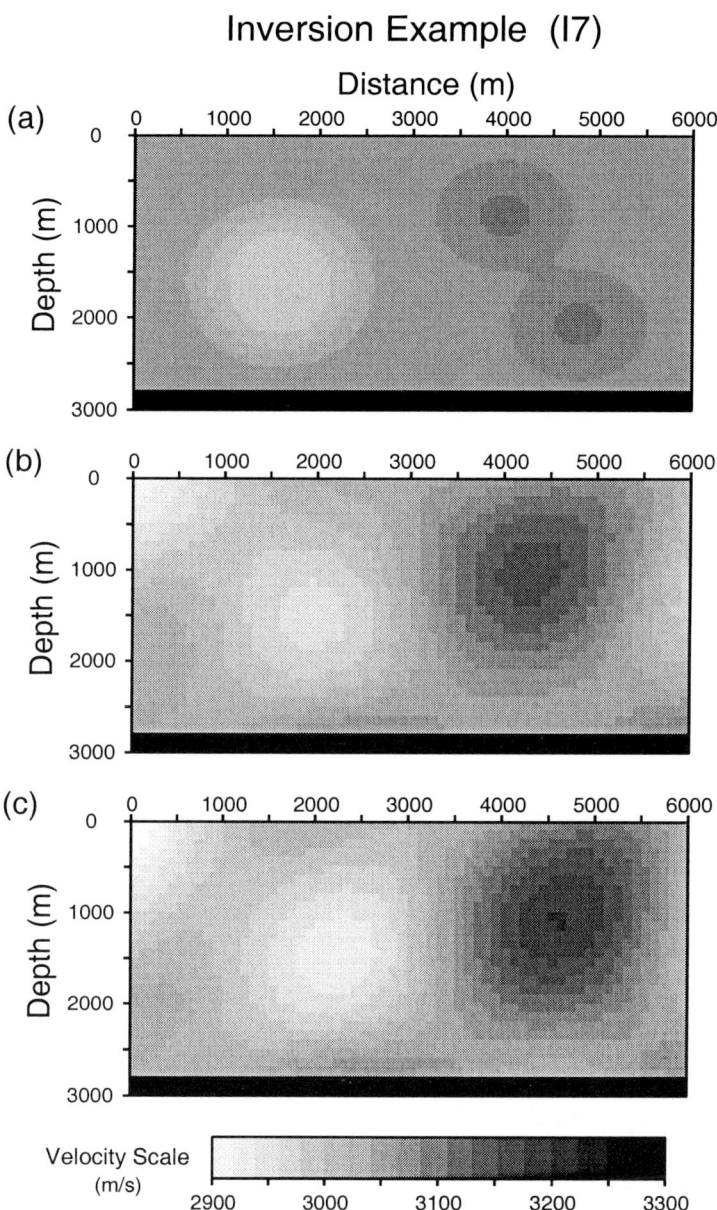

Figure 5.12. Inversion example I7: (a) the synthetic model with three arbitrary localized velocity anomalies over constant background velocity distribution; (b) the inversion solutions after five iterations and (c) after 20 iterations.

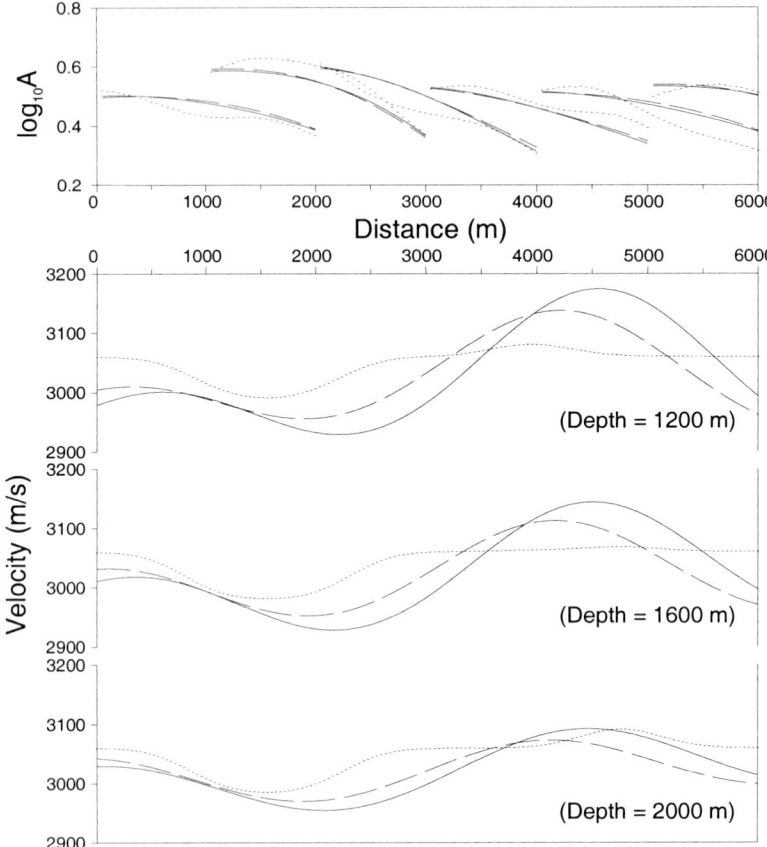

Figure 5.13. Bottom: comparison of velocity profiles for the synthetic model (dotted) and model estimates after five (dashed) and 20 (solid) iterations, at 1200 m, 1600 m and 2000 m. Top: comparison of the corresponding synthetic (dotted) and estimated amplitude curves (dashed and solid).

Finally we consider a synthetic model with an arbitrary velocity distribution and more structure at high wavenumber. Figure 5.12a shows a model (I7) with a negative velocity anomaly and two smaller positive velocity anomalies. As in the inversion of I7, we set $N = 2$ and $\Delta k = 0.25$ km^{-1}. The inversion solutions after 5 and 20 iterations are shown in Figures 5.12b and 5.12c. Comparing the inversion models with the synthetic model, we see that the main anomaly is approximately resolved, but the inversion fails to separate the two smaller positive velocity anomalies. As in Figure 5.1, the Fourier components in the z-direction are poorly resolved. Figure 5.13 shows the velocity profiles at depths 1200 m, 1600 m and 2000 m, respectively, and the corresponding amplitude curves. We conclude from these two last examples that the inversion method is stable when high wavenumber structure is present, and even if this structure remains unresolved, the solution provides a reasonable smoothed approximation to the actual structure.

5.8. Discussion

We have demonstrated the amplitude inversion procedure to estimate velocity variation in 2-D structures. Although the problem is relatively poorly constrained because of the reflection configuration, we have still obtained some quite encouraging results using a 2-D Fourier series model parameterization of the slowness distribution instead of the commonly used cellular parameterization. Certainly, the choice of the number of terms in the 2-D Fourier series and the fundamental wavenumber Δk are important factors influencing the inversion processing that need further study. In this chapter we have focused primarily on establishing that reflection seismogram amplitudes contain sufficient information to permit inversion for an unknown velocity distribution.

In the use of subspace inversion method, we partition model parameters into separate subspaces on the basis of sensitivity of the amplitude response with respect to model parameters (the coefficients of 2-D Fourier series). The inversion converges steadily in the case of models with smoothly varying velocity and velocity gradient. Even where the model velocity cannot be exactly represented by the inversion parameterization, the long wavelength components of the model are satisfactorily recovered. The relatively less successful behaviour of the inversion in the presence of data noise, or in the case of systematic errors in the reflector geometry, may indicate significant limitations on the use of amplitude data in tomographic inversion. Nevertheless, the amplitudes of the reflected signals are sensitive to the location of the velocity anomalies and the inversions provide an approximate image of velocity variation, thus demonstrating that amplitude data contain information that can constrain unknown velocity variation.

Chapter 6

Sensitivities of traveltimes and amplitudes in joint inversion

Abstract   To understand the information content of seismic traveltimes and amplitudes, we investigate the sensitivities of traveltimes and ray-amplitudes with respect to interface geometry and interval velocity variations, respectively. We analyse the difference in the sensitivities based on the eigenvalues and eigenvectors of the Hessian matrix, which is defined in terms of the Fréchet matrix and its adjoint associated with different norms chosen in the model space. We find that for reflection tomographic inversion, traveltime and amplitude data do indeed contain complementary information. Both for reflector geometry and for interval velocity variations, the traveltimes are sensitive to the model components with small wavenumbers, whereas the amplitudes are more sensitive to the components with high wavenumbers. The model resolution matrices, after the rejection of eigenvectors corresponding to small eigenvalues, may also provide some insight into how the addition of amplitude information can potentially contribute to the recovery of physical parameters.

6.1.   Introduction

In the previous two chapters, we have seen that ray-amplitude data alone can be used to constrain interface geometry and velocity variation, respectively. In this chapter, we focus our investigation on the difference in sensitivities of traveltimes and amplitudes in the reflection seismic configuration.

A sensitivity analysis is carried out by linearizing the problem in the vicinity of a computed solution. Given a model perturbation around the current estimate, we can calculate the variation of the predicted observations. The partial derivatives of the predicted observations with respect to the model parameters, $\mathbf{F}$, comprise the Fréchet matrix, which is sometimes called the sensitivity matrix in forward and inverse problems. Evaluating the eigenvalues of the matrix $\mathbf{F}^T\mathbf{F}$ allows us to analyse the sensitivity of the observations to the model parameters, since the eigenvalues give a measure of the sensitivity of the corresponding eigenvectors.

Most previous work on sensitivity has been carried out by means of singular value analysis. The singular value of the Fréchet matrix $\mathbf{F}$ is commonly defined as the positive square-root of the eigenvalue of the matrix $\mathbf{F}^T\mathbf{F}$ (or $\mathbf{F}\mathbf{F}^T$). In the context of crosshole tomography, singular value analysis results were shown by Bregman *et al.* (1989). Pratt and Chapman (1992) and Farra and Le Bégat (1995) applied singular value analysis to anisotropic crosshole traveltime inversion. For reflection tomography, Farra and Madariaga (1988) and Stork (1992b) showed singular value analysis results for traveltime inversion. In the previous two chapters, we also applied singular value analysis to amplitude inversion. However, Delprat-Jannaud and Lailly (1992) and Pratt and Chapman (1992) showed that, when using the singular value of the Fréchet matrix $\mathbf{F}$ (i.e. the eigenvalue of $\mathbf{F}^T\mathbf{F}$), data sensitivities are influenced strongly by the way in which the model is parameterized.

In iterative linearized inversion, it is the Hessian matrix (and not simply the Fréchet matrix) that measures the perturbation of the data misfit function resulting from a model perturbation applied to the current solution. Therefore, the sensitivity analysis can be performed by studying the Hessian matrix, which has different forms, depending on the norm chosen in model space. The matrix $\mathbf{F}^T\mathbf{F}$ above is, in fact, only one form of the Hessian, which follows from the use of the L^2-norm (the Euclidean norm) in the model space. Delprat-Jannaud and Lailly (1992) showed that the eigen-solution of the Hessian with L^2-norm in model space is influenced by the chosen model discretization interval rather than by the physical problem under consideration, when they explored a case of traveltime inversion with a model defined by B-spline functions.

To ensure the convergence of discrete eigenvalues and eigenvectors, towards the solution of a continuum spectral problem (i.e. stabilization), we need to introduce a different norm in the model space (assumed to be a Hilbertian norm associated with a scalar product of the model components). For instance, we can define a usual L^2-norm in the Sobolev space H^1, which is the space of L^2-functions such that the spatial derivative of the function is L^2 (Tarantola, 1987b).

Comparing the traveltime inversion using the H^1-norm with that using the L^2-norm, Delprat-Jannaud and Lailly (1992) showed that using the H^1-norm the influence of the model discretization interval is negligible (provided the discretization is sufficiently fine) and the model components are determined intrinsically by the traveltime data. In the following sensitivity studies we apply different norms in model space to the cases of traveltime inversion and amplitude inversion. Traveltime and amplitude data show different sensitivities to different Fourier components of the model.

As described above, a continuum spectrum of the Hessian may be obtained by using different norms in the model space. A norm in the model space is defined in terms of a scalar product, which is written explicitly in this chapter as a weighted sum of two correlation matrices. One is the correlation of model parameters and the other is the correlation of their first spatial derivatives. Alternatively, this scalar product can be understood as a working definition of the model covariance matrix. The latter is often used in the linearized inverse problem for expressing expected correlations between different model parameters. Therefore, we can physically understand the scalar product, such as the H^1-norm, as adding terms that penalize large spatial derivatives to the objective function. The weighting factor for the scalar product in the model space allows emphasis on the model perturbations or on their spatial derivatives.

We are interested in understanding which model parameters most influence reflection traveltimes and amplitudes. The Hessian represents the local curvature of the data misfit function. Its eigenvalue distribution indicates how the different parameters contribute to the information content of the traveltime and amplitude data, and thus is important for predicting the performance of iterative inversion techniques. In order to invert seismic traveltimes and amplitudes simultaneously, we propose an empirical definition of the data covariance matrix which balances the relative sensitivities of different types of data.

We consider the joint use of both data types for interface geometry and for 2-D interval velocity variations. In both cases, we find that the joint inversions can provide better solutions than that using traveltimes alone. The potential benefit of including amplitude data constraints in seismic reflection traveltime tomography is, therefore, that it may be possible to obtain a better resolution of the known ambiguity between reflector depth uncertainty and interval velocity uncertainty.

6.2. The Hessian and the norm in model space

We conduct the sensitivity analysis by means of eigen-analysis of the Hessian matrix, which relates the influence of each model perturbation to the perturbation of the objective function. We now discuss the linearized inverse problem in terms of the least-squares formulation, and form the Hessian matrix in terms of the Fréchet matrix and its adjoint (associated with a chosen norm in the model space).

6.2.1   The linearized problem and the Hessian

Given a set of observed data, we want to find a model that best matches the data. The least-squares formulation of the problem is to find a model that minimizes the objective function (Tarantola, 1987b),

$$J(\mathbf{m}) = \frac{1}{2}\|f(\mathbf{m}) - \mathbf{d}_{\mathrm{obs}}\|_{\mathrm{D}}^2 \,, \tag{6.1}$$

where $\mathbf{d}_{\mathrm{obs}}$ is the observed data set and $f(\mathbf{m})$ is the forward prediction. The objective function $J(\mathbf{m})$ measures the misfit between observed and calculated data. Its definition calls for the choice of a norm $\|\;\|_{\mathrm{D}}^2$ in the data space $\mathcal{D}$. This norm is associated with a scalar product $(\;,\;)_{\mathrm{D}}$, given by

$$(\mathbf{a}, \mathbf{b})_{\mathrm{D}} = \mathbf{a}^{\mathrm{T}}\mathbf{C}_{\mathrm{D}}^{-1}\mathbf{b} \,, \qquad \forall (\mathbf{a}, \mathbf{b}) \in \mathcal{D}^2 \,, \tag{6.2}$$

where $\mathbf{C}_{\mathrm{D}}$ is a symmetric positive definite matrix normally chosen to describe the covariances of the elements of $\mathcal{D}$. In the following numerical analysis, for inversions using one type of data (traveltimes or amplitudes) performed on modelled data, free of noise, we choose $\mathbf{C}_{\mathrm{D}}$ as an identity matrix with dimension $(data)^2$. In joint inversions in which we include both types of data simultaneously, we use a balancing factor, controlling the contribution to the objective function of observations with different physical dimensions, in order to manipulate the corresponding elements in the identity matrix.

The inverse problem, being non-linear, can be solved by successive linearizations. We generally assume that $f(\mathbf{m})$ is differentiable around a current estimate $\mathbf{m}_0$. Let us denote the Fréchet derivative matrix $\mathbf{F} = \nabla_{\mathbf{m}}f$, which is a linear map such that the approximate equality

$$f(\mathbf{m}_0 + \delta\mathbf{m}) \approx f(\mathbf{m}_0) + \mathbf{F}\,\delta\mathbf{m} \tag{6.3}$$

is valid for small perturbations, $\delta\mathbf{m}$. The linearized inverse problem consists of minimizing the objective function

$$J(\delta\mathbf{m}) = \frac{1}{2}\|\mathbf{F}\,\delta\mathbf{m} - \delta\mathbf{d}\|_{\mathrm{D}}^2 \,, \tag{6.4}$$

where the data residual, $\delta\mathbf{d} = \mathbf{d}_{\mathrm{obs}} - f(\mathbf{m}_0)$, is the difference between data and prediction of the current model and the parameter perturbation $\delta\mathbf{m}$ is the "model" to be solved. We may make a quadratic approximation to this objective function at point $\delta\mathbf{m}_0$ in the form of a Taylor series,

$$J(\delta\mathbf{m}) \approx J(\delta\mathbf{m}_0) + \hat{\mathbf{g}}^{\mathrm{T}}(\delta\mathbf{m} - \delta\mathbf{m}_0) + \frac{1}{2}(\delta\mathbf{m} - \delta\mathbf{m}_0)^{\mathrm{T}}\mathbf{H}\,(\delta\mathbf{m} - \delta\mathbf{m}_0) \,, \tag{6.5}$$

in terms of the gradient vector $\hat{\mathbf{g}}$ and the Hessian matrix $\mathbf{H}$.

We now introduce the adjoint $\mathbf{F}^\dagger$ of $\mathbf{F}$ associated with the scalar product $(\ ,\ )_M$ in the model space $\mathcal{M}$ as follows:

$$(\mathbf{C}_D^{-1}\delta\mathbf{d},\ \mathbf{F}\,\delta\mathbf{m})_D = (\mathbf{F}^\dagger\mathbf{C}_D^{-1}\delta\mathbf{d},\ \delta\mathbf{m})_M\ ,\quad \forall(\delta\mathbf{m},\ \delta\mathbf{d})\in\mathcal{M}\times\mathcal{D}.$$

(6.6)

This allows the gradient vector to be expressed as

$$\hat{\mathbf{g}} = \mathbf{F}^\dagger\mathbf{C}_D^{-1}(\mathbf{F}\,\delta\mathbf{m} - \delta\mathbf{d})$$

(6.7)

and the Hessian matrix as

$$\mathbf{H} = \mathbf{F}^\dagger\mathbf{C}_D^{-1}\mathbf{F}\ .$$

(6.8)

This Hessian matrix measures the influence of a perturbation about the solution model on the objective function (6.4). The solution of a linearized inverse problem can be obtained from $\hat{\mathbf{g}}(\delta\mathbf{m}) = 0$, i.e.

$$\mathbf{F}^\dagger\mathbf{C}_D^{-1}\mathbf{F}\,\delta\mathbf{m} = \mathbf{F}^\dagger\mathbf{C}_D^{-1}\delta\mathbf{d}\ ,$$

(6.9)

and hence the perturbation $\delta\mathbf{m} - \delta\mathbf{m}_0$ makes a change to the objective function (see equation 6.5) equal to

$$\Delta J = J(\delta\mathbf{m}) - J(\delta\mathbf{m}_0) = \frac{1}{2}(\delta\mathbf{m} - \delta\mathbf{m}_0)^{\mathrm{T}}\mathbf{H}\,(\delta\mathbf{m} - \delta\mathbf{m}_0)\ .$$

(6.10)

If a model perturbation gives rise to a large perturbation of the objective function, it is a sensitive component in the model and is well determined in the solution. Similarly, a perturbation that has either a small or no effect on the objective function is a less sensitive or insensitive component, and is correspondingly poorly determined or not determined in the solution.

Ideally such a sensitivity analysis can be evaluated at each iteration of the linearized inversion, where an optimum solution from the linearized problem (6.9) is $\delta\mathbf{m} = \delta\mathbf{m}_0$ so that ΔJ in equation (6.10) is equal to zero. In the following numerical analysis, however, we evaluate the Hessian matrix at the true solution point, $\mathbf{m}$, the optimal model found at the global minimum of the non-linear misfit function (6.1). In the vicinity of the solution point, the curvature of the misfit function, defined by the Hessian, shows the sensitivity of the objective function to the traveltime and amplitude data as follows:

$$\Delta J(\mathbf{m}) = J(\mathbf{m} + \Delta\mathbf{m}) - J(\mathbf{m}) = \frac{1}{2}\Delta\mathbf{m}^{\mathrm{T}}\mathbf{H}\,\Delta\mathbf{m}\ .$$

(6.11)

Since the misfit function (6.1) is not completely linear for the model, the relative sensitivities evaluated at the global minimum may not apply for all iterations of the linearized inversion, if the initial estimate is remote from the solution $\mathbf{m}$.

The Hessian, i.e. the local curvature of the data misfit function, describes all possible combinations of parameter perturbations (or model parameters) that give rise to the same change ΔJ as seen in equation (6.10) [or in equation (6.11)], and represents an M-dimensional ellipsoid (where M is the total number of model parameters). An eigen-analysis of $\mathbf{H}$ will give the lengths and the directions of the principal axes of the ellipsoid,

$$(\mathbf{H} - \lambda_i \mathbf{I})\mathbf{v}_i = 0 , \qquad (6.12)$$

where λ_i is the eigenvalue and $\mathbf{v}_i$ is the associated eigenvector. Eigenvectors corresponding to large and small eigenvalues give, respectively, well and poorly determined parameter combinations.

We now rewrite equation (6.12) using matrix form, i.e.

$$\mathbf{H} = \mathbf{V}\mathbf{\Lambda}\mathbf{V}^{\mathrm{T}} ,$$

where $\mathbf{\Lambda}$ is a diagonal matrix consisting of the eigenvalues λ_i and $\mathbf{V}$ is the matrix of associated eigenvectors. The solution of a linearized inverse problem (equation 6.9) can now be expressed as

$$\delta\tilde{\mathbf{m}} = (\mathbf{F}^{\dagger}\mathbf{C}_{\mathrm{D}}^{-1}\mathbf{F})^{-1}\mathbf{F}^{\dagger}\mathbf{C}_{\mathrm{D}}^{-1}\delta\mathbf{d} = \mathbf{H}^{-1}\mathbf{H}\,\delta\mathbf{m} = \mathbf{V}\,\mathbf{V}^{\mathrm{T}}\delta\mathbf{m} , \qquad (6.13)$$

where the matrix $\mathbf{V}\,\mathbf{V}^{\mathrm{T}}$ is called the model resolution matrix of the problem. The calculated model resolution matrix, after the rejection of eigenvectors corresponding to small eigenvalues, can provide some insight into how the relative sensitivity actually translates into accuracy and efficiency in a real inversion problem.

6.2.2   The Hessian and the norm in model space

We now discuss the form of the Hessian $\mathbf{H}$, which depends on the definition of the scalar product $(\,,\,)_{\mathrm{M}}$ in the model space $\mathcal{M}$. The scalar product is, however, related to the choice of the norm $\|\ \|_{\mathrm{M}'}$ in the dual space $\mathcal{M}'$.

Following Delprat-Jannaud and Lailly (1992, 1993), we define the Hilbertian norm for $\|\ \|_{\mathrm{M}'}$, where $\mathcal{M}'$ is thus a Hilbert space. Let us consider two model perturbations $\delta\mathbf{U}^{(1)}(\mathbf{x})$ and $\delta\mathbf{U}^{(2)}(\mathbf{x})$, which are the perturbations of reflector geometry, or slowness variation, or their combination (in general, not simply the perturbations of the model parameters $\delta\mathbf{m}$). The norm $\|\delta\mathbf{U}(\mathbf{x})\|_{\mathrm{M}'}$ in the dual space $\mathcal{M}'$ is defined as

$$(\delta\mathbf{U}^{(1)}(\mathbf{x}), \delta\mathbf{U}^{(2)}(\mathbf{x}))_{\alpha} = (1 - \alpha) \int_{\Omega} \delta\mathbf{U}^{(1)}(\mathbf{x})\delta\mathbf{U}^{(2)}(\mathbf{x})d\Omega$$

$$+\alpha \int_{\Omega} \nabla_{\mathbf{x}}\delta\mathbf{U}^{(1)}(\mathbf{x})\nabla_{\mathbf{x}}\delta\mathbf{U}^{(2)}(\mathbf{x})d\Omega , \qquad (6.14)$$

$$\forall(\delta\mathbf{U}^{(1)}, \delta\mathbf{U}^{(2)}) \in \mathcal{M}' \times \mathcal{M}' ,$$

where Ω is a generic term consisting of the reflector geometry component and the slowness component of the model, and α is a weighting parameter which can take values within $[0, 1)$.

We now adopt the model parameterization defined by

$$\mathbf{U}(\mathbf{x}) = \sum_{i=1}^{M} c_i \beta_i(\mathbf{x}) \, , \tag{6.15}$$

where c_i is the amplitude coefficient of the ith basis function $\beta_i(\mathbf{x})$. We denote the projections on the basis functions $\{\beta_i(\mathbf{x}), \forall i (1 \leq i \leq M)\}$ as vectors (of the model parameters),

$$\begin{aligned} \delta\mathbf{m}^{(1)} &= (\delta c_i^{(1)})_{(i=1,M)} \, , \\ \delta\mathbf{m}^{(2)} &= (\delta c_i^{(2)})_{(i=1,M)} \, , \end{aligned} \tag{6.16}$$

for $\delta\mathbf{U}^{(1)}(\mathbf{x})$ and $\delta\mathbf{U}^{(2)}(\mathbf{x})$, respectively. The scalar product equation (6.14) is then given by

$$\left(\delta\mathbf{U}^{(1)}(\mathbf{x}), \ \delta\mathbf{U}^{(2)}(\mathbf{x}) \right)_\alpha = \langle \mathbf{D}_\alpha \delta\mathbf{m}^{(1)}, \ \delta\mathbf{m}^{(2)} \rangle \, , \tag{6.17}$$

$$\forall (\delta\mathbf{U}, \ \delta\mathbf{m}) \in \mathcal{M}' \times \mathcal{M} \, ,$$

where $\mathbf{D}_\alpha$ is an $M \times M$ symmetric positive definite matrix with tensors defined as

$$[D_{ij}]_\alpha = (1 - \alpha) \int_\Omega \beta_i(\mathbf{x}) \beta_j(\mathbf{x}) \, d\Omega + \alpha \int_\Omega \nabla_\mathbf{x} \beta_i(\mathbf{x}) \nabla_\mathbf{x} \beta_j(\mathbf{x}) \, d\Omega \, , \tag{6.18}$$

$$\forall i (1 \leq i \leq M) \, , \ \ \forall j (1 \leq j \leq M) \, .$$

In the equation (6.18), we define the norm $\| \ \|_{\mathbf{M}'}$ in $\mathcal{M}'$ by a weighted sum of two integral terms, in which one term is the sum of the L^2 scalar product of the model components and the other is the sum of the L^2 scalar product of the first spatial derivatives of the model components. Note that the correlation matrices in the definition of the operator $\mathbf{D}_\alpha$ are scaled in application, so that the quantities are dimensionless and the matrices are balanced with respect to each other. In the particular case of model parameterization using regular discrete grids, the vector $\delta\mathbf{U}$ and the vector $\delta\mathbf{m}$ in equation (6.17) would be identical.

The scalar product depends on the weighting parameter α which can take values within $[0, 1)$. When $\alpha = 0$ it defines the L^2-norm. When $\alpha = 0.5$ it corresponds to the usual norm in the Sobolev space H^1, which is equal to the sum of the usual L^2-norm of the function and of the L^2-norm of its spatial derivative (Tarantola, 1987b). This process can be physically understood as penalizing both the data misfit and the first derivatives in the inversion, i.e. we search for model perturbations with small spatial derivatives. This is consistent with ray theory which requires a locally smooth model.

We denote the transpose of matrix $\mathbf{F}$ associated with the inner product $\langle\,,\,\rangle$ as $\mathbf{F}^T$, defined by

$$(\mathbf{C}_D^{-1}\delta\mathbf{d},\ \mathbf{F}\,\delta\mathbf{m})_D = \langle\mathbf{F}^T\mathbf{C}_D^{-1}\delta\mathbf{d},\ \delta\mathbf{m}\rangle\,, \tag{6.19}$$

$$\forall(\delta\mathbf{m},\ \delta\mathbf{d}) \in \mathcal{M} \times \mathcal{D}\,.$$

Comparing it with equation (6.6), we obtain

$$\langle\mathbf{F}^T\mathbf{C}_D^{-1}\delta\mathbf{d},\ \delta\mathbf{m}\rangle = (\mathbf{F}^\dagger\mathbf{C}_D^{-1}\delta\mathbf{d},\ \delta\mathbf{m})_M\,, \tag{6.20}$$

$$\forall(\delta\mathbf{m},\ \delta\mathbf{d}) \in \mathcal{M} \times \mathcal{D}\,.$$

We now define the scalar product $(\,,\,)_M$ in the model space $\mathcal{M}$ as

$$(\delta\mathbf{m}^{(1)},\ \delta\mathbf{m}^{(2)})_\alpha = \langle\mathbf{D}_\alpha\delta\mathbf{m}^{(1)},\ \delta\mathbf{m}^{(2)}\rangle\,, \tag{6.21}$$

$$\forall(\delta\mathbf{m}^{(1)},\ \delta\mathbf{m}^{(2)}) \in \mathcal{M}^2\,,$$

where $\langle\,,\,\rangle$ is the duality product between $\mathcal{M}$ and $\mathcal{M}'$ (Delprat-Jannaud and Lailly, 1992). We then have

$$\langle\mathbf{F}^T\mathbf{C}_D^{-1}\delta\mathbf{d},\ \delta\mathbf{m}\rangle = \langle\mathbf{D}_\alpha\mathbf{F}^\dagger\mathbf{C}_D^{-1}\delta\mathbf{d},\ \delta\mathbf{m}\rangle\,, \tag{6.22}$$

$$\forall(\delta\mathbf{m},\ \delta\mathbf{d}) \in \mathcal{M} \times \mathcal{D}$$

and the following equality

$$\mathbf{F}^\dagger = \mathbf{D}_\alpha^{-1}\mathbf{F}^T\,. \tag{6.23}$$

Thus we can express the Hessian matrix as

$$\mathbf{H} = \mathbf{F}^\dagger\mathbf{C}_D^{-1}\mathbf{F} = \mathbf{D}_\alpha^{-1}\mathbf{F}^T\mathbf{C}_D^{-1}\mathbf{F}\,. \tag{6.24}$$

In this chapter, we assume that the matrix $\mathbf{C}_D$ is equal to the identity matrix, for one type of data without noise.

We cannot define a continuum spectral problem associated with operator $\mathbf{F}^T\mathbf{F}$, and the numerical solutions are strongly dependent on the discretization and do not converge as the discretization is refined. However, we can define a continuum spectral problem for the Hessian $\mathbf{F}^\dagger\mathbf{F}$ associated with the scalar product $(\,,\,)_\alpha$, $(0 < \alpha < 1)$, and obtain accurate numerical solutions of this continuum spectral problem. Delprat-Jannaud and Lailly (1992) showed the results of eigen-analysis of $\mathbf{F}^\dagger\mathbf{F}$ with the H^1-norm, compared to the one with the

L^2-norm, in their traveltime inversion with a model parameterized as B-spline interpolation of discretized points. In the following sections we also apply those two norms in the case of traveltime inversion, but with a different model parameterization, and we extend the method to the case of amplitude inversion for the analysis of traveltime and amplitude sensitivities to interface and slowness distribution.

6.3.   Sensitivities to interface geometry

6.3.1   Interface parameterization and the Hessian operator

We now study the sensitivities of reflection traveltime and ray-amplitude data to the geometry of a reflecting interface. The relative sensitivity is measured by the magnitude of the eigenvalues of the Hessian matrix for the linearized equation. The analyses for traveltimes and amplitudes are initially considered separately. The relative sensitivity of the amplitudes compared to that of the traveltimes is then evaluated by means of a joint inversion including both types of data. The model resolution matrix, for the pseudo-inverse using a truncation of the SVD, is also given in each case.

We consider a 2-D stratified velocity structure consisting of variable-thickness layers, and assume that the depth of the interface dividing the layers varies continuously. A continuous interface, band limited in wavenumber, may be approximated by the Fourier series for its periodic continuation, i.e.

$$Z(x) = \sum_{n=0}^{N} [a_n \cos(n\pi\Delta kx) + b_n \sin(n\pi\Delta kx)] , \qquad (6.25)$$

where a_n and b_n are amplitude coefficients of the nth harmonic term, a basis function used in equation (6.15), with a wavenumber equal to an integer multiple of the fundamental wavenumber Δk (the reciprocal of the fundamental wavelength), and N is the number of harmonic terms. This parameterization enables an accurate study of the scale dependence of the traveltime and amplitude variation.

To illustrate the calculation of the Fréchet matrix for the study of sensitivities, we consider a model with a constant velocity (2500 m/s) and a single flat reflector at depth 2000 m as a reference model for the following perturbation analysis. The velocity within the layer below the reflector is set equal to 2800 m/s. (We shall return to this geometry and this reference model for a number of synthetic examples, shown in section 6.5). We generate a synthetic experiment with a realistic reflection acquisition geometry in which 10 shots are located on the surface at intervals of 1000 m, with data recorded at 25 receivers for each shot.

The minimum and maximum shot-receiver offsets are equal to 100 m and 2500 m respectively. Thus, we have 250 traveltime and 250 amplitude observations. The Hessian matrix, determined with the model estimate equal to the solution, is calculated using equation (6.15), in terms of the Fréchet matrix and the operator $\mathbf{D}_\alpha$, relating to different norms in the model space.

Following the definition of the scalar product in the model space, the scalar product of the model perturbations can be written in terms of the correlation matrix of the perturbations of interface depth and the correlation matrix of their slopes, correlating along the spatial direction. Considering a reflector parameterized by equation (6.25), these two correlation matrices can be explicitly written in terms of basis functions as

$$(\hat{\mathbf{B}}_0)_{ij} = \int_{I_\mathbf{x}} \beta_i(x)\beta_j(x)\,dx \qquad (6.26a)$$

and

$$(\hat{\mathbf{B}}_x)_{ij} = \int_{I_\mathbf{x}} \frac{d\beta_i(x)}{dx}\frac{d\beta_j(x)}{dx}\,dx\,, \qquad (6.26b)$$

respectively, where $I_\mathbf{x}$ is that part of the interface from which rays are reflected. In the following analysis, we show only the results associated with the coefficients of cosine terms, and thus set $\beta(x) = \cos(n\pi\Delta kx)$ in equation (6.26). Because a sine function is a phase-shifted cosine function, the sensitivity of observations to the coefficients of sine terms is similar to that of observations to the coefficients of cosine terms. The question of the independence of the sine and cosine terms, related to the phase (i.e. the lateral position of the anomalies) will be discussed later.

In the following numerical experiments the two correlation matrices in equation (6.26) are scaled as follows:

$$\mathbf{B}_0 = \frac{\varepsilon}{\text{trace}(\hat{\mathbf{B}}_0)}\hat{\mathbf{B}}_0 \qquad \text{and} \qquad \mathbf{B}_x = \frac{\varepsilon}{\text{trace}(\hat{\mathbf{B}}_x)}\hat{\mathbf{B}}_x\,, \qquad (6.27)$$

where ε can be arbitrarily set in the range of 1.0 to M, for the $M \times M$ matrix, to avoid floating point overflow in the matrix inverse calculation. The effect of the scaling operation in equation (6.27) is to normalize the two matrices, so that they are relatively balanced and dimensionless. Thus, the operator $\mathbf{D}_\alpha$ is built as

$$\mathbf{D}_\alpha = (1 - \alpha)\,\mathbf{B}_0 + \alpha\,\mathbf{B}_x\,. \qquad (6.28)$$

The first of the two terms in equation (6.28) penalizes large perturbations, the second penalizes solutions with large gradients (slopes). The scalar product used in equations (6.26)-(6.28) can alternatively be understood as a working definition of the model covariance matrix, $\mathbf{C}_M^{-1}$, if we set $\mathbf{C}_M^{-1}$ as $\mathbf{D}_\alpha$ times an identity matrix with dimension *(model parameter)*$^{-2}$ (see section 6.5 for a further discussion of this point). Penalizing solutions with a non-zero value of α implies that a near horizontal solution is to be preferred and non-horizontal slopes are undesirable. The choice of parameter α for the scalar product in model space allows emphasis on the model perturbations or on their spatial derivatives.

Note that the model parameterization in equation (6.25) should include high order components so that the discretized solution approximates the continuum solution. This differs from a possible inversion practice in which an attempt is made to obtain a long wavelength solution by truncating the high wavenumber components in the parameterization. In the following sensitivity analysis, we use wavenumbers up to $k_{max} = 5$ km^{-1}. The lateral resolving power of any reflection method based on ray theory will be limited to the first Fresnel zone width, given by $(\frac{1}{2}\lambda d)^{1/2}$, where λ is the seismic wavelength and d is the depth of the reflector (Sheriff and Geldart 1995). Assuming a dominant wavelength of 100 m (for realistic exploration frequencies), the Fresnel zone width will be of the order of 300 m for this example. Thus, a wavenumber of 5 km^{-1}, which corresponds to an interface wavelength of 200 m, is an adequate high wavenumber limit.

6.3.2   Interface inversion with the L^2-norm chosen in model space

We consider an interface with wavenumbers ranging from zero to 5 km^{-1} and discretized by equation (6.25) with the number of harmonic terms $N = 80$, corresponding to a discretization interval Δk equal to $1/16$ km^{-1}. The eigenvalues and associated eigenvectors of the Hessian matrix, with the L^2-norm chosen in the model space, and the model resolution matrix of reflection traveltime and amplitude inversions for the interface geometry are shown in Figure 6.1. The resultant eigenvalues shown in the upper part of the figure are normalized relative to the maximum value and ordered in decreasing size. Each associated eigenvector shown in the middle part of the figure is represented as a column of pixels, ranging from zero wavenumber at the top to maximum wavenumber at the bottom. The eigenvector is normalized for each eigenvalue, with darker shades representing positive values (up to +1) of the element in an eigenvector and lighter shades representing negative values (down to -0.6). The model resolution matrices of traveltime and amplitude inversion are shown in the bottom of the figure.

We first examine the traveltime sensitivity to the reflection interface (lefthand side of Figure 6.1). Since we only have a finite number of data, the solution will be underdetermined. The sharp decrease in the magnitude of eigenvalues of the Hessian (after the 53rd) clearly indicates that a null space exists for such

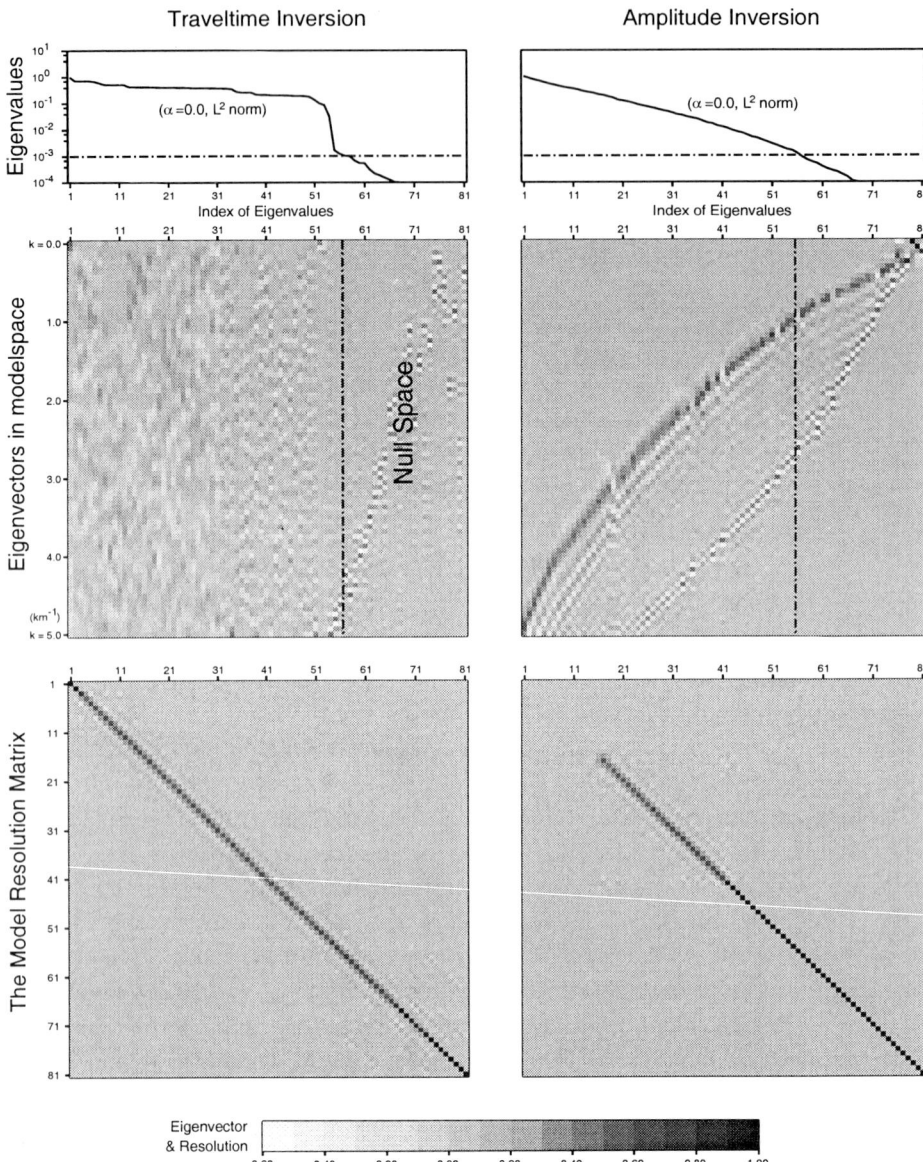

Figure 6.1. Independent sensitivities of reflection traveltimes and amplitudes to interface wavenumber components, in terms of the eigenvalues and eigenvectors of the Hessian matrices and the model resolution matrices in linearized inversion, where the L^2-norm ($\alpha = 0$) in model space is chosen. The eigenvalues, in order of decreasing magnitude, are normalized relative to the maximum value. Each eigenvector is represented as a vertical column, below the associated eigenvalue. In each eigenvector column, wavenumbers of the model components increase from top to bottom. The model resolution matrices for the problems are shown in the lower half of the figure. Integers along the sides refer to the wavenumber indices.

an over-fine discretization. To determine the geometry of a reflection interface with such densely sampled wavenumber components, more observations
would naturally be helpful for a traveltime inversion. However, for a fixed experimental geometry, we have little choice but to deal with this null space in
some other manner. Reparameterization using a coarser set of basis functions
would be one choice; alternatively we could simply ignore the null space, as in
a pseudo-matrix inversion by SVD with truncation.

If we ignore the null space by truncated SVD, biases in the solution are
inevitable (Ory and Pratt, 1995). The model resolution matrix indicates the
reliability of the solution.

To obtain the model resolution matrix, let us arbitrarily ignore the eigenvectors with relative eigenvalues smaller than 10^{-3}, set approximately corresponding to the obvious discontinuity appearing in the eigenvalue curve. The
diagonal elements of the model resolution matrix, shown in Figure 6.1, for
traveltime are

$$
\left\{
\begin{array}{cccccccccc}
1.00; & 0.98; & 0.81; & 0.63; & \cdots; & 0.67; & 0.75\text{--}0.88; & 0.90\text{--}0.97; & 0.98 \\
(1) & (2) & (3) & (4) & (5\text{-}65) & (66) & (67\text{-}73) & (74\text{-}80) & (81)
\end{array}
\right\}
$$

and the magnitude for elements 5-65 is around 0.63-0.67. Each component
$\delta\tilde{m}_j$ is the combination of its true solution δm_j and neighbouring components.
The magnitudes show that traveltime inversion can determine relatively better
both the interface components with wavenumbers equal to or close to zero and
the components with very short wavelengths (step changes causing reflections),
compared to the rest of the components.

From the eigenvalues and the associated eigenvectors we can see that the objective function in traveltime inversion is most sensitive to the components
with wavenumber of zero and Δk. The first eigenvector for traveltime is
given as a normalized vector of $\{0.52, 0.38, \cdots, [\text{with the remaining elements}$
$< \mathcal{O}(0.01)]\}$. The most significant elements are the first and second elements.
The 53rd-55th eigenvectors (representing higher wavenumber components)
have small eigenvalues (close to 10^{-3}) and indicate that high wavenumber
components have relatively weak sensitivities in traveltime inversion. For the
remaining eigenvalues, there is no clear dependence of reflection traveltime on
the wavenumber.

In the case of amplitude inversion, the ray-amplitudes are significantly more
sensitive to interface components with higher wavenumbers (shorter wavelengths), as demonstrated by the distribution of the eigenvalues and associated
eigenvectors of the Hessian matrix shown in the right-hand side of Figure 6.1.
The eigenvector pattern shows that the amplitude sensitivity decreases quasi-
exponentially with wavenumber. This may be expected as the curvature of the
interface has a significant effect on the recorded amplitude because of the con-

sequent focusing or defocusing of the beam. However, in the case of traveltime inversion with the L^2-norm chosen, there is no such dependence.

In amplitude inversion, the eigenvalues decrease smoothly without the obvious discontinuity that appears in traveltime inversion. We follow the traveltime inversion case and truncate the eigenvectors $\mathbf{v}_i$ for $i \geq 56$ associated with the smallest eigenvalues, representing those components with the weakest sensitivities, in the calculation of the model resolution matrix. The diagonal elements of the model resolution matrix are

$$\left\{ \begin{array}{cccccccccc} 0.00; & 0.01; & 0.02; & 0.16; & 0.58; & \ldots; & 0.70; & 0.86; & 0.98; & 1.00 \\ \text{(1-13)} & \text{(14)} & \text{(15)} & \text{(16)} & \text{(17)} & \text{(...)} & \text{(41)} & \text{(42)} & \text{(43)} & \text{(44-81)} \end{array} \right\},$$

where the values for elements 18-40 gradually increase from 0.60 to 0.70. The model resolution matrix clearly shows that amplitude inversion with the L^2-norm can well constrain all wavenumber components except those components with small wavenumbers.

6.3.3   Interface inversion with constraint of spatial derivatives

We now consider Figure 6.2, which shows the same sensitivity analysis but with the H^1-norm (as $\alpha = 0.5$ in equation 6.28) chosen in the model space. This takes into account not only the depth variation but also the slope of the reflector.

In the previous subsection we saw that, if we choose the L^2-norm in the model space for the calculation of the Hessian matrix for traveltime inversion, the objective function is most sensitive to the mean depth of a reflection horizon and the component with fundamental wavenumber Δk, but there is no simple dependence on the other non-zero wavenumbers. The clear dependence of the amplitude L^2 objective function sensitivity on wavenumber is seen in traveltime inversion with the choice of the H^1-norm in the model space, as shown in Figure 6.2.

Although the choice of a norm in model space is essential to the process of traveltime inversion, as can be seen by comparing the trend of eigenvectors of the Hessian with the L^2-norm (Figure 6.1) and the H^1-norm (Figure 6.2), for the case of amplitude inversion we see that the value of α has only a slight influence on the trend of eigenvalues. Even with the L^2-norm in model space, a strong dependence of reflection ray-amplitude on the wavenumber of interface components exists. Therefore, the constraint term of the reflector slope is not necessarily required in the objective function for amplitude inversion, although the introduction of a H^1-norm in the model space slightly improves the condition number, as seen from the comparison of eigenvalue curves. In the case of amplitude inversion the condition number of the Hessian matrix is 10^7 for the L^2-norm and 10^6 for the H^1-norm with $\alpha = 0.5$.

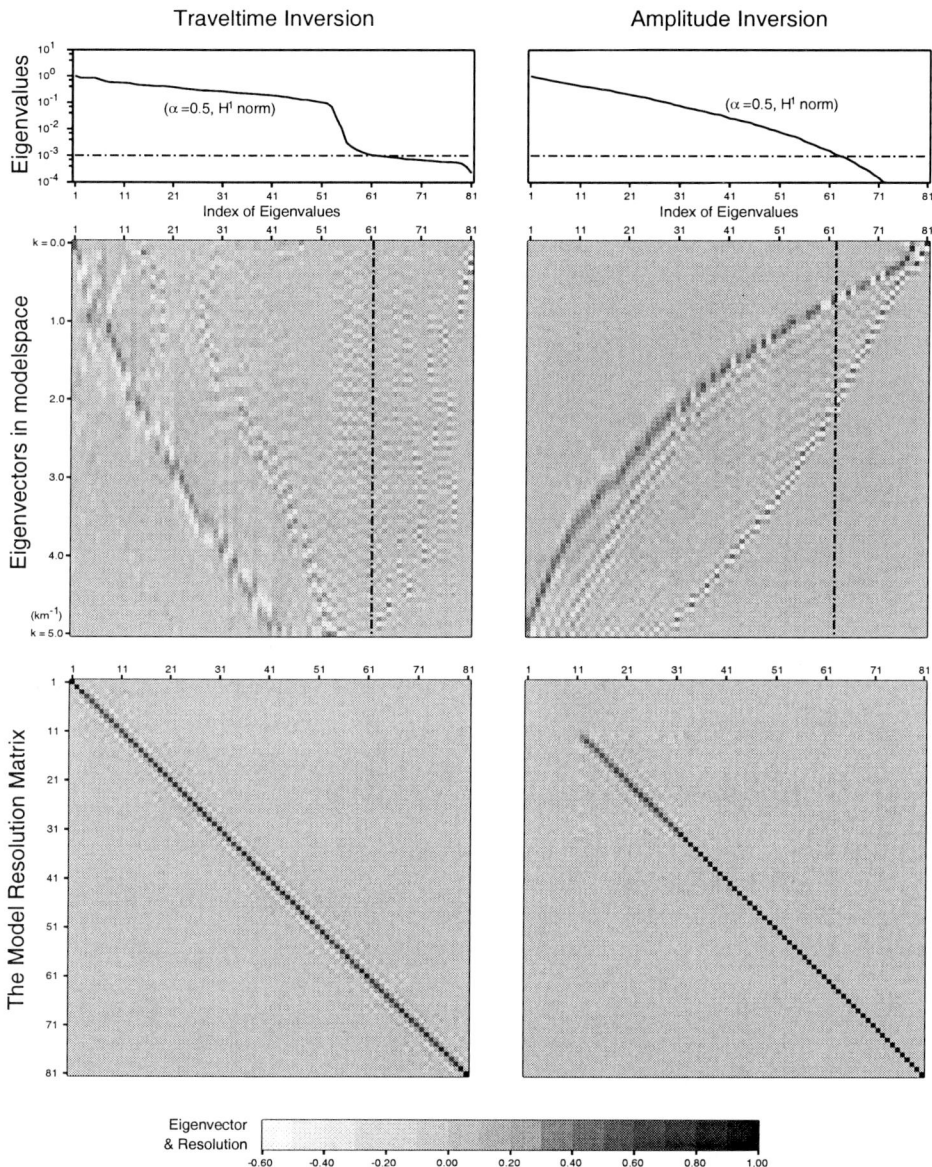

Figure 6.2. Independent sensitivities of reflection traveltimes and amplitudes to the interface wavenumber components, where the H^1-norm ($\alpha = 0.50$) in model space is chosen.

Figures 6.1 and 6.2 also depict the model resolution matrix for each problem. These matrices show the extent to which each model parameter (i.e. each Fourier coefficient) is uniquely resolved; a perfectly resolved parameter has a single non-zero value of 1.0 on the main diagonal of the corresponding row

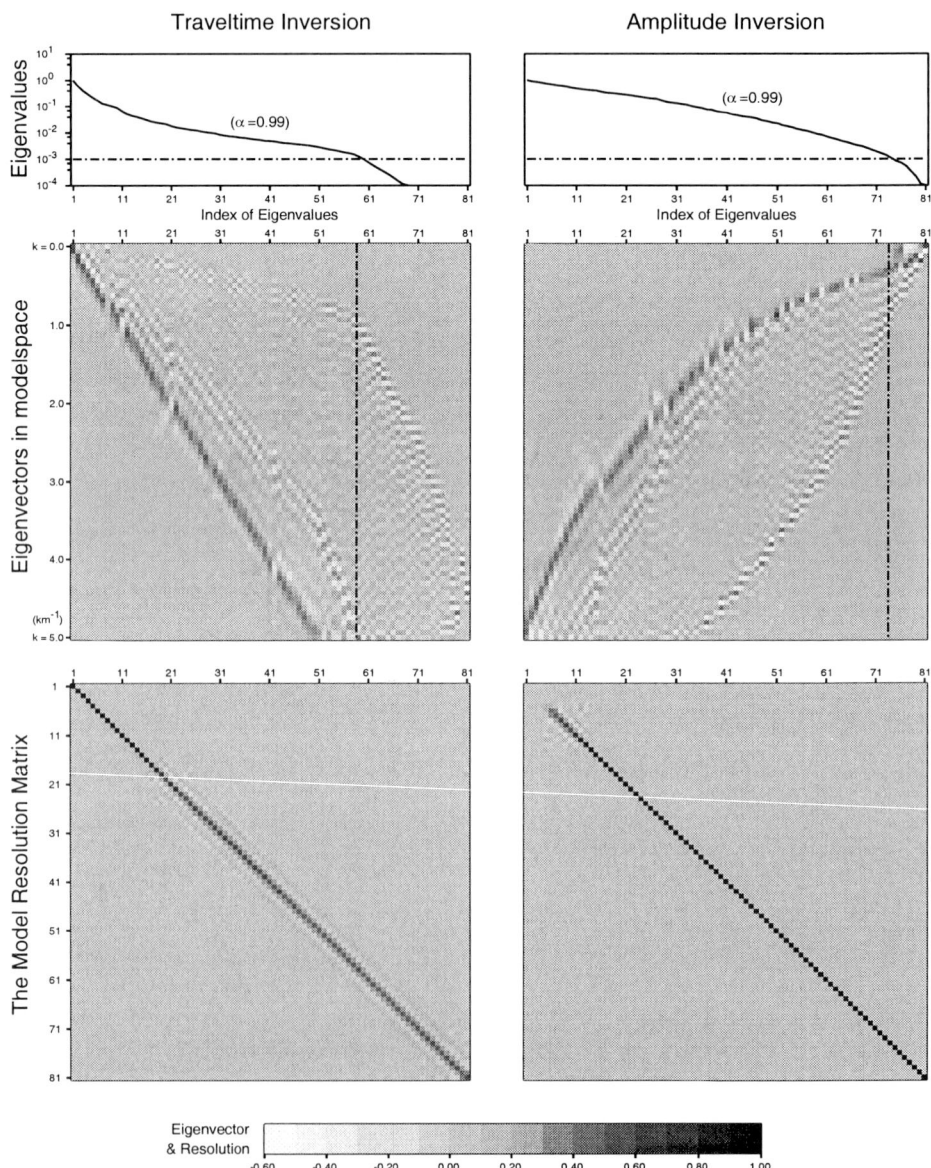

Figure 6.3. Independent sensitivities of reflection traveltimes and amplitudes to the interface wavenumber components, where $\alpha = 0.99$ is set in the scalar product in the model space.

of the resolution matrix, and off-diagonal values show the extent to which the parameter is non-unique with respect to other model parameters. The model resolution matrices show some apparent improvement in resolution when the H^1-norm is introduced, when compared with the results from the L^2-norm. Traveltime inversion using the H^1-norm appears to resolve the first two components, as the value of the first two diagonal elements of the model resolution matrix is equal to 1.0. In the case of amplitude inversion the model resolution matrix is also improved for components with intermediate and small wavenumber. These improvements are due to the use of a gradient penalization term in the inversion; they indicate that amplitude inversion, coupled with a constraint on the gradient of the solution, will apparently resolve the low wavenumbers. It should be noted, however, that the additional information does not come from the data, but from the constraints.

In practice, the choice of α is perhaps arbitrary, depending only on the need to minimize the magnitude of the perturbations, or their slopes. We now consider a case of interface inversion in which the interface gradients are more heavily penalized in the inversion. Note that $\alpha = 1.0$ does not constrain the magnitude of the model perturbations at all, and therefore results in a singular matrix. For the results shown in Figure 6.3, we set $\alpha = 0.99$ in equation (6.28). [Experiments have shown that if α takes the value 0.9, the result is similar to Figure 6.3; a choice of $\alpha = 0.999$ results in a singular matrix.] With this level of penalization on the spatial derivatives, the small wavenumber components are emphasized and the corresponding eigenvalues (at small indices for traveltime and at large indices for amplitude) are increased relative to the rest. In this case, the eigenvector pattern of the reflection traveltime clearly shows linear dependence on the wavenumber. Traveltime inversion can now apparently determine the 17 lowest wavenumber components, as the first 17 diagonal elements of the model resolution matrix are equal to 1.0. Also evident is the suppression of the high wavenumber components. In the case of amplitude inversion, the model resolution matrix is almost an identity matrix, except for the three components with the lowest wavenumber. Once again, the high level of penalization on the interface gradient will possibly cause models with no gradient to be generated, in spite of the data.

In summary, with an L^2-norm in the model space, reflected traveltimes are most sensitive to the mean depth of a reflector, whereas reflected amplitudes are most sensitive to the shorter wavelengths. By introducing the constraint of spatial derivatives in the definition of the Hessian ($\alpha \neq 0.0$), the traveltime sensitivities will be dependent on the wavenumber of the interface components. This dependence is in the opposite direction to the sensitivity of ray-amplitudes to the interface wavenumber components. Therefore, the information content in reflection traveltimes and reflection amplitudes are indeed complementary in linearized inversion, being sensitive to different components of the interface.

6.3.4   Joint inversion for interface geometry

In the numerical examples shown above, we have treated the two types of data separately. Another point that we shall address in this chapter is the relative value of including amplitude and traveltime data in a joint inversion. We now consider what will happen if we take the examples of Figure 6.1 and Figure 6.2 and use both types of data simultaneously, so that

$$\mathbf{d} = \begin{bmatrix} \mathbf{d}_{ampl} \\ \mathbf{d}_{time} \end{bmatrix} ,$$

in the definition of the objective function.

If both physical data types are included, the relative effects will be determined, at least partly, by the statistical reliability of the two data types. Ideally, we would seek to incorporate such statistical information. In practice, such information is usually unavailable or unreliable. As a working solution, we characterize these statistics using a covariance matrix defined as a diagonal matrix,

$$\mathbf{C}_D = \text{diag}\{\sigma_j^2\} ,$$

in terms of the estimated uncertainty of the jth measurement σ_j. Without data noise, we first set the standard deviation σ_j as the rms values of the observation data, denoted by σ_{ampl} and σ_{time}, two constants corresponding to amplitudes and traveltimes respectively. In physical terms, this procedure can be understood to remove the physical dimensions so that both types of data make a similar contribution in a joint inversion. With this model we performed a number of experiments (not shown) which revealed that reflection amplitudes absolutely dominate the inversion for interface geometry, and traveltimes only weakly influence the mean depth of interfaces. Based on this observation, we now propose an alternative to the definition for the data covariance matrix, which balances the data contributions not directly by the data magnitudes but by the data sensitivities.

Denoting the data covariance matrices for amplitudes and traveltimes as $\mathbf{C}_{ampl}$ and $\mathbf{C}_{time}$, respectively, the combined covariance matrix, which includes both types of data, is then defined as

$$\mathbf{C}_D^{-1} = \begin{bmatrix} \kappa^2 \, \mathbf{C}_{ampl}^{-1} & \mathbf{0} \\ \mathbf{0} & \mathbf{C}_{time}^{-1} \end{bmatrix} , \tag{6.29}$$

where κ is a dimensionless balancing factor. If we normalize $\mathbf{C}_{ampl}^{-1}$ and $\mathbf{C}_{time}^{-1}$ to the maximum value, or set them to identity matrices after suppressing data errors (see the real data examples shown in Chapter 7), the dimensionless balancing factor κ is given by

$$\kappa = \left(\frac{\text{trace } (\mathbf{F}_{\text{time}})}{\text{trace } (\mathbf{F}_{\text{ampl}})} \right)^g, \qquad (6.30)$$

where the ratio of matrix traces is an empirical quantity indicating the relative sensitivities of the traveltimes compared to the amplitudes, and g can be understood as a second, somewhat arbitrary, weighting factor. The trace of a matrix is defined as the sum of its eigenvalues. In our numerical example, $\kappa = 1.685 \times 10^{-2}$, which indicates that the amplitudes are much more sensitive to variations in the interface model than are the traveltimes. Therefore, the larger g is, the greater the influence of the traveltimes. Experiments have shown that the optimal value of g is in the range of 0.25 to 0.75 for the interface inversion.

Figure 6.4 shows the eigen-analysis results of the Hessian for the joint inverse problem using both data types, in which we chose $p = 0.75$. The sensitivity patterns for both the L^2-norm and the H^1-norm are modified from those observed in Figure 6.1 and Figure 6.2. Comparing the patterns of eigenvectors obtained for the joint inverse problem with those observed for the individual traveltime and amplitude problems, we see that the amplitudes appear to dominate the eigenvector patterns for the components with high wavenumbers and the traveltimes are dominant for those with low wavenumbers. With the L^2-norm the components with small wavenumber have less sensitivity than those using the H^1-norm in the model space, shown by the comparison of the eigenvalues corresponding to the combination of those small wavenumber components. The model resolution matrices for the joint inversion appear to be a combination of those in the previous (individual) inversions. With the H^1-norm the resolution of the inversion to model components of small wavenumber is enhanced, since the penalization of spatial derivatives emphasizes equally the model components with small wavenumber in the inversion. Therefore, if we choose $\alpha = 0.99$, as in Figure 6.3, the model resolution matrix of the joint inversion will be close to an identity matrix.

In the case of joint inversion including amplitude data and traveltime data, or in the case of inversion using only reflection amplitudes, the largest eigenvalue of the Hessian matrix corresponds to the interface component with the highest wavenumber. A situation in which the maximum sensitivity is for the shortest wavelength would not necessarily lead to a stable inversion, using iterative matrix inversion methods. A requirement of stability is that it is possible to truncate the parameterization at an arbitrary high wavenumber without significantly changing the answer. If maximum sensitivity is at the short wavelength parameter, the answer will depend to a high degree on the choice of truncation wavenumber. The amplitude inversion examples given

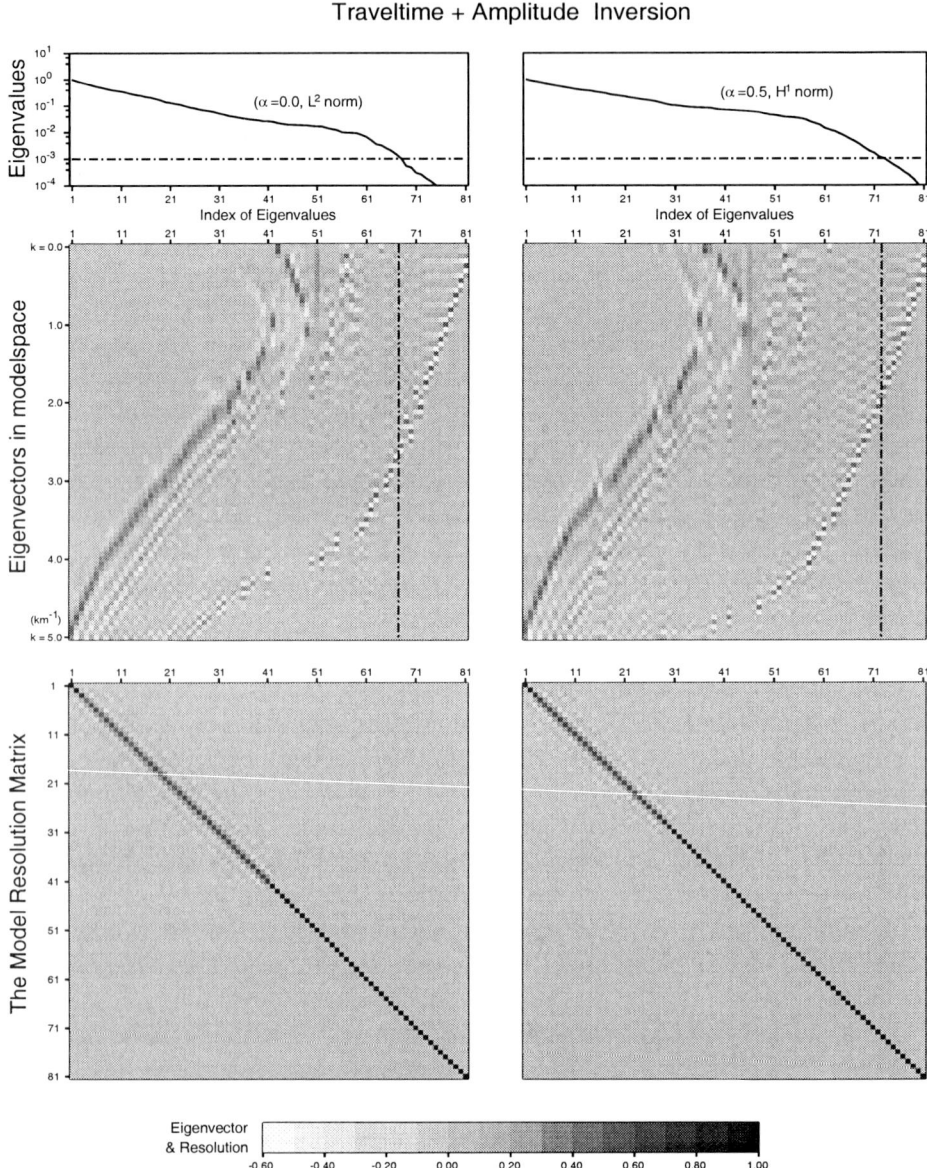

Figure 6.4. Joint sensitivity analysis of reflection traveltimes and amplitudes to the interface wavenumber components.

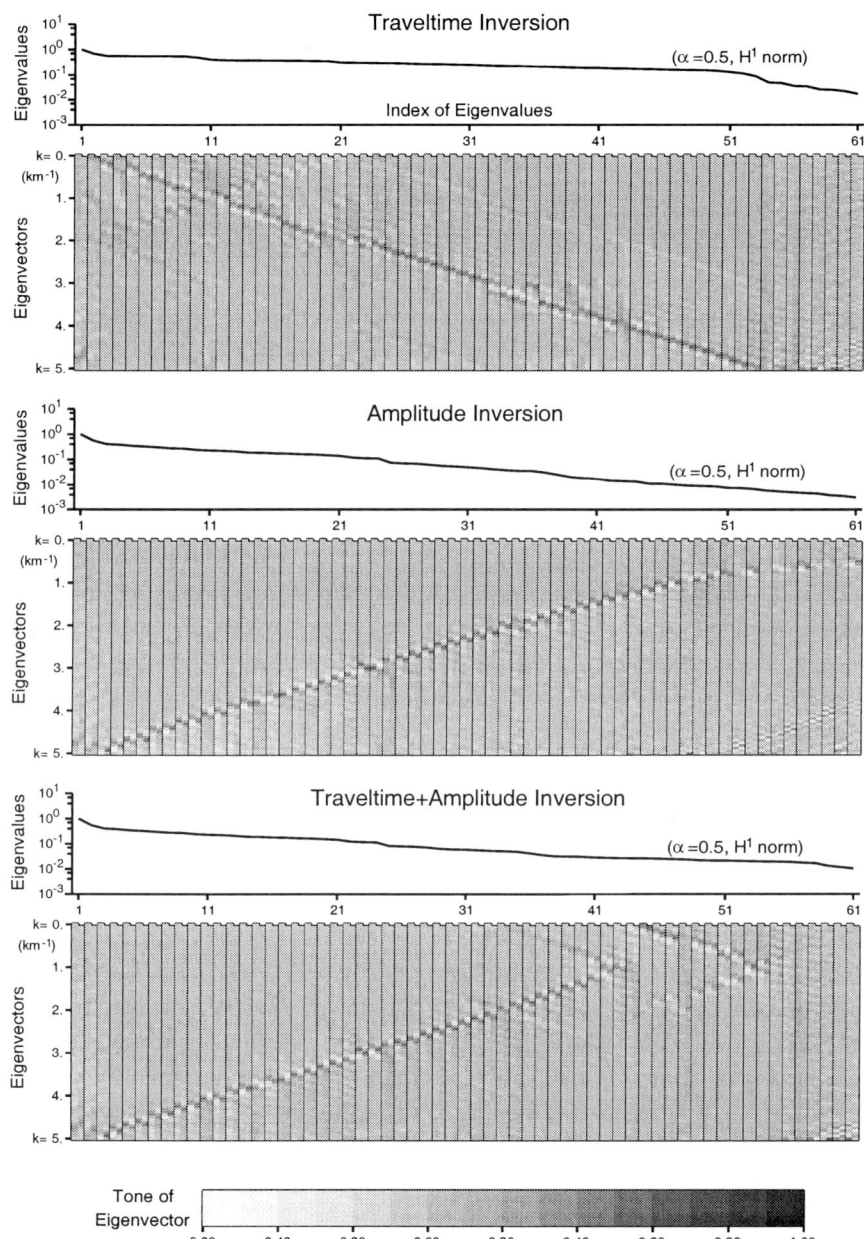

Figure 6.5. Eigen-results of the Hessian matrices ($\alpha = 0.5$, H^1-norm) in traveltime inversion, amplitude inversion and the joint inversion for interface geometry in which both sine and cosine terms are included. Each eigenvector, including the 81 cosine coefficients $\{a_n, (n = 0, 80)\}$ and the 80 sine coefficients $\{b_n, (n = 1, 80)\}$ in the interface parameterization, is diagrammatically shown as two columns, the left column showing the cosine coefficients and the right showing the corresponding the sine terms. Only 61 of the largest eigenvalues and associated eigenvectors are shown.

in Chapter 4 have shown, however, that putting different groups of interface components with different ranges of sensitivity into separate subspaces can effectively stabilize the inversion procedure, where it is assumed that components with sufficiently high wavenumbers to fit the continuous interface (up to the limit of the Fresnel zone width) are included in the model parameterization. An alternative approach is also possible; a traveltime inversion is first performed to reconstruct the longer wavelength components, and then amplitude data are used in the inversion to constrain high wavenumber components of the model.

6.3.5   Determination of the phase

In the parameterization of equation (6.25), we did not explicitly express the phases of the harmonic terms, which control the lateral position of any anomalies on a reflector. Naturally, the smaller the coefficient, the less important is the phase, defined by $\varphi_i = \tan^{-1}(b_i/a_i)$, in describing the overall reflector geometry. If both a_i and b_i are independently sensitive components of the model, then the corresponding phase φ_i would be well determined by inversion. Figure 6.5 shows the eigen-analysis of the Hessian matrix with $\alpha = 0.5$ (the H^1-norm), in which the coefficients of both sine and cosine terms (total 161 parameters) have been included. Only 61 of the largest eigenvalues and the corresponding eigenvectors of the Hessian are shown in the figure. This plot is similar to previous figures, but in this case each eigenvector is plotted with two columns; the left-hand column shows the coefficients of cosine terms and the right-hand column shows the coefficients of the sine terms (note that the zero wavenumber has no sine coefficient). This result was obtained using the uniform single layer model with a single reflector.

From Figure 6.5 we can see that the eigenvalues and associated eigenvectors of the sine functions have the same trend as those of the cosine terms, in each of the three cases of traveltime inversion, amplitude inversion and joint inversion using both types of data. The eigenvectors for traveltime data contain a natural progression from small wavenumbers (associated with large eigenvalues) to large wavenumbers (associated with small eigenvalues). The trend is reversed for amplitude data, and for joint amplitude and traveltime inversion. In the case of traveltime inversion, the cosine and sine terms do not appear to be independently resolved. For the amplitude data, there is some mixing of sine and cosine terms at the largest eigenvalues, associated with the short wavelengths. As the eigenvalues decrease, the cosine and sine terms become better resolved from each other. Also, in the case of joint inversion, at low wavenumbers the model components are not distinct. However, we see in sections 6.6 and 6.7 that the joint inversion can better resolve the interface structure.

6.4.   Sensitivities to 2-D slowness variation

6.4.1   *Slowness representation and the Hessian operator*

We now investigate the sensitivities of seismic reflection traveltimes and amplitudes to 2-D, smoothly varying slowness variations within the layers, again by analysing the eigenvalues in the model space of the Hessian described by different norms in the model space. We represent the slowness distribution within a layer by a truncated 2-D Fourier series,

$$u(\mathbf{r}) = a_{00} + \sum_{m=1}^{N} [a_{m0}\cos(\pi\mathbf{k}\cdot\mathbf{r}) + b_{m0}\sin(\pi\mathbf{k}\cdot\mathbf{r})]$$

$$+ \sum_{m=-N}^{N} \sum_{n=1}^{N} [a_{mn}\cos(\pi\mathbf{k}\cdot\mathbf{r}) + b_{mn}\sin(\pi\mathbf{k}\cdot\mathbf{r})] , \qquad (6.31)$$

where the location vector $\mathbf{r} = x\mathbf{i} + z\mathbf{j}$, the wavenumber vector $\mathbf{k} = m\Delta k\mathbf{i} + n\Delta k\mathbf{j}$, and a_{mn} and b_{mn} are amplitude coefficients of the (m, n)th harmonic term. We assume that slowness varies continuously, and that any discontinuity is explicitly represented by a smooth interface. For the calculation of the Fréchet derivatives that follows, we use the uniform single layer model defined in the previous section.

Considering only 2-D slowness variation, the correlation matrices of the model variation and its horizontal and vertical derivatives are

$$(\hat{\mathbf{B}}_0)_{ij} = \int_{I_\mathbf{x}\times I_\mathbf{z}} \beta_i(x, z)\beta_j(x, z)\, dx\, dz ,$$

$$(\hat{\mathbf{B}}_x)_{ij} = \int_{I_\mathbf{x}\times I_\mathbf{z}} \frac{\partial\beta_i(x, z)}{\partial x} \frac{\partial\beta_j(x, z)}{\partial x}\, dx\, dz , \qquad (6.32)$$

$$(\hat{\mathbf{B}}_z)_{ij} = \int_{I_\mathbf{x}\times I_\mathbf{z}} \frac{\partial\beta_i(x, z)}{\partial z} \frac{\partial\beta_j(x, z)}{\partial z}\, dx\, dz ,$$

where $I_\mathbf{x} \times I_\mathbf{z}$ is the rectangular area through which rays propagate. The operator $\mathbf{D}_\alpha$ can then be expressed as

$$\mathbf{D}_\alpha = (1 - \alpha)\mathbf{B}_0 + \alpha(\mathbf{B}_x + \mathbf{B}_z) , \qquad (6.33)$$

where $\mathbf{B}_0 = (1/w_0)\hat{\mathbf{B}}_0$, $\mathbf{B}_x = (1/w_x)\hat{\mathbf{B}}_x$ and $\mathbf{B}_z = (1/w_z)\hat{\mathbf{B}}_z$, with the scale factors, w, defined by the trace of the matrices as in equation (6.27). As in the analysis of the interface sensitivities in the previous section, in the following numerical experiments we show only the results associated with the coefficients of cosine terms. The symmetry inherent in the model parameterization of 2-D Fourier series implies that only 1/4 of the k-plane ($k_z \geq 0, k_x \geq 0$) is required. However, we show the result on half the k-plane ($k_z \geq 0$) in the following analysis and attempt to gain an insight into the dependence on the phase (spatial location) of the slowness variation.

6.4.2   Sensitivity analysis with the L^2-norm

Figure 6.6 shows the eigenvalues and associated eigenvectors of the Hessian matrix and the model resolution matrices for traveltime and amplitude inversions. The L^2-norm ($\alpha = 0.0$) is chosen in the model space. For the slowness model, the maximum wavenumber under consideration is 5.0 km^{-1}. The discretization interval of wavenumber $\Delta k = 1.0$ km^{-1} ($N = 5$), is set in equation (6.31). Therefore, 61 ($= 5 \times 5 + 6 \times 6$) cosine coefficients a_{mn} are determined in the eigen-analysis. In each eigenvector square, the horizontal wavenumbers of model components increase from left to right (with the wavenumber ranging from -5 to 5 km^{-1}) and the vertical wavenumbers increase from top to bottom (from zero to 5 km^{-1}). Only those 20 eigenvectors associated with the largest eigenvalues of the Hessian are shown in the figure. These eigenvectors are the linear combinations of model parameters which have the most influence on the traveltime and amplitude data. The model resolution matrices for each inversion are displayed as a 61 $\times$ 61 matrix split into several panels, each corresponding to a constant horizontal wavenumber k_x. The diagonal elements of the resolution matrices are displayed in an inset in these figures.

From the eigenvalues and associated eigenvectors of the Hessian shown in Figure 6.6 we see that in a traveltime inversion the objective function is most sensitive to the model components with smaller wavenumber, typically to the zero-valued component. However, in amplitude inversion the most sensitive components seem to prefer large k_x values, but mid-range k_z values, as seen from the first and second eigenvectors associated with the largest eigenvalues. For the surface geometry that was employed, both mid-range k_z and large k_x cause large transverse slowness derivatives, which focus or defocus energy. Neele *et al.* (1993a) showed explicitly the dependence of amplitudes on the higher transverse derivatives of the slowness field and, therefore, on the higher wavenumber components. This dependence was also illustrated in their inversions using real data (Neele *et al.*, 1993b).

To calculate the model resolution matrix of a pseudo-matrix inversion, we truncate eigenvalues at 10^{-3}, as in the previous section with interface inversion, indicated by a dash-dotted line in the figure. The model resolution matrices shown in the bottom of the figure clearly indicate which slowness component

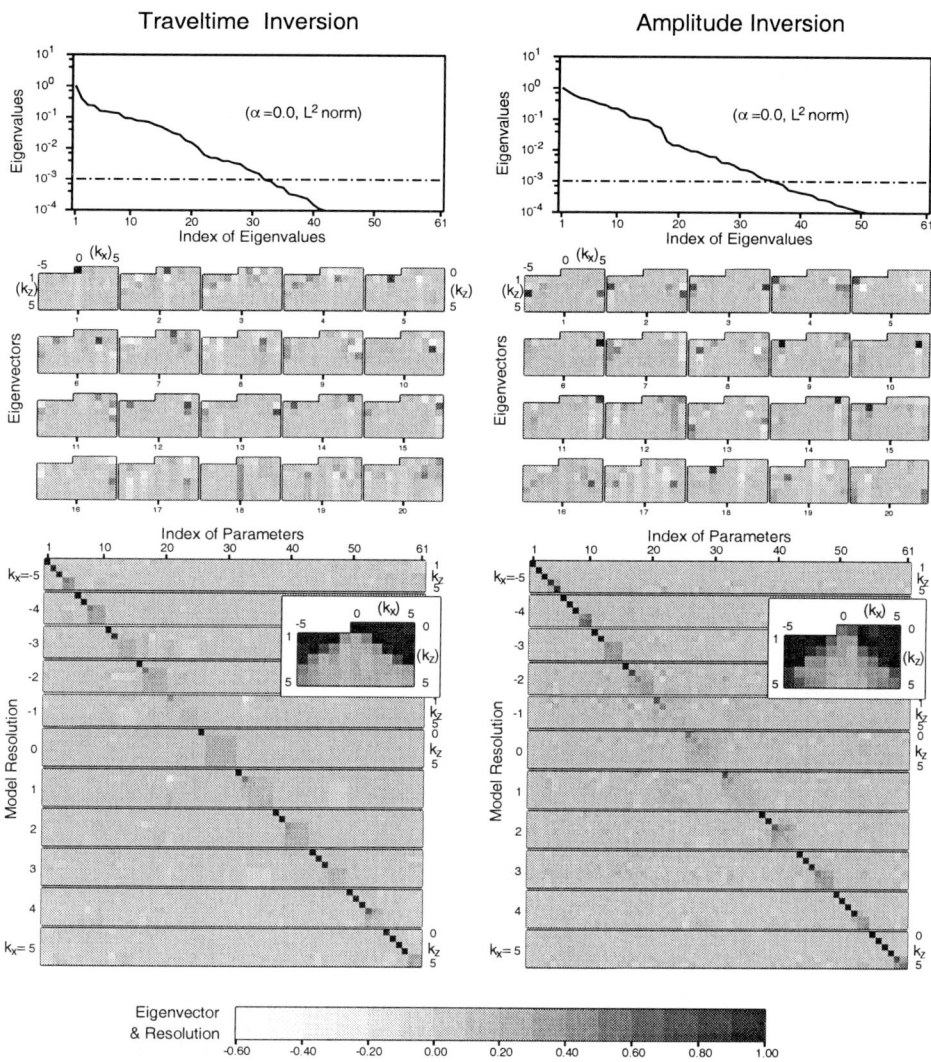

Figure 6.6. Eigenvalues and eigenvectors of the Hessian matrix and the model resolution matrices for traveltime and amplitude inversions for slowness variation. The L^2-norm is chosen in the model space. In each eigenvector image, horizontal wavenumbers of the model components increase from left to right (with the wavenumber ranging from -5 to 5 km^{-1}) and the vertical wavenumbers increase from top to bottom (from zero to 5 km^{-1}). Only those 20 eigenvectors associated with the largest eigenvalues of the Hessian are shown. The diagonal elements of each resolution matrix, computed using a pseudo-inverse by SVD truncation, are shown in the matrix-like rectangular inset.

can or cannot be constrained by the pseudo-inverse. Slowness components of vertical variation at higher (absolute) horizontal wavenumbers k_x can be resolved by both traveltime and amplitude inversions with the L^2-norm chosen in model space. When the (absolute) horizontal wavenumber decreases, the constraint on the vertical variation becomes gradually weaker. When $k_x = 0$, neither inversion can constrain the vertical variation.

These observations are summarized by the 6×11 matrix-like rectangle shown in the inset on each model resolution matrix, consisting of the diagonal elements of the model resolution matrix. When the diagonal element is equal to 1, the corresponding component of the solution is the true model parameter. If the element is zero, the inversion cannot constrain the model component at all. The diagonal elements clearly show that neither inversion can resolve wavenumbers in the purely vertical direction, that traveltimes cannot constrain any vertical variation with high wavenumber k_z, but that amplitude inversion can apparently constrain the vertical variation of slowness at a high horizontal wavenumber k_x.

6.4.3   *Sensitivities with penalization of spatial derivatives*

We now examine the sensitivities shown in Figure 6.7, in which the H^1-norm was chosen in the model space. In the traveltime inversion, the sensitivities to the wavenumber appear almost identical to the case with the L^2-norm shown in Figure 6.6. For the amplitudes, the eigenvectors corresponding to the largest eigenvalues show that they are still sensitive to the combination of the components with high wavenumber **k**, as with the L^2-norm, but that additional sensitivity to the components of intermediate wavenumber has been created. The sensitivity to the intermediate components is due to the use of the smoothing regularization. If the degree of smoothing increases, the inversion is still more sensitive to the small wavenumber components, as can be seen from the result with $\alpha = 0.99$ shown in Figure 6.8.

In a reflection amplitude inversion, once we set $\alpha = 0.99$ in the scalar product of model perturbation, the objective function is sensitive to perturbations in the background slowness a_{00}, as seen from the first and second eigenvectors. This apparent sensitivity is due to the constraints. The amplitudes are still sensitive, however, to the slowness components with larger (absolute) wavenumber (shorter wavelength), as shown in eigenvectors 1 and 2. For the traveltime inversion, the absolute value of the largest eigenvalue is increased, due to the stronger penalization of the spatial derivatives. The model resolution matrices, after the rejection of eigenvectors associated with small eigenvalues (Figure 6.8), show that the traveltime inversion can only constrain the slowness components defining horizontal variation with $n = 0$ and $n = 1$. However, the amplitudes provide information on the high wavenumber components. The model resolution ma-

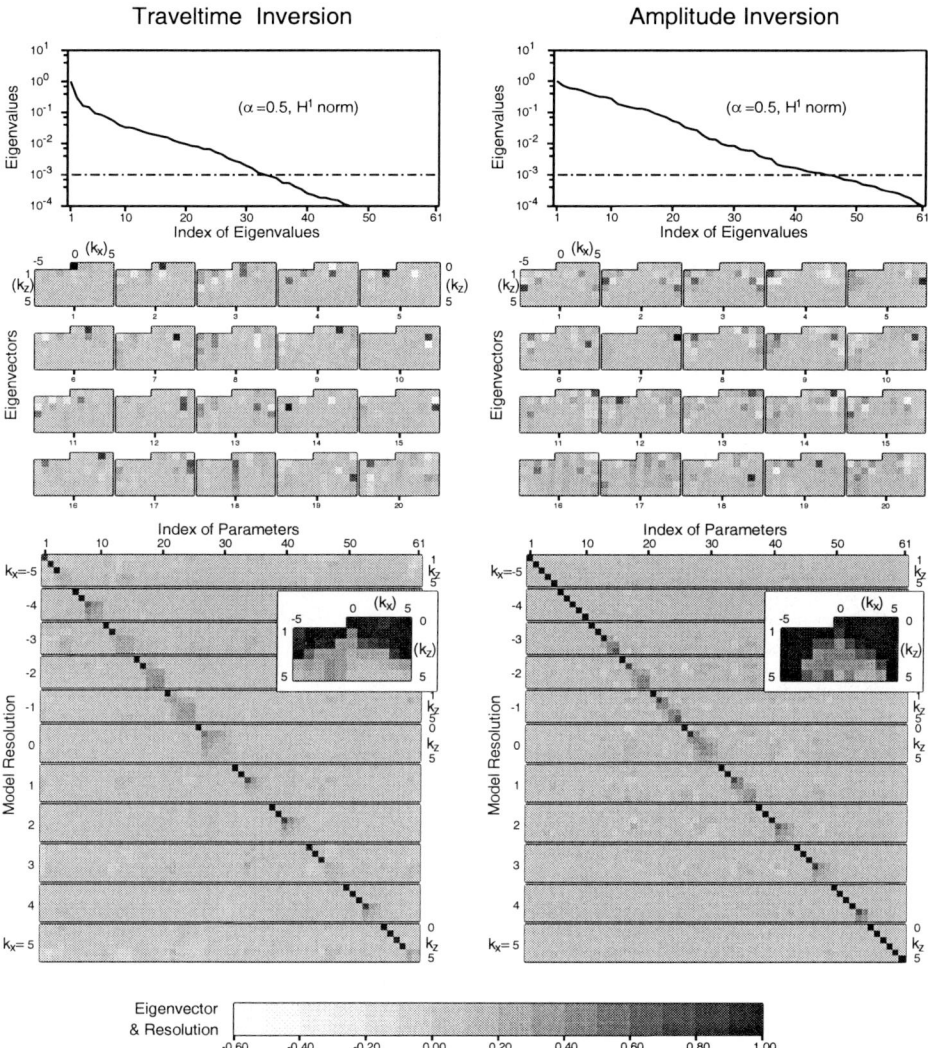

Figure 6.7. Eigenvalues, associated eigenvectors of the Hessian matrix and the model resolution matrices for traveltime and amplitude inversions for slowness variation. The H^1-norm ($\alpha = 0.50$) is chosen in the model space.

trices for the amplitude inversion, in both the cases of $\alpha = 0.50$ and $\alpha = 0.99$, are almost an identity matrix. The different information content of amplitudes and traveltimes is again evident.

From the sensitivities shown in Figures 6.6, 6.7 and 6.8, we see that the eigenvectors clearly show a symmetrical pattern for $k_x > 0$ and $k_x < 0$ in each case. Therefore, ambiguities between different harmonic components inevitably exist in the inversion.

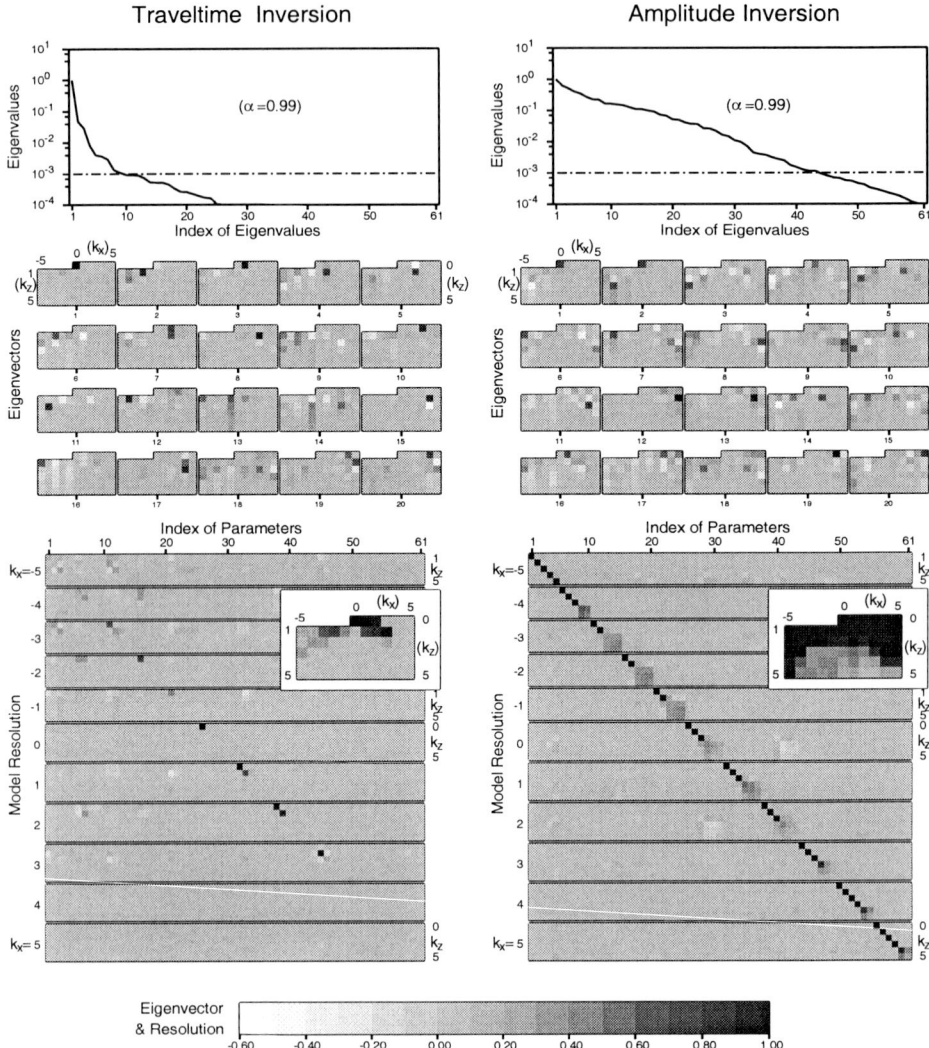

Figure 6.8.   Eigenvalues, associated eigenvectors of the Hessian matrix and the model resolution matrices for traveltime and amplitude inversions for slowness variation, where $\alpha = 0.99$ is used.

6.4.4   Joint inversion for slowness variations

For a joint inversion using both amplitude data and traveltime data simultaneously for slowness variation, we use the same definition of the data covariance matrix as equation (6.29). In equation (6.29) the parameter κ is used to balance the relative contributions of the traveltime and the amplitude data, and is determined by using the ratio of the traces of the two sensitivity matrices, $\mathbf{F}_{time}$ and $\mathbf{F}_{ampl}$.

For the joint inversion of both traveltimes and amplitudes for interface geometry, we determined a value of $\kappa = 1.685 \times 10^{-2}$; the traveltimes were not as sensitive to the geometry as the amplitudes were. In contrast, for this example in which we are inverting the same data for the interval slowness variation, we find the ratio of traces of the sensitivity matrices is $\kappa = 16.81$; the traveltimes are significantly more sensitive to the slowness model, than are the amplitudes. Once more, this is an indication that amplitudes and traveltimes are sensitive to different physical parameters. This is an important result, with a potential benefit that it may be possible to better resolve the known ambiguity between reflector depth uncertainty and interval velocity uncertainty by including amplitude data in seismic reflection traveltime tomography.

The experiments we carried out that the range of 0.75 to 1.25 for the second balancing factor g is appropriate in this case. For $g \leq 1.0$, the contributions of traveltimes and amplitudes when the L^2-norm is used are balanced, whereas traveltimes dominate the inversion with the H^1-norm. For $g > 1.0$, amplitudes dominate the inversion with the L^2-norm, whereas traveltimes and amplitudes are balanced in the case of inversion with the H^1-norm. The result with the balancing factor $g = 1.0$ is shown in Figure 6.9.

From Figure 6.9 we see that, with the L^2-norm used in the model space, the objective function is most sensitive to the background slowness a_{00}, dominated by traveltime data, and then to components with high wavenumber $\mathbf{k}$, dominated by amplitude data. With the H^1-norm in the inversion, the most significant parameters in the slowness model are the background slowness a_{00} and those components defining the horizontal variation of the slowness. The sensitivities to the variation with high wavenumber $\mathbf{k}$ are reduced relative to the background slowness.

In this subsection and in subsection 6.3.4, we have considered a joint inversion using both traveltimes and amplitudes. In the following sections, we show some synthetic examples of inversion for interface geometry and velocity variation separately, and explore how the inclusion of amplitudes in the tomography improves the solution of traveltime inversion.

6.5.   Inversion formula

The solution for the linearized inverse problem (equation 6.9) can be given by

$$\delta\mathbf{m} = (\mathbf{F}^{\dagger}\mathbf{C}_D^{-1}\mathbf{F} + \mu\,\mathbf{I})^{-1}\mathbf{F}^{\dagger}\mathbf{C}_D^{-1}\delta\mathbf{d} \,, \tag{6.34}$$

where μ is a dimensionless damping factor introduced to stabilize the inverse procedure, $\mathbf{I}$ is the identity matrix with units of *(model parameter)*$^{-2}$, and $\mathbf{C}_D^{-1}$ is the data covariance matrix containing the two weighting parameters, κ and

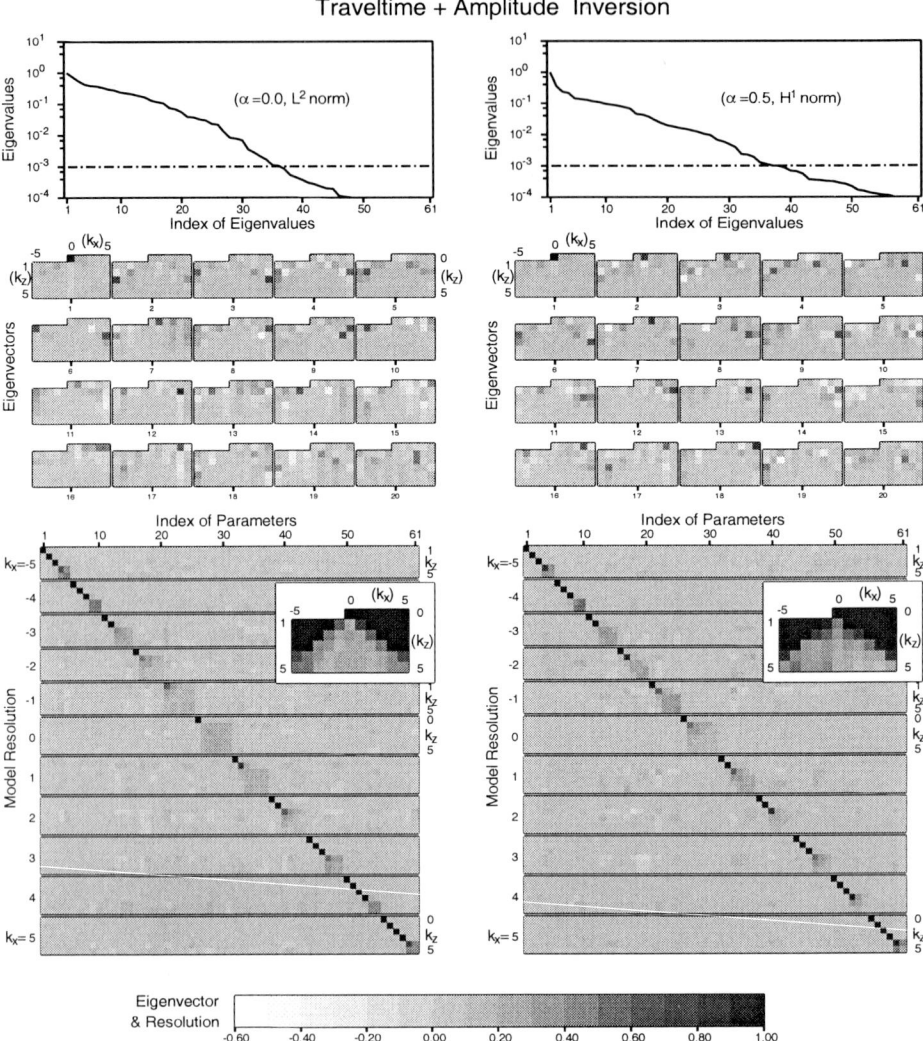

Figure 6.9. Eigenvalues and eigenvectors of the Hessian matrix and the model resolution matrices in the case of a joint inversion of traveltime and amplitude data for slowness variation.

g discussed earlier. Equation (6.34) contains the full Hessian matrix

$$\mathbf{H} = \mathbf{F}^{\dagger}\mathbf{C}_{\mathrm{D}}^{-1}\mathbf{F} = \mathbf{D}_{\alpha}^{-1}\mathbf{F}^{T}\mathbf{C}_{\mathrm{D}}^{-1}\mathbf{F}$$

(equation 6.24). In spite of the use of the additional constraint $\mathbf{D}_{\alpha}$ ($\alpha \neq 0$), we still require a strategy for selecting the damping factor, μ; as we have seen, even with the H^{1}-norm, the Hessian is a singular matrix. The choice of μ is addressed below.

The adjoint $\mathbf{F}^{\dagger}$ of the matrix $\mathbf{F}$ being given by $\mathbf{F}^{\dagger} = \mathbf{D}_{\alpha}^{-1}\mathbf{F}^{T}$ (equation 6.23), equation (6.34) can be expressed as

$$\delta\mathbf{m} = (\mathbf{F}^{T}\mathbf{C}_{D}^{-1}\mathbf{F} + \mu\,\mathbf{D}_{\alpha}\,\mathbf{I})^{-1}\mathbf{F}^{T}\mathbf{C}_{D}^{-1}\delta\mathbf{d}\,, \qquad (6.35)$$

where the operator $\mathbf{D}_{\alpha}$ is introduced to penalize first spatial derivatives of the model, and then can be alternatively understood as a working definition of the model covariance matrix, $\mathbf{C}_{M}^{-1}$, expressing expected correlations between different model parameters.

In fact, if we set $\mathbf{C}_{M}^{-1} = \mathbf{D}_{\alpha}\,\mathbf{I}$, equation (6.35) will be the solution of the well-known stochastic inversion (Franklin, 1970; Jackson, 1979; Tarantola and Val-lette, 1982; Tarantola, 1987b), which stabilizes the inverse problem by adding a term to the data misfit function that depends on the *a priori* model, $\mathbf{m}_{ref}$, and its covariance matrix,

$$J(\mathbf{m}) = \frac{1}{2}\{[f(\mathbf{m}) - \mathbf{d}_{obs}]^{T}\mathbf{C}_{D}^{-1}[f(\mathbf{m}) - \mathbf{d}_{obs}]$$

$$+\mu\,(\mathbf{m} - \mathbf{m}_{ref})^{T}\mathbf{C}_{M}^{-1}(\mathbf{m} - \mathbf{m}_{ref})\}\,, \qquad (6.36)$$

where the scalar μ acts as a trade-off parameter that controls the relative weights of *a priori* information and the observed data. The model covariance matrix has been extensively used to remove numerical instabilities by damping poorly constrained parameters towards the reference model, and allowing only well-constrained parameters to be controlled by the data. However, the dependence of the second term in equation (6.36) on the reference model $\mathbf{m}_{ref}$ may result in unwarranted structure. In contrast, the model regularization given by the operator $\mathbf{D}_{\alpha}$ is defined by the scalar product of the basis functions $\{\beta_{i}(\mathbf{x}),\ \forall\,i(1\leq i \leq M)\}$ (and not by the model $\mathbf{m}$ or the perturbation $\delta\mathbf{m}$, see equation 6.14). Thus the form of the constraint used in equation (6.35) does not depend on the starting model or on any arbitrary reference model. This would appear to be more appropriate for examining lateral variations in seismic structure, especially when no strong *a priori* model exists for the region (Constable *et al.*, 1987; Sambridge, 1990).

In the following synthetic examples, we use the inversion formula (equation 6.35), in which $\alpha = 0.50$ (the H^{1}-norm) in the scalar product $\mathbf{D}_{\alpha}$ is arbitrarily chosen. $\mathbf{D}_{\alpha}$ contains an intermediate degree of penalization for the spatial derivatives. The parameters κ and g used for $\mathbf{C}_{D}^{-1}$ are the same as in the sensitivity analysis. The trade-off parameter μ is determined using the relation $\mu = \rho/r$, where

$$r = \frac{\mathrm{trace}(\mathbf{D}_{\alpha}\,\mathbf{I})}{\mathrm{trace}(\mathbf{F}^{T}\mathbf{C}_{D}^{-1}\mathbf{F})} \qquad (6.37)$$

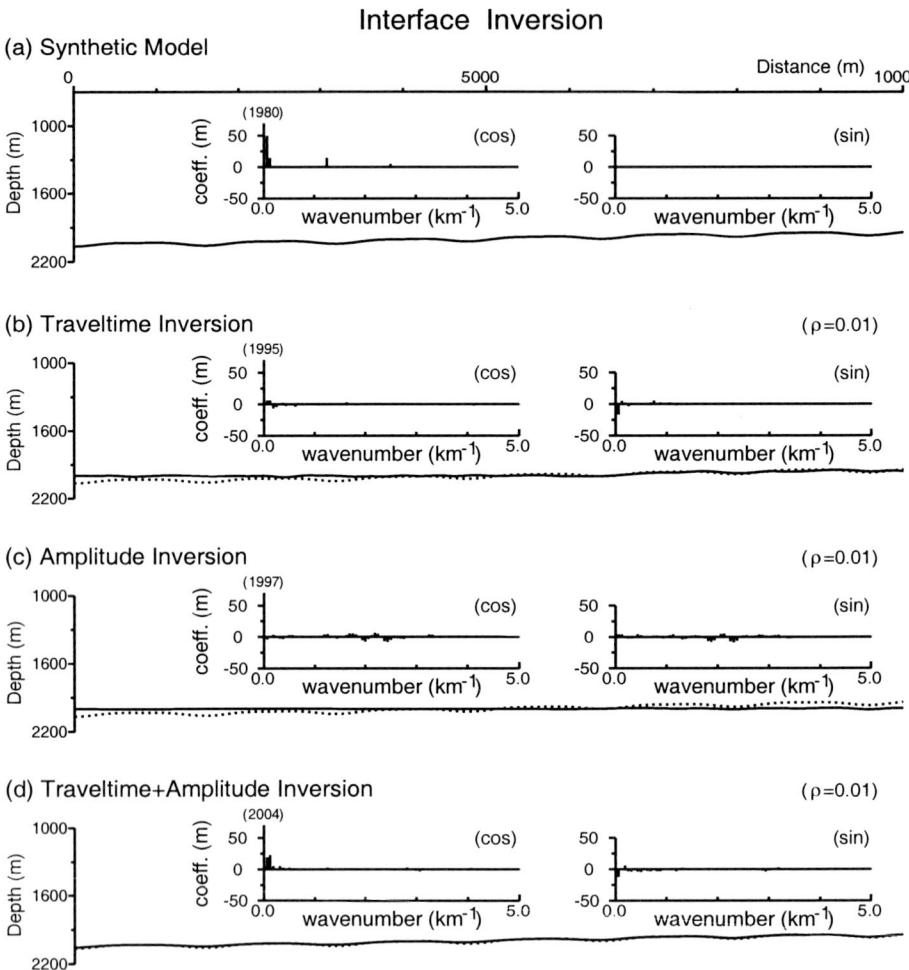

Figure 6.10.   An example of interface inversion: (a) synthetic model; (b) traveltime inversion; (c) amplitude inversion; (d) the joint inversion using both traveltime and amplitude data simultaneously. Dotted lines in (b), (c) and (d) show the true interface, which is compared with the inversion solution (solid lines), after the first iteration. Model parameters, i.e. the amplitude coefficients of the Fourier components, are also depicted in the figure.

is an empirical quantity used to normalize the matrices. Equation (6.35) is thus rewritten as

$$\delta \mathbf{m} = r \left(r \mathbf{F}^{\mathrm{T}} \mathbf{C}_{\mathrm{D}}^{-1} \mathbf{F} + \rho \mathbf{D}_\alpha \mathbf{I} \right)^{-1} \mathbf{F}^{\mathrm{T}} \mathbf{C}_{\mathrm{D}}^{-1} \delta \mathbf{d} \,, \tag{6.38}$$

where the dimensionless factor ρ easily controls the total amount of regularization in the inversion.

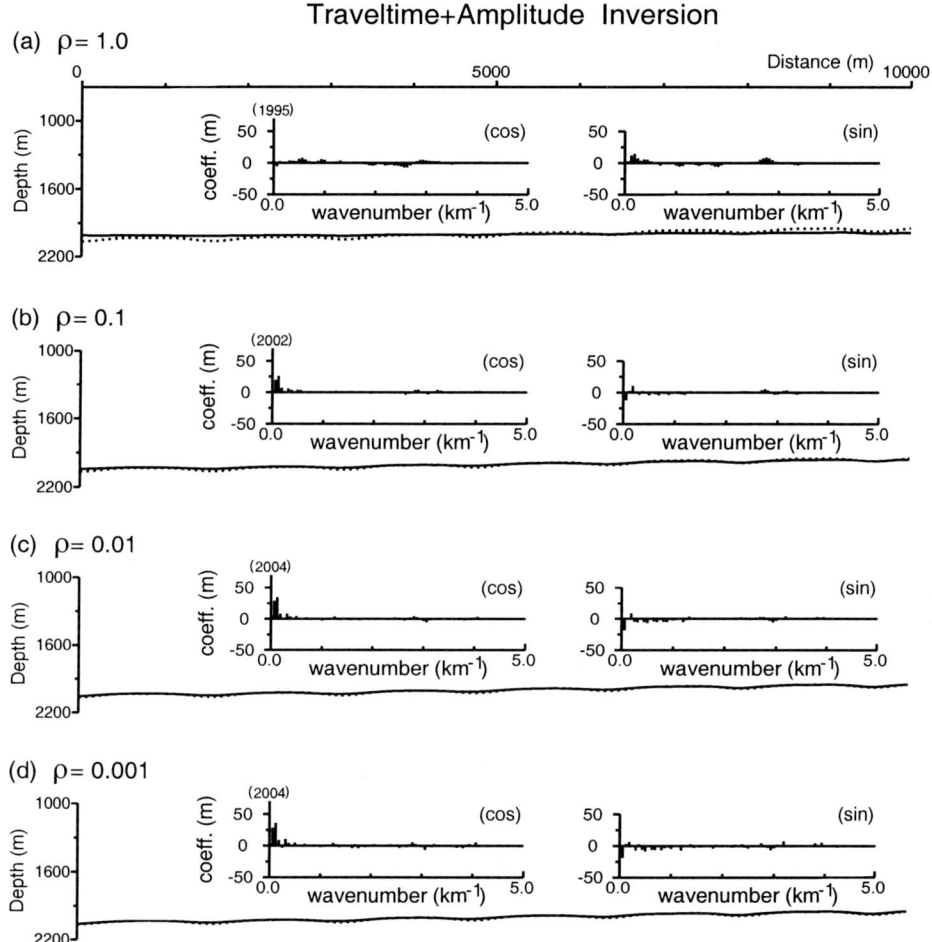

Figure 6.11. Four results of joint inversion using both traveltime and amplitude data, for interface geometry. The values of ρ used in the inversion formula (equation 6.38) are (a) 1.0; (b) 0.1; (c) 0.01; (d) 0.001.

6.6. Joint inversion for an interface

Let us consider an interface (shown in Figure 6.10a) defined by equation (6.25), in which the amplitude coefficients a_0, a_1, a_2, a_{20} and a_{40} are 1980, 50, 15, 15 and 5 m, respectively, the remaining coefficients are zero, and the fundamental wavenumber Δk is 1/16 km^{-1}. The amplitude coefficients, a_n of the cosine terms and b_n of the sine terms, are also shown, plotted against wavenumbers in the figure. For the calculation of synthetic data, we use the same observation configuration as in the previous sections. In the inversion, the starting model is given by a flat straight reflector, at a depth of 2000 m (the same uniform structure used in sensitivity analysis). The number of harmonics in the parameterization is set at $N = 80$ in equation (6.25). We therefore have 161 parameters (81 cosine and 80 sine terms) in the inversion.

The solutions of traveltime inversion, amplitude inversion and the joint inversion using both types of data are shown in Figures 6.10b–d, respectively. Although the problem would normally be solved iteratively, the results after only one iteration are sufficient to show the dramatic differences between the inversions. These inversions were run with the factor ρ in equation (6.38) set equal to 0.01.

In the case of traveltime inversion with the H^1-norm, the zero and low wavenumber components were shown to be the most sensitive components (see Figure 6.5a). However, from Figure 6.10b we see that the solution after the first iteration for the components with zero wavenumber and low wavenumber are compromised. In practice, more iterations may be needed, and the Fréchet matrix for each iteration will need to be recalculated.

In the case of the amplitude inversion shown in Figure 6.10c, we see that slight changes to interface components with intermediate wavenumber have been obtained and that the long wavelength components were not obtained.

In the joint inversion using both types of data, however, the solution (Figure 6.10d) is dramatically improved, in a single iteration. The significance of these figures is that neither inversion alone (Figures 6.10b and 6.10c) is successful in resolving the two smallest wavenumbers, but that the joint inversion (Figure 6.10d) does begin to resolve them.

Figure 6.11 shows a comparison of joint inversions with different values of ρ applied. Penalizing first derivatives, introduced by $\mathbf{C}_M^{-1}$ or the operator $\mathbf{D}_\alpha$ ($\alpha = 0.5$, the H^1-norm) in equation (6.38), implies that a near horizontal solution is to be preferred. As the value of ρ decreases from 1.0 (Figure 6.11a) to 0.001 (Figure 6.11d), the inversion gradually allows the inclusion of larger slopes. In this case, the slope of the interface is dominated by the amplitude coefficients of harmonic terms with small wavenumber. It is clearly seen from the figure that these amplitude coefficients gradually increase as ρ decreases.

6.7.   Joint inversion for slowness

To demonstrate the inversion for velocity variation, we give a synthetic model, shown in Figure 6.12a and defined by equation (6.31), in which we set

$$a_{00} = 0.4, \quad a_{10} = -0.001, \quad a_{20} = 0.005, \quad a_{13} = -0.004, \quad a_{23} = -0.005,$$

$$b_{10} = 0.004, \quad b_{20} = -0.006, \quad b_{12} = -0.008, \quad b_{22} = 0.002 \ (s/km)$$

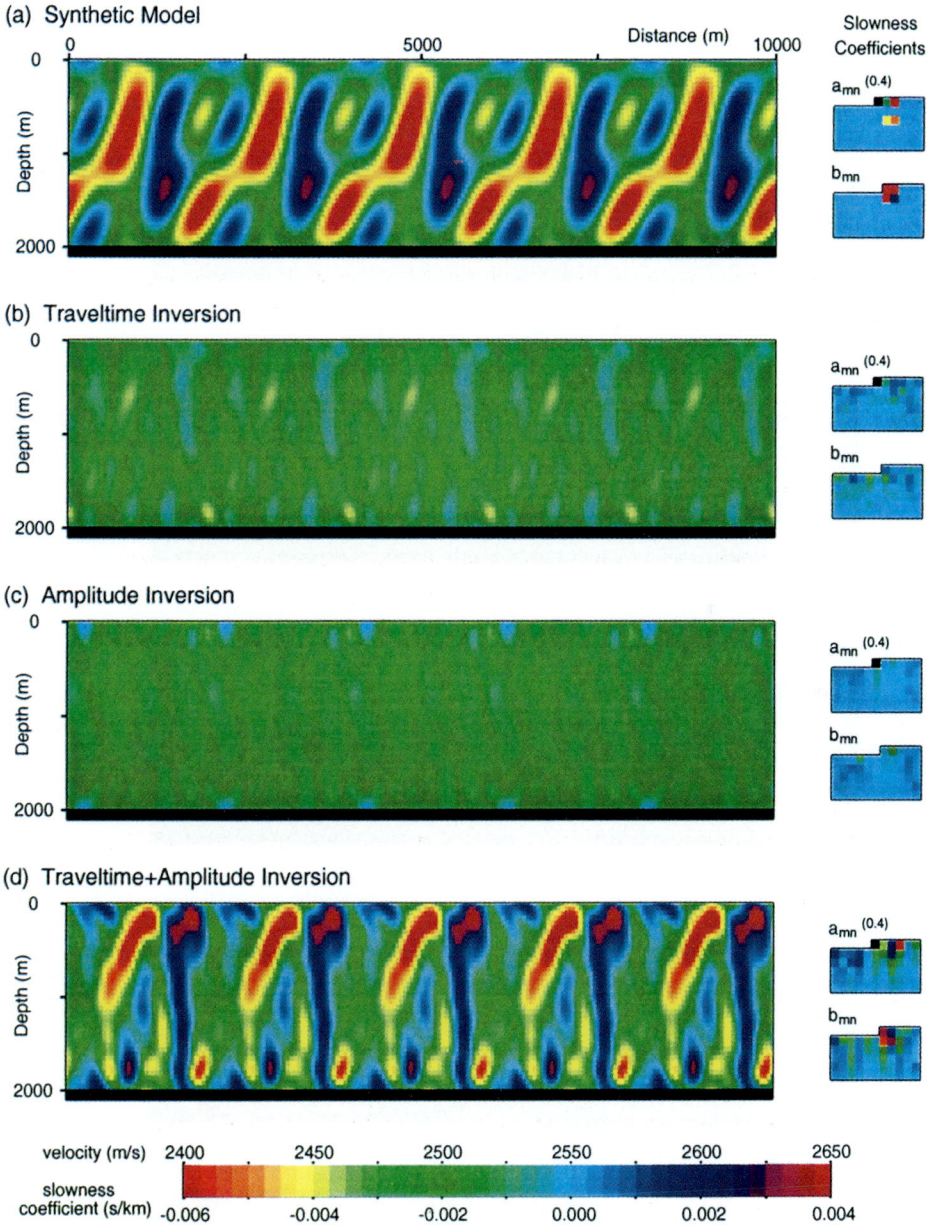

Figure 6.12. An example of velocity inversion: (a) synthetic model; (b) traveltime inversion; (c) amplitude inversion; (d) the joint inversion using both traveltime and amplitude data simultaneously, where $\rho = 0.01$ is used in the inversion (equation 6.38) to control the amount of regularization in the inversion. Model parameters, i.e. the coefficients of the Fourier components, are graphically represented as images on the right side of the figure.

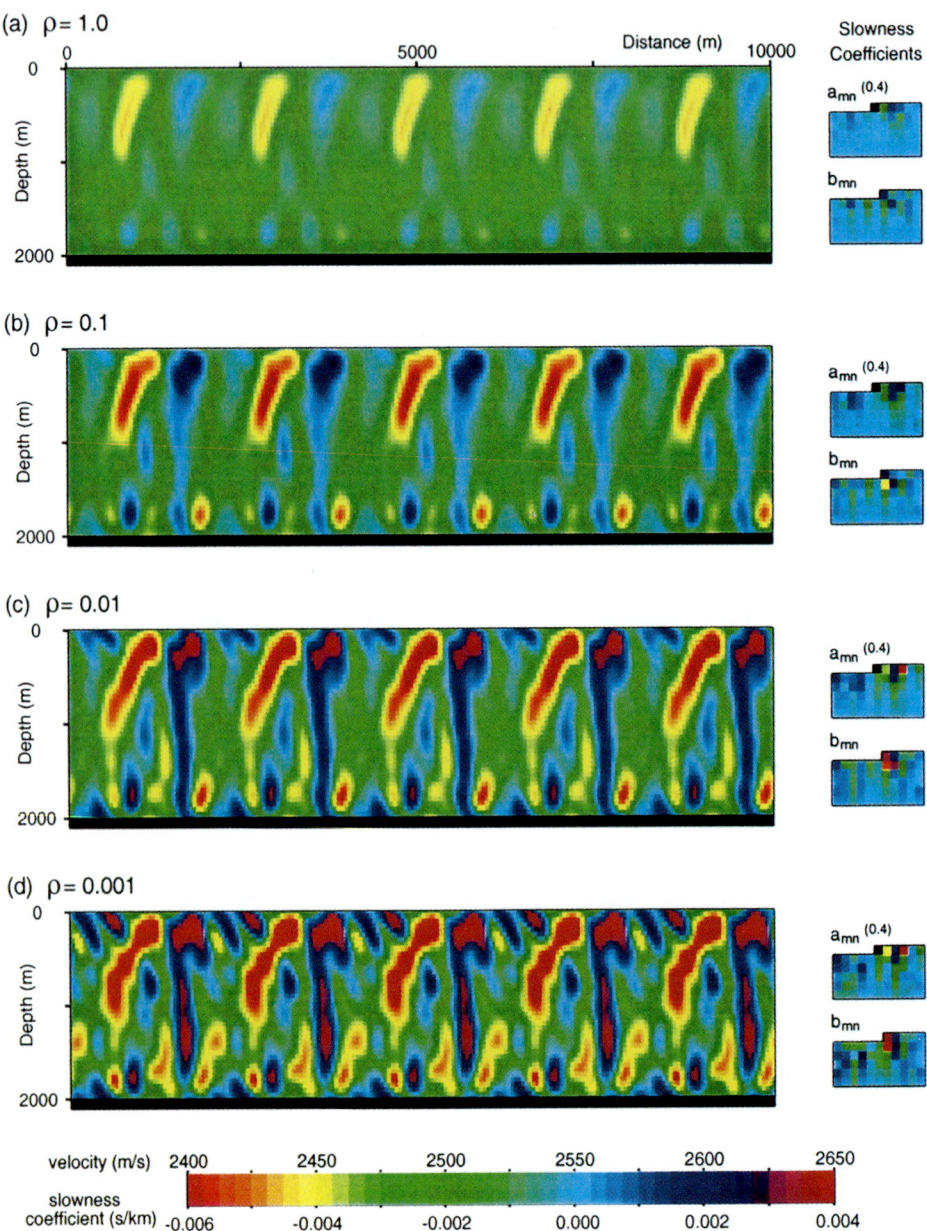

Figure 6.13.   Four results of joint inversion using both traveltime and amplitude data for velocity variation. The values of ρ used in the inversion formula (equation 6.38) are (a) 1.0; (b) 0.1; (c) 0.01; (d) 0.001.

with the remaining coefficients being zero-valued, and $\Delta k = 1.0$ km^{-1}. A graphical representation of thsse amplitudea coefficients a_{mn} and b_{mn} is also shown in the figure. In the inversion we set $N = 5$ in the model parameterization, therefore, a total of 121 parameters (61 cosine coefficients a_{mn} and 60 sine coefficients b_{mn}) are determined. The starting model is given by the constant background slowness (a_{00}). The Fréchet matrix based on this starting model was used in the previous section for the sensitivity analysis.

The results of traveltime and amplitude inversions and the joint inversion, after the first iteration, are shown in Figures 6.12b–d, respectively. In those inversions, $\rho = 0.01$ is used. From Figure 6.12 we see that both traveltime and amplitude inversions show some image of velocity variation, but far from the true model. It is interesting to note that the amplitude inversion appears to recover anomalies close to the surface and close to the interface, whereas the traveltime inversion appears to recover the anomalies in between. The joint inversion shows an encouraging result, in which the basic features of the velocity variation are reconstructed.

Figure 6.13 presents four solutions of the joint inversion using both travel-time and amplitude data simultaneously. These solutions differ in the value of ρ used. Comparing Figure 6.13 with the synthetic model (Figure 6.12a), we see that the anomalous structure is clearly reconstructed by using the constraint of both types of data, when we set $\rho = 0.1$ (Figure 6.13b) or $\rho = 0.01$ (Figure 6.13c). When $\rho = 1.0$ (Figure 6.13a) the inversion gives the smoothest model (in the sense of the first derivatives of the slowness), whereas $\rho = 0.001$ (Figure 6.13d) gives the roughest structure. Clearly, the value of ρ cannot be reduced indefinitely; it is required in order to stabilize the inversion in the presence of noise.

In this section we have considered a simultaneous inversion for the slowness variation, using both types of data (traveltimes and amplitudes). However, the numerical analysis performed by Neele *et al.* (1993a) suggests that, when amplitudes and traveltimes are combined in an inversion for the velocity structure, a non-linear traveltime inversion must be performed before amplitude data are included, ensuring that the non-linear behaviour of amplitudes due to ray shift induced by slowness perturbations is minimized. This is because traveltimes show a more linear dependence on slowness perturbations than do amplitudes, whereas amplitudes depend on the spatial derivatives of the slowness distribution. The success in the simultaneous inversion shown in this chapter is partially due to the application of operator $\mathbf{D}_\alpha$, which penalizes the first derivatives of slowness variation, and partially due to the data covariance matrix we proposed, which balances the contribution, in terms of relative sensitivities, of different types of data.

6.8.   Discussion

We have compared the sensitivities of reflection seismic amplitudes and traveltimes to the variation in interface geometry and slowness distributions. In general, the information content of the traveltimes and the amplitudes are complementary, being sensitive to different features of the model. Traveltimes are more sensitive to model components with smaller wavenumbers, but amplitudes are more sensitive to the components defined by the larger wavenumbers.

The ray theory used in this chapter, however, does not account for the finite extent of the Fresnel zone. Under the ray approximation, the sensitivity of amplitudes to lateral slowness or interface variations is always dominated by the high-end k_x values. The amplitude depends on the second derivatives of the traveltime function; short-wavelength structure has the largest effect on these derivatives, causing large amplitude fluctuations. The way around this is to include the Fresnel zone in the computation of the Fréchet derivatives, removing the local character of ray theory. Snieder and Lomax (1996) presented a procedure for doing this in an intuitive fashion, but more exact methods are possible, although expensive. In this procedure, the sensitivity of amplitudes becomes a maximum for scale lengths comparable to the size of the Fresnel zone, with lower sensitivity for both larger and smaller scale lengths, without the need for elaborate weighting schemes or roughness penalization terms in the misfit function. Our solution of cutting off the parameterization of the model at scale lengths comparable to the size of the Fresnel zone is a first step towards a more complete and accurate description of the dependence of amplitudes on slowness and discontinuity structure and, therefore, the main conclusions of the present investigation would be unaltered if a more exact forward method were used. However, the size of the amplitude Fréchet derivatives would naturally be different.

The investigation into amplitude sensitivity confirms the observations from the inversion examples given in Chapters 4 and 5. Although the previous chapters showed that seismic amplitude data contain information which independently constrains aspects both of reflector geometry and of the slowness variation, the amplitude inversion should not be viewed as an alternative to reflection traveltime tomography. It should be viewed as a complementary set of constraints on the inversion problem for reflector geometry and interval slowness. For a joint inversion using both traveltime and amplitude data, an empirical definition of the data covariance matrix which balances the relative sensitivities of different types of data has been proposed. The inversion results suggest that in this manner a joint inversion can provide a much better solution than that using one type of data alone.

Amplitudes and traveltimes are also sensitive to different physical parameters. The ratio of traces of the sensitivity matrices has shown that traveltimes are more sensitive to variations in the slowness model than are amplitudes, whereas amplitudes are more sensitive to variations in interface geometry than are traveltimes. Therefore, we may design a cost-effective joint inversion for slowness variation and interface geometry, using both types of data, in which traveltimes dominate the determination of the slowness variation and amplitudes dominate the determination of the interface geometry. Because amplitudes can provide better constraints on model components with shorter wavelength and traveltimes are more sensitive to longer wavelength components, the joint inversion may better resolve the known ambiguity between reflector depth uncertainty and interval slowness uncertainty in seismic reflection traveltime tomography. This prospect needs to be investigated by further inversion tests.

Chapter 7

Amplitude inversion of a multi-layered structure

Abstract   In this chapter we focus our investigation on amplitude inversion for constraining the interface geometry of, typically, a multi-layered structure. We find that the two main difficulties in the amplitude inversion are the high wavenumber oscillation in the shallower interface and the failure to reconstruct long wavelength components of deep, curved interfaces. A damped subspace method may overcome the first difficulty, and a multi-scale scheme may be used to combat the second problem. To enhance the resolution of the low wavenumber components and, simultaneously, to reduce the instability of high wavenumber components in the amplitude inversion, we then develop a so-called *multi-stage damped, subspace method*, a combination of these two schemes. We see from the inversion experiments that the multi-stage damping scheme is an appropriate inversion scheme for obtaining a geologically realistic model.

7.1.   Introduction

When a ray beam illuminates a curved interface, the curvature of the interface will cause focusing or defocusing of the ray tube propagating by transmission or reflection, and will significantly influence the seismic amplitude recorded at the surface. That is why the reflection amplitude data can be used, separately or jointly with traveltime data in reflection seismic tomography, to constrain the geometry of subsurface reflectors. In Chapter 4 we saw an investigation on this topic but it was limited to a model consisting of only one reflecting interface. We now extend this investigation to a multi-layered structure.

As the number of interfaces increases, the difficulty in reconstructing curved interfaces will increase accordingly. We use a realistic multi-layered structure model to demonstrate the difficulties arising from the amplitude inversion and the capabilities of variant inversion algorithms in different realistic circumstances. In the inversion, as in the previous chapters, we parameterize each interface by a set of Fourier series with different wavenumbers. We see that the main difficulty in inverting a deep interface is the lack of constraint on the long wavelength components of a curved interface. Another problem is the high wavenumber oscillation appearing in the inversion of shallow interfaces.

To overcome these difficulties, we test a subspace method with a damping scheme. The subspace method restricts the local minimization of the misfit function to a relatively small number of subspaces of the model parameters, as seen in Chapters 4 and 5, whereas the damping factors can be used to balance different sensitivities of the model parameters. We see that such a damped subspace method, in which the most sensitive components (those with high wavenumbers, as seen in Chapter 6) are most strongly damped, is an efficient procedure, compared to the subspace method without damping. For a multi-layered structure, however, this method is still unable to recover deep interfaces with a high degree of curvature.

To overcome the latter problem, we then test a multi-scale inversion scheme, starting with long wavelength components only and gradually adding short wavelength components into the inversion. This multi-scale scheme was verified to be efficient in the context of traveltime inversion (Williamson, 1990; Lutter and Nowack, 1990). However, since seismic amplitudes are significantly more sensitive to the short wavelength components, completely rejecting these components in the earlier stages of a multi-scale inversion may cause instability in the amplitude inversion.

A combination of the multi-scale scheme and the damped subspace scheme, therefore, leads to the development of the so-called *multi-stage damped, subspace method*. We see from the investigation in this chapter that the multi-stage damping scheme is appropriate for constraining the interface geometries of a multi-layered structure when using reflection amplitudes in seismic inversion.

7.2.   Forward calculation and inverse method

Let us first summarize briefly the forward calculation and the inverse method used in the inversion process.

7.2.1   Ray tracing

We consider a two-dimensional stratified velocity structure consisting of variable-thickness, homogeneous, isotropic layers, in which the interface between two layers is parameterized by a truncated Fourier series (equation 6.25). The ray path is determined by Fermat's principle (equation 2.7). For a constant velocity layered model, we may write it explicitly as a system of linear equations:

$$\left[(x_i - x_{i+1}) + (z_i - z_{i+1})\frac{\partial z_i}{\partial x_i} \right] \frac{1}{v_{i+1}d\ell_{i+1}}$$

$$+ \left[(x_i - x_{i-1}) + (z_i - z_{i-1})\frac{\partial z_i}{\partial x_i} \right] \frac{1}{v_i d\ell_i} = 0 , \qquad (7.1)$$

$$\text{for} \quad i = 1, \cdots, 2K - 1 ,$$

where (x_i, z_i) is the ith intersection point between the ray and an interface, $d\ell_i$ is the length of the ith ray segment, given by

$$d\ell_i = \sqrt{(x_i - x_{i-1})^2 + (z_i - z_{i-1})^2} , \qquad (7.2)$$

and v_i is the corresponding velocity. It is a tridiagonal linear system,

$$\mathbf{T}\mathbf{x} = \mathbf{b} , \qquad (7.3)$$

where $\mathbf{T}$ is the tridiagonal matrix and $\mathbf{x}$ is the vector of unknowns $\{x_i\}$. Linking intersection points with straight lines, we obtain a two-point ray between the source (x_0, z_0) and the receiver (x_{2K}, z_{2K}).

7.2.2 Ray-amplitude and its perturbation

As a seismic pulse propagates through a layered model consisting of homogeneous layers, its amplitude generally decreases with increasing length of the ray path. With the high-frequency assumption, the ray amplitude is determined by the geometrical spreading function along the ray path, and by reflection and transmission coefficients at each interface:

$$A(\ell) = A_0 C_Q C / L(\ell) , \qquad (7.4)$$

where A_0 is the amplitude at the source point, C_Q is the inelastic attenuation factor, C is the product of all the relevant reflection or transmission coefficients, calculated using Zöppritz's amplitude equations (Chapter 3), and $L(\ell)$ is the explicit expression of the geometrical spreading function, which depends on the velocity contrasts, the angle of incidence, and the local slope and curvature of the interfaces (Chapter 1).

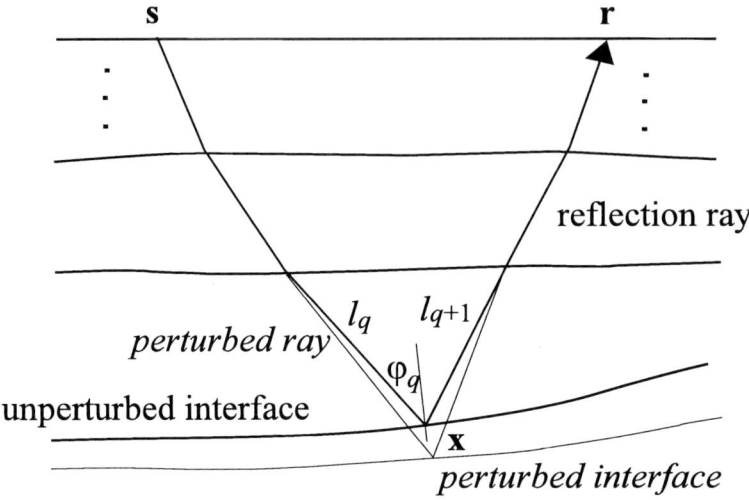

Figure 7.1. Approximating the perturbation of a reflection ray, due to the perturbation of the interface, connecting the source point **s** and the receiver **r**, where ℓ_q and ℓ_{q+1} are the lengths of the qth and $(q + 1)$th ray segments, and φ_q is the angle of incidence at the qth interface.

The inelastic attenuation is given by

$$C_Q(\ell) = \exp[-\eta(z)\,\ell(z)]\,, \tag{7.5}$$

where $\eta(z)$ is the average absorption coefficient, averaged from the sea-bottom to the depth z, and $\ell(z)$ is the ray length of the ray reflected from the target interface at depth z. As the depth z increases, the average absorption $\eta(z)$ generally decreases. Other amplitude effects, such as those caused by fine layering (the layer thickness being less than the seismic wavelength), may be assumed to be multiplicative scaling factors applied to the observed amplitude data (Almoghrabi and Lange, 1986; Juhlin and Young, 1993). The logarithmic amplitude is used in the inversion (Nowack and Lutter, 1988). Using log variables converts the multiplicative structure to an additive data structure so as to reduce the effect of multiplicative scaling in the least-squares inversion.

The matrix of Fréchet derivatives of the geophysical response with respect to the model parameters is often calculated based on the difference between the perturbed and unperturbed responses. To obtain the partial derivatives in this fashion one would need to retrace rays through the model for each parameter perturbation, because changing the boundary or the velocity distribution may significantly change the ray path. Retracing rays however is the most time-consuming part in any iterative inversion procedure. To speed up the calculation, we introduce the approximation (see Figure 7.1) that the ray path perturbation caused by the perturbation of the interface can be determined *solely* from the two adjacent ray segments intersecting the perturbed interface. The perturbed ray segments are determined according to Fermat's principle. For the calculation of the perturbation of the transmission/reflection coefficient δC, and the perturbation of the geometrical spreading δL, all other ray segments are left unperturbed.

This first-order approximation to the perturbation of amplitude (and travel-time) speeds up the inversion dramatically, especially for a multi-layered model. The computation times for amplitude inversion and traveltime inversion are now similar. The precision of this non-standard approach as opposed to an approach in which the entire trajectory is modified by a perturbation of an interface (Farra *et al.*, 1989; Nowack and Lyslo, 1989) is also justified because, in reflection geometry with limited source-receiver offsets, the focusing effect of the downward ray and the defocusing effect of the upward ray travelling through an interface in the overburden are largely cancelled (as the two piercing points are close to each other), and in addition the amplitude perturbation (due to the perturbation of the reflector) depends mainly upon the perturbations in the vicinity of the perturbed reflection point. In the following inversion examples, we retrace rays and recalculate the Fréchet matrix at each iteration.

7.2.3   The subspace inversion method

If the data residuals have a Gaussian distribution, we can adopt a least-squares objective, defined as the misfit function between the observed data $\mathbf{d}_{obs}$ and the predicted data $\mathbf{d} = f(\mathbf{m})$. We solve the non-linear optimization problem of minimizing the data misfit iteratively. At each iteration, the model update $\delta\mathbf{m}$ is estimated using a subspace method,

$$\delta\mathbf{m} = -\mathbf{A}(\mathbf{A}^{\mathrm{T}}\mathbf{H}\mathbf{A})^{-1}\mathbf{A}^{\mathrm{T}}\hat{\mathbf{g}}, \qquad (7.6)$$

where $\hat{\mathbf{g}}$ is the gradient vector, $\mathbf{H}$ is the Hessian matrix and $\mathbf{A}$ is the subspace projection matrix, given by

$$\mathbf{A} = [\mathbf{a}^{(1)}, \mathbf{a}^{(2)}, \cdots, \mathbf{a}^{(q)}],$$

in which the basis vectors $\{\mathbf{a}^{(j)}\}$ are built up using the gradient vector. Assuming the model parameters can be classified as q different groups, we partition the gradient vector into q vectors, so that

$$
\mathbf{A} =
\begin{bmatrix}
\mathbf{g}^{(1)} & 0 & \cdots & 0 \\
0 & \mathbf{g}^{(2)} & & \vdots \\
\vdots & 0 & & \vdots \\
\vdots & \vdots & & \vdots \\
0 & 0 & \cdots & \mathbf{g}^{(q)}
\end{bmatrix},
\tag{7.7}
$$

where $\mathbf{g}^{(j)}$ is the normalized vector of the gradient component $\hat{\mathbf{g}}^{(j)}$, corresponding to the jth parameter group. Each column in $\mathbf{A}$ is a basis vector $\mathbf{a}^{(j)}$, spanning the full model space. In this manner, all q basis vectors in $\mathbf{A}$ are orthogonal.

In the interface inversion presented below, we partition the parameters defining short, intermediate and long wavelength components of the interface geometry into separate subspaces. Inversion tests shown in Chapter 4 suggest that this partitioning scheme, based on the dependence of amplitudes on the interface wavenumbers, is effective in obtaining an unbiased solution for interface geometry, where the example models include a single interface. When further considering the case of multi-layered structure, tests described in the next section reveal, however, that the interface geometry of a deep reflector buried beneath a sequence of layers is not well reconstructed by using the inversion method described above.

7.3.   Preliminary inversion test

7.3.1   *A multi-layered velocity model*

A multi-layered velocity model is shown in Figure 7.2, in which five interfaces separate constant velocity layers. The top layer of the model is assumed to be a sea-water layer, and the velocities for remaining layers are shown in the figure. The model consists of two synclinal interfaces (interfaces 3 and 4), which lie beneath two approximately flat interfaces (interfaces 1 and 2), and are above another flat interface (interface 5). This example model, chosen to illustrate the problems in the reconstruction of a deep and curved interface, is built with reference to an existing seismic profile from the North Sea (Chapter 8).

Synthetic amplitudes for 201 CDP gathers reflected from interfaces 2-5 are calculated, where we simulate the loss of data due to muting in a practical data process. The interval between CDPs is 50 m. In each gather, the minimum offset is 199.0 m with a trace interval of 33.3 m. In the synthetic calculation, the average absorption coefficients $\eta(z)$, averaged from the sea-bottom to the target interfaces 2-5, are set equal to 1.60, 0.61, 0.55 and 0.51

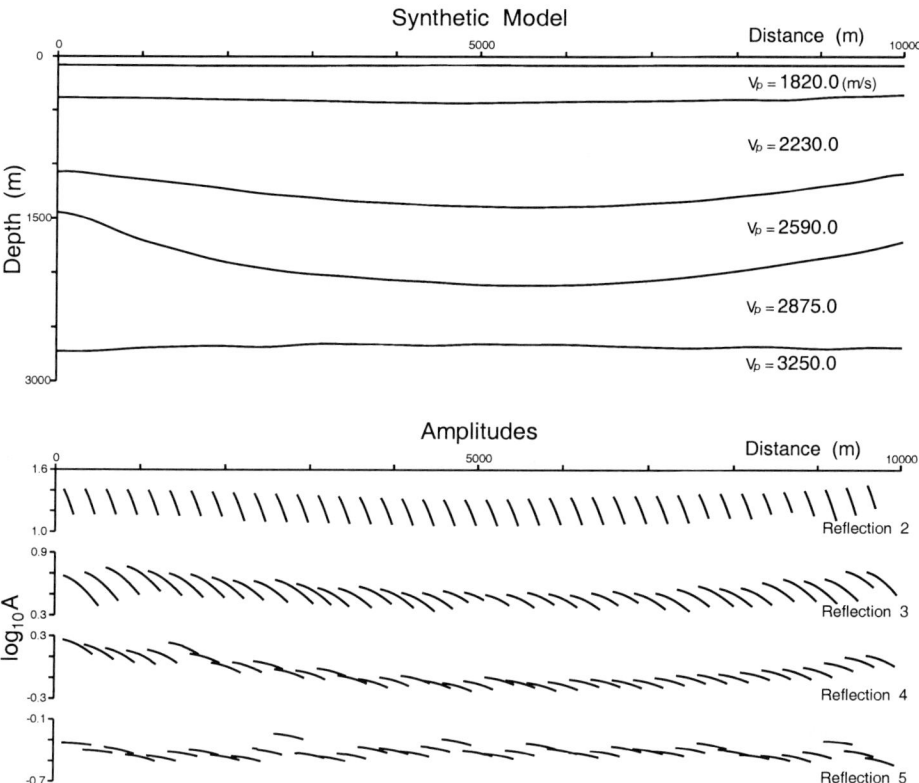

Figure 7.2. A synthetic model of a multi-layered structure, and the seismic amplitudes of four reflections.

(all $\times$ 10^{-3}), respectively. Calculated amplitude samples are also depicted in Figure 7.2.

In the following inversion experiments, aimed at investigating the information content of seismic amplitudes with respect to the interface geometry, we assume that the absorption coefficients $\eta(z)$ are known *a priori*. Note that Nowack and Matheney (1997) used instantaneous frequency data (i.e. pulse broadening) to help constrain the attenuation structure.

7.3.2 Inversion test

A layer-stripping approach is used in the following inversion tests. In order to gain some insight into the capability of amplitude inversion for the reconstruction of interface geometry in a multi-layered structure, we assume that all interfaces above the target are known *exactly* and invert for one interface at a time. The subspace formula (7.6) is used in this test, which is referred to as Inversion-I.

Reflection amplitudes can only constrain the velocity contrasts at interfaces, not the absolute values of velocities (which may be determined by traveltime data). In the inversion for the qth interface, we assume a value for v_q (the layer above interface q) and invert for v_{q+1}. The number of harmonic terms in the model parameterization (equation 6.16) is set using $N = 20$. Thus, there are 42 model parameters for each interface inversion. These 42 parameters are classified into six groups:

$$
\begin{aligned}
\mathbf{m}^{(1)} &= \{v_{q+1}\} \,, \\
\mathbf{m}^{(2)} &= \{a_0, \, a_n, \, b_n, \, (n = 1, \, 2)\} \,, \\
\mathbf{m}^{(3)} &= \{a_n, \, b_n, \, (3 \leq n \leq 5)\} \,, \\
\mathbf{m}^{(4)} &= \{a_n, \, b_n, \, (6 \leq n \leq 10)\} \,, \\
\mathbf{m}^{(5)} &= \{a_n, \, b_n, \, (11 \leq n \leq 15)\} \,, \\
\mathbf{m}^{(6)} &= \{a_n, \, b_n, \, (16 \leq n \leq 20)\} \,.
\end{aligned}
\tag{7.8}
$$

The velocity parameter v_{q+1} is allocated its own group, separate from interface parameters. The latter are grouped according to model sensitivity. The sensitivity tendency of amplitudes with respect to interface perturbations has a pseudo-linear relationship with interface wavelengths (as seen in the previous chapter).

Inversion results for individual interfaces are shown in Figure 7.3: the starting models for all inversions are flat (dashed lines); the inversion results are plotted in solid lines and compared to the true interfaces (dotted lines). In the figure, v_p is the true velocity, v_0 is the initial guess and v_a is the velocity obtained from the amplitude inversion. The data fit is also shown in Figure 7.3, in which the solid and dotted lines are, respectively, predicted amplitudes and the input of the inversion.

In this inversion the shallower layers are far better constrained than the deeper layers. The starting models for both interface 2 and interface 5 are 100 m deeper than the mean depths of the true model. The inversion can reconstruct the mean depth for interface 2 but not for interface 5. There is also a trade-off between the mean depth of the interface 5 and the velocity contrast at the reflector: the mean depth is deeper than the true model, while the velocity v_6 (velocity below interface 5) is greater than its true value. Whether the mean depth of the interface is well determined essentially depends on the ray angle coverage. In practice, increasing the maximum offset of the observations may help to determine the deeper interfaces better. Once the mean depth of the

Amplitude Inversion (I)

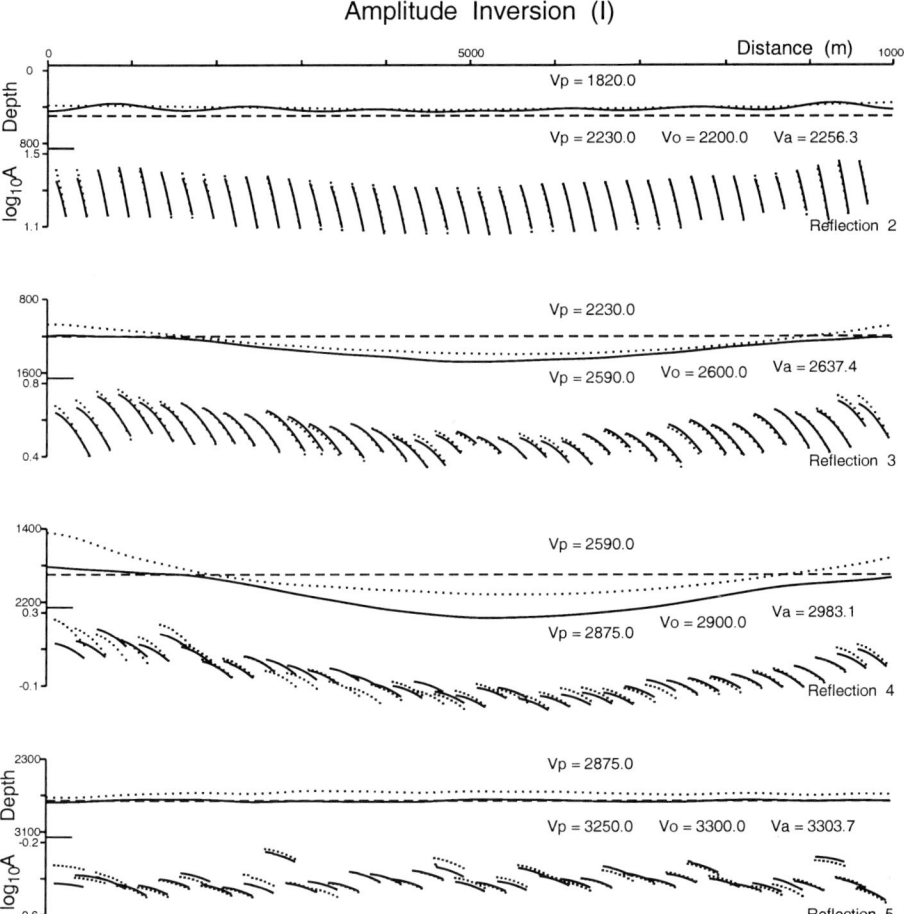

Figure 7.3. Amplitude inversion-I: the subspace method. Dotted lines, dashed lines and solid lines are the true model, the initial estimate and the inversion solution for each individual interface 2-5. The true velocity, the initial guess and the estimate from amplitude inversion are denoted by v_p, v_0 and v_a, respectively.

interface is well determined, the velocity contrast should be well determined as a consequence.

Comparing the results of the two synclinal interfaces, we see that interface 3 is better resolved than interface 4. The data fitting and the interface solution for reflection 4 are not acceptable.

An additional problem is the appearance of a short wavelength oscillation on interface 2, not present in the original model.

If we compare the inversion result with $N = 20$ (the number of harmonic terms) with the result using $N = 40$ in model parameterization (not shown), we would observe minimal differences (in both the data match and the model match).

7.3.3   Summary

In summary, there are two major difficulties in the amplitude inversion for multi-layered structures:

1) The high wavenumber oscillatory phenomenon appearing on shallow interfaces.

2) The failure to recover the long wavelength components of deep, curved interfaces.

These two difficulties are the consequence of the fact that the seismic amplitudes are extremely sensitive to the interface components with short wavelengths (Chapter 6). To combat the difficulties arising from amplitude inversion, we investigate a damped subspace method and a multi-scale inversion scheme in the following sections.

7.4.   Damped subspace method

7.4.1   Methodology

To stabilize the inversion procedure by moderating the influence of short wavelength components, we now introduce a wavelength-dependent damping scheme to balance the sensitivities of different model components. The subspace formula (7.6) is then modified as

$$\delta\mathbf{m} = -\mathbf{A}[\mathbf{A}^{\mathrm{T}}(\mathbf{H} + \mathbf{D})\mathbf{A}]^{-1}\mathbf{A}^{\mathrm{T}}\hat{\mathbf{g}} , \qquad (7.9)$$

where the damping matrix is given by $\mathbf{D} = \mathrm{diag}\{\mu_j\}\,\mathbf{I}$, expressed in terms of damping factors μ_j with units of (*model parameter*)$^{-2}$, and the identity matrix $\mathbf{I}$. Separate values of μ_j can be applied to different model components or groups. In this scheme, damping the most sensitive components (i.e. the shortest wavelength components) of the interface model helps the inversion to converge to the global minimum.

If a constant damping factor μ were used in equation (7.9), it could then be interpreted as the Lagrange parameter as used in the Levenberg-Marquardt method. As in the Levenberg-Marquardt method, therefore, we can initially set μ_j to large positive values so that the good initial convergence properties of the steepest descent method can come into play. Subsequently we can gradually reduce the μ_j synchronously so that a kind of Gauss-Newton method may dominate in the region close to the solution point, where the data residuals are small and may be nearly linear in the parameter changes. Note that in Bayesian approaches the μ_j are interpreted as priors and therefore need to be kept constant.

An advantage of equation (7.9) lies in the fact that the inverse of matrix $\mathbf{A}^T\mathbf{H}\mathbf{A}$ in equation (7.6) may not exist. When $\mathbf{H}$ is nearly singular, the parameter update $\delta\mathbf{m}$ tends to be large. Adding positive constants to the diagonal of $\mathbf{H}$ improves the condition number of matrix $\mathbf{H}+\mathbf{D}$, and the inverse of $\mathbf{A}^T(\mathbf{H}+\mathbf{D})\mathbf{A}$ in equation (7.9) could be guaranteed. When using equation (7.6) directly, the inverse of $\mathbf{A}^T\mathbf{H}\mathbf{A}$ is calculated by using the SVD method, in which normalized eigenvalues less than 10^{-6} are truncated.

7.4.2   Inversion result

The inversion result from the damped subspace method (which is referred to as Inversion-II) is shown in Figure 7.4. In this experiment (and those following), the model parameters in the overburden are the solution values of the inversion, and no longer the exact values of the true model (used in Inversion-I). The damping factors used in equation (7.9) are

$$
\begin{aligned}
\mu_1 &= 10^{-4}, & \mu_2 &= 10^{-4}, \\
\mu_3 &= 10^{-3}, & \mu_4 &= 10^{-2}, \\
\mu_5 &= 10^{-1}, & \mu_6 &= 1.0,
\end{aligned}
\tag{7.10}
$$

corresponding to the different parameter groups $\{\mathbf{m}^{(j)}, j = 1, \cdots, 6$, allocated in expression (7.8). The choice of damping parameters is somewhat arbitrary, but is based on a limited number of experiments. We found that the damping factors μ_j should be chosen in the interval [0.0, 1.0]. The damping factors listed in equation (7.10) are based on the sensitivities of the corresponding parameter groups, i.e. the relative magnitudes of the eigenvalues of the Hessian matrix.

The most significant improvement in Inversion-II is that the high wavenumber oscillation shown on interface 2 of Inversion-I is not evident in this inversion. Interface 2 is now better constrained. The data fit for reflection 2 is also improved relative to Inversion-I. For interfaces 3 and 4, however, the inversion result is similar to the previous inversion without the damping. The structure of the long wavelength syncline of interface 4 is still not resolved.

Even though the structure of interface 4 is not well reconstructed, the geometry of interface 5 is well determined by the amplitude inversion. Since a reflection ray travels through interfaces in the overburden twice (going down to the reflector and back to the surface), the transmission focusing of the downward ray is partly cancelled by the defocusing of the upward ray, provided

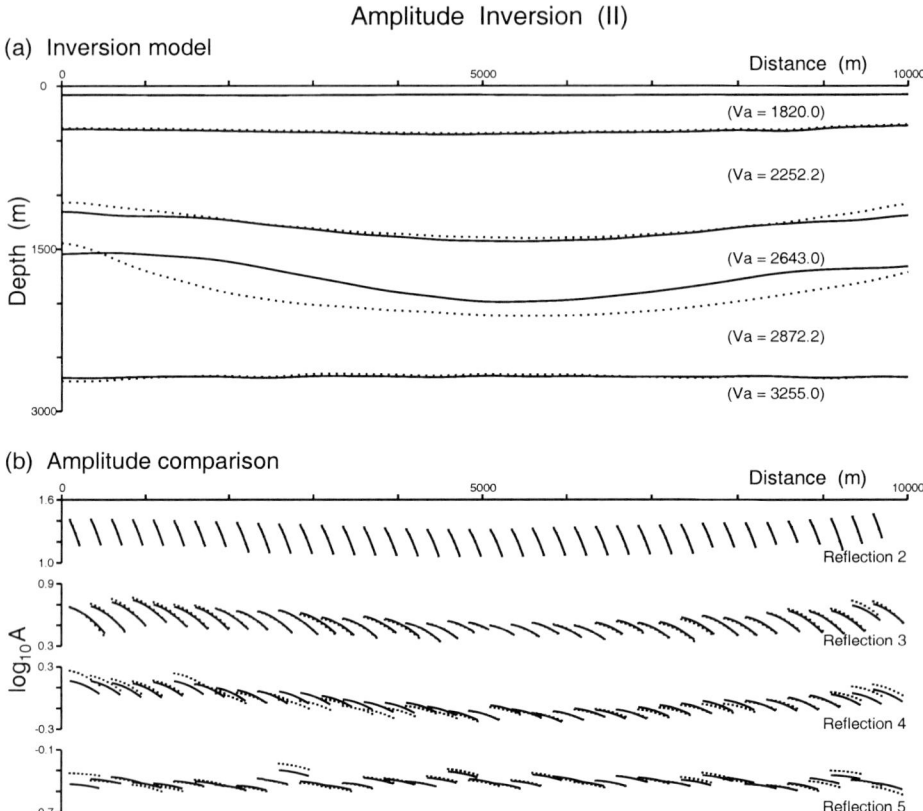

Figure 7.4. Amplitude inversion-II: damped subspace method. The solid lines represent the inversion result, the dotted lines represent the true model.

that the curvature of the interface does not change too rapidly between the first and second piercing points of the ray within the limited offset. Although the perturbation of an interface in the overburden causes the perturbation of ray path, a given reflection amplitude contains scant information that constrains the curvature of the interface in the overburden.

The valid reconstruction of interface 5 supports the contention that errors in the specification of interfaces in the overburden are not important in inverting for the current interface, thereby validating a layer-stripping approach. The failure in the reconstruction of interface 4 (shown in Figures 7.3 and 7.4) is thus not due to errors in the estimate of the overlying interfaces and further efforts are required to combat the problem.

7.5.   Multi-scale scheme

7.5.1   Methodology

For the reconstruction of the long wavelength components of the interfaces (using wavelengths comparable to the total length of observations in a particular experiment), we next consider a multi-scale scheme as an alternative approach to the amplitude inversion. In this inversion processing, we first restrict the model solution to include only low wavenumbers and then gradually increase N, the number of harmonic terms in the model parameterization, until a satisfactory result is obtained, beyond which the data misfit ceases to decrease and the model no longer changes.

This multi-scale approach was used by Williamson (1990) and Lutter and Nowack (1990) in traveltime inversion, and by Bunks *et al.* (1995) in waveform inversion. Nowack and Matheney (1997) also used an iterative inversion starting with a smooth model, to invert amplitudes and traveltimes.

The subspace inversion formula equation (7.6), without damping, is used iteratively, beginning with a limited number of low wavenumber harmonic terms. The iteration is stopped when the misfit function does not decrease any more: the reduction of the data misfit S is less than 1% of S. We then move to the next stage in which more harmonic terms are used, i.e. the next subset(s) of parameters in expression (7.8) are included.

7.5.2   Inversion result

We refer to this inversion experiment as Inversion-III. The initial model is a straight line and N is initially set equal to 2. The solution is used as a starting model for a subsequent inversion with $N = 5$. Doubling N to 10, and then to 20, we try to obtain a final solution that includes high wavenumber components. The inversion convergence of interfaces 3-5 is shown in Figure 7.5. In parts (a) and (b), the dotted lines are the starting models for each inversion stage and the solid lines are the inversion solutions. Part (c), the convergence for reflector 5, shows clearly how important the higher wavenumber components are (for the data fit). Increasing the number of harmonic terms from $N = 10$ to $N = 20$ does not cause any significant change in the interface geometry, but the data fit is improved. In this experiment we see that $N = 20$ is sufficient for the modelling. The inversion with $N = 40$ (not shown) only slightly decreased the data misfit.

The comparison between the final inverted interfaces (solid lines) and the true subsurface geometries (dotted lines) is shown in Figure 7.6, which establishes that the main features of the structure, especially the synclinal curvature of interfaces 3 and 4, is reasonably reconstructed. The predicted amplitudes (solid lines) and the observed amplitudes (dotted lines) for all four reflections match each other much better than in previous inversions. Of course, a perfect

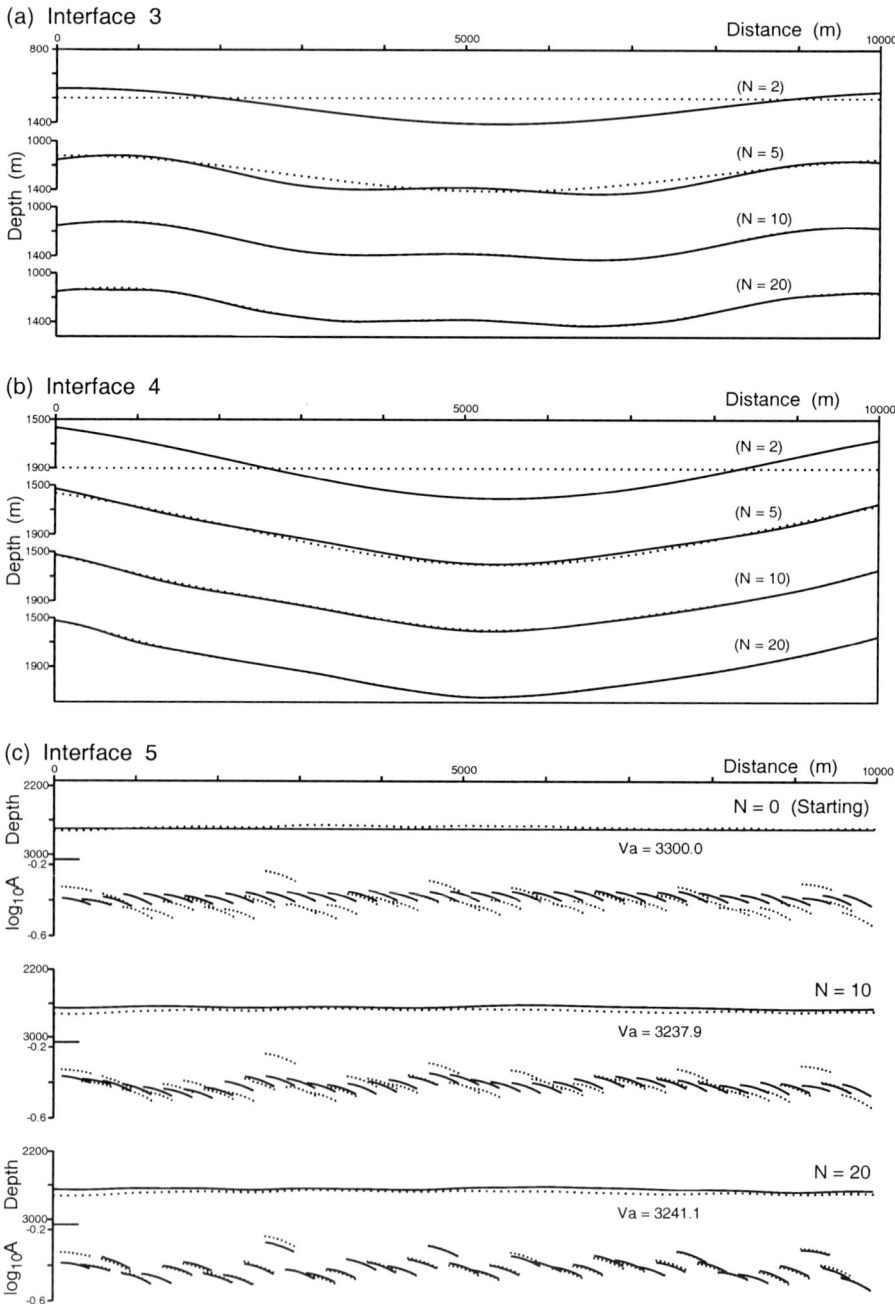

Figure 7.5. Convergence of interface inversions, using a multi-scale scheme, for interfaces 3-5. The dotted lines are the result of the preceding inversion (above, solid line). In (c) solid lines also represent predicted amplitudes, dotted lines represent the input of inversion.

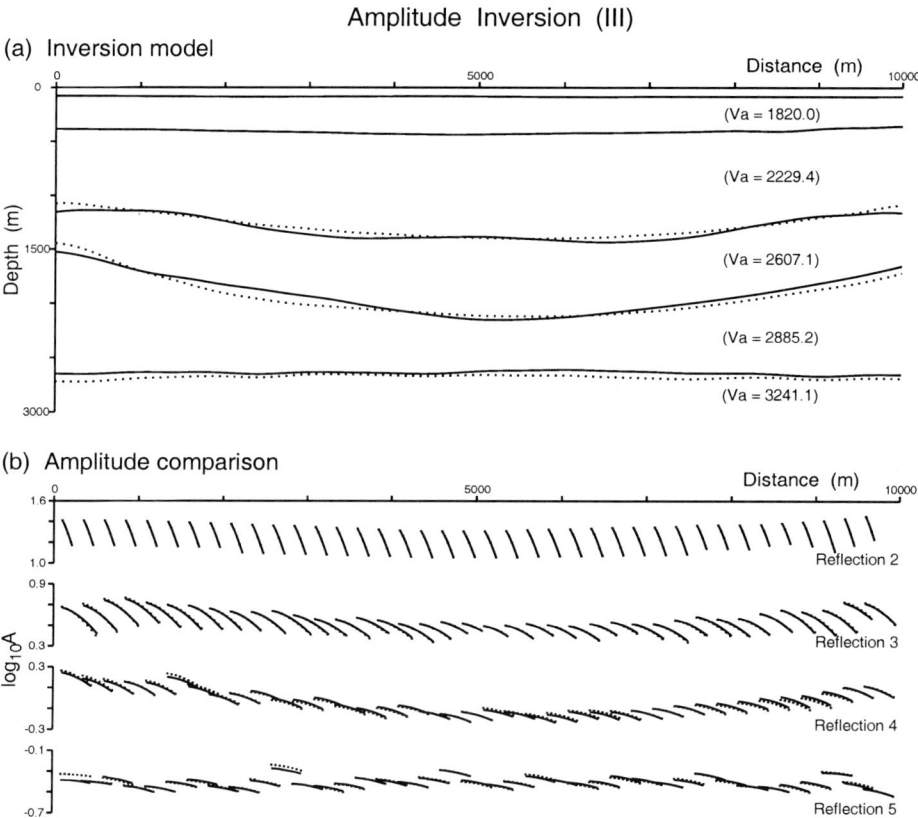

Figure 7.6. Amplitude inversion-III: the multi-scale scheme. The solid lines represent the inversion result, the dotted lines represent the true model.

fit of predicted and observed data sets does not necessarily mean that the inversion solution is the true model. The inversion result shown in Figure 7.6 is contaminated with a low amplitude but high wavenumber oscillation. This observation suggests that completely rejecting high wavenumber components in the earlier stages of the multi-scale inversion prevents the amplitude inversion from converging to a true solution.

7.6. Multi-stage damped subspace method

7.6.1 Methodology

We recall that Inversion-I reveals two major problems in the amplitude inversion. The damped subspace scheme in Inversion-II attempts to reduce the possibility of high wavenumber instability, i.e. to overcome the first problem

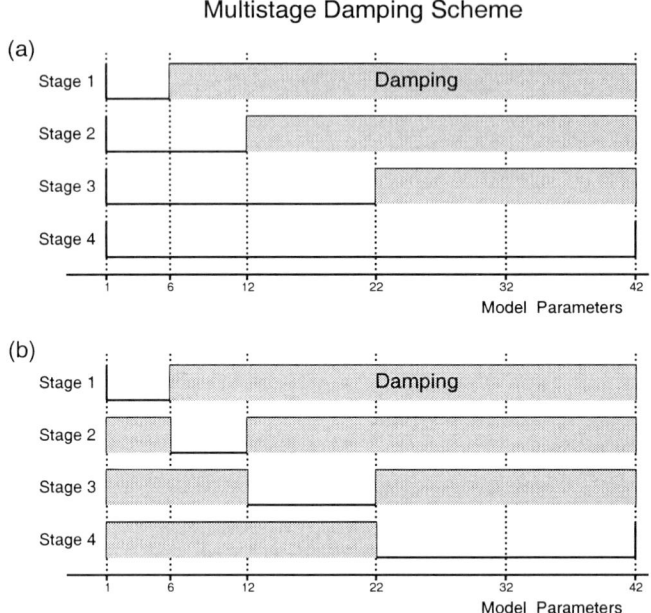

Figure 7.7. Schematic presentation of two multi-stage damping schemes. The horizontal axis corresponds to the six groups of parameters [defined in expression (7.8), here groups 1 and 2 are combined]. The damping factor is equal to 1 in the shaded region and equal to zero in the unshaded region.

in the inversion, but if a large range of wavenumbers is included in the model parameterization, the convergence of the low wavenumber components may be either prevented or occur too slowly to be useful, as seen in both Inversion-I and Inversion-II. The multi-scale scheme in Inversion-III then attempts to enhance the low wavenumber components by concentrating on only the low wavenumber components in interfaces in the first few stages, i.e. to overcome the second problem. However, since amplitudes are most sensitive to high wavenumber interface components, the final result of Inversion-III could depend to a high degree on the choice of truncation wavenumber in the first stage. These considerations lead to the use of a combination of the damping scheme with the multi-scale scheme—a multi-stage damped subspace method, which is shown diagrammatically in Figure 7.7.

In this method, the maximum wavenumber components for which the calculation is stable are included in the parameterization at the start. Instead of the gradually increasing series μ_j in equation (7.10) for all iterations, we now set μ equal to 0 or 1, according to the plan shown in Figure 7.7. The choice of damping factors at each stage follows the design of the multi-scale scheme. Two designs for the multi-stage damping scheme are shown in Figure 7.7, where (a)

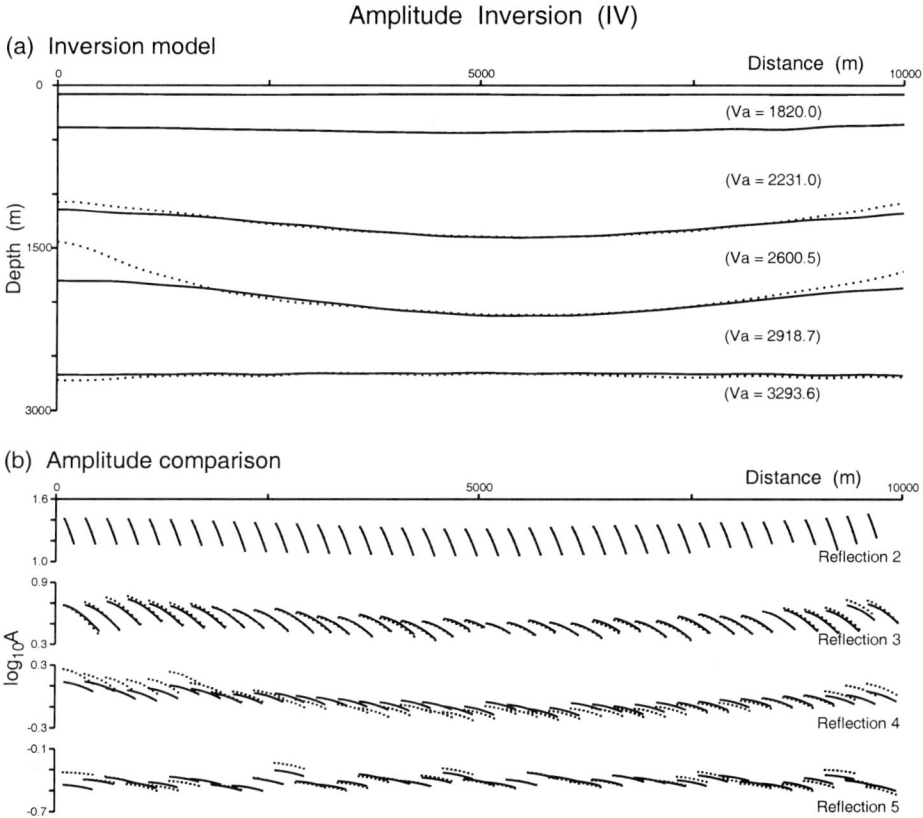

Figure 7.8. Amplitude inversion-IV: multi-stage damped subspace method. The solid lines represent the inversion result, the dotted lines represent the true model.

is similar to the multi-scale scheme discussed above, with the difference that here $\mu = 1$ (previously $\mu \to \infty$, a complete rejection defined in the multi-scale scheme), and (b) is a modification in which one parameter group is emphasized at each stage. Numerical experiments demonstrate that in the case of amplitude inversion, these two schemes display slight difference both in the interface solutions and the predicted amplitude curves, and that the second scheme does have some advantages when used in the traveltime inversion. This is because traveltimes and amplitudes are sensitive to different features in the model interface. In the following inversion tests, we use the second scheme.

7.6.2   Inversion result

Figure 7.8 shows the result of the inversion with the multi-stage damping scheme (Inversion-IV). As expected, the result is an improvement on both

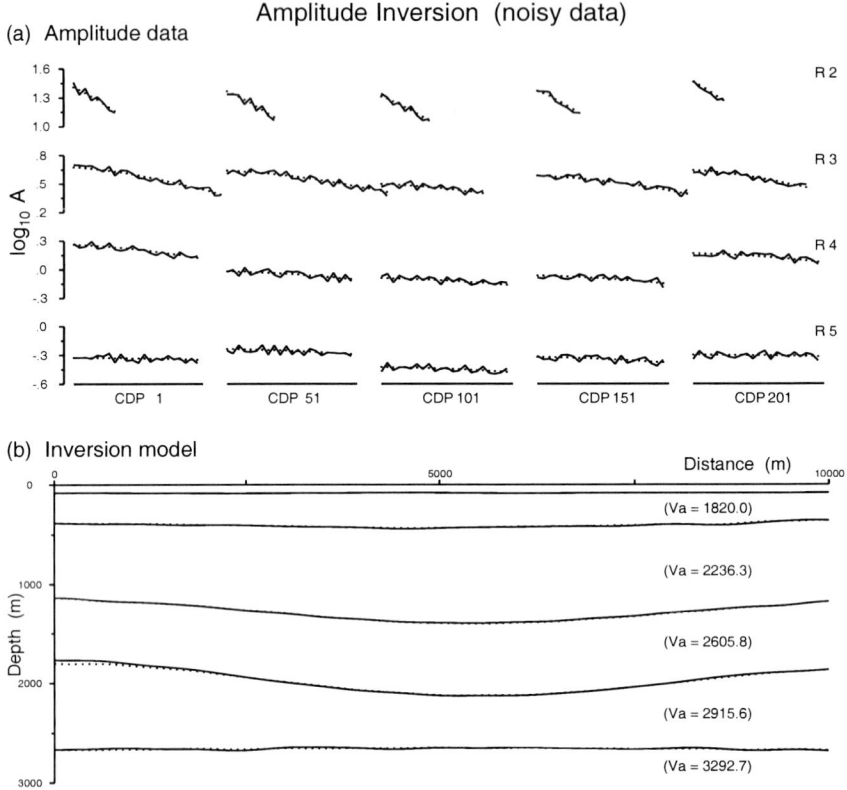

Figure 7.9. Amplitude inversion in the presence of data noise: (a) depicts the noisy amplitude curves (solid lines) and noise-free data (dotted lines); (b) shows the comparison of inversion results from noisy data (solid lines) and noise-free data (dotted lines).

Inversion-II and Inversion-III. The interface solutions generally match the true model interfaces, except for the mismatches at both ends of interfaces 3 and 4. The data fit for reflections 2, 3 and 5 is quite good, whereas reflection 4 retains relatively large residuals. It appears, however, that the layer-stripping approach does not obscure lateral variations in deeper interfaces, as the amplitudes and geometry of reflection 5 are well matched.

To test the robustness of the method in the presence of data noise, we repeat the previous example, with the synthetic amplitude data modified by the addition of 5 per cent white noise. The test result is shown in Figure 7.9, in which (a) depicts both the noisy data (solid lines) and the noise-free data (dotted lines) and (b) shows the comparison between inversion models from the noisy data (solid lines) and from the noise-free data (dotted lines). Because the multi-stage damped subspace method suppresses the "high-frequency" random noise in the

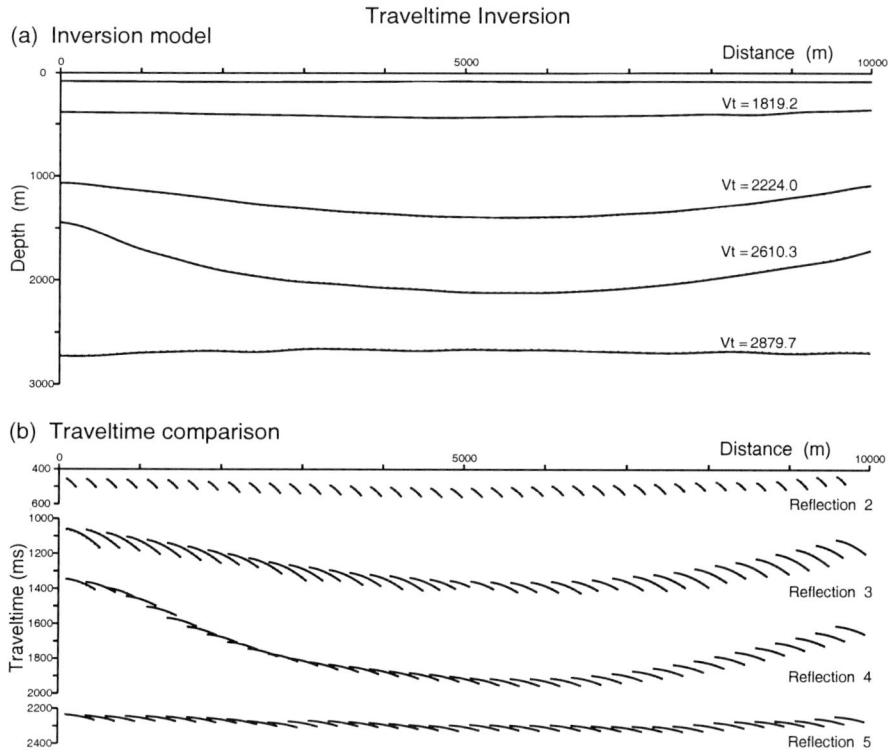

Figure 7.10. Traveltime inversion using multi-stage damped subspace method. The dotted lines (for the true model) here plot on the top of solid lines (the inversion result) and are hardly visible.

first two stages, the inversion procedure shows itself to be robust in the presence of noise.

To show the validity of the multi-stage damped subspace method, we also apply the same scheme to the traveltime inversion. We again assign parameters, according to the different ranges of the interface wavenumbers, into separate subspaces, simply following the scheme of expression (7.8), this time with v_q, instead of v_{q+1}, as an unknown parameter. The starting model for the traveltime inversion is a set of flat interfaces, each of which is 40 m different from the mean depth of the true interface. The inversion procedure converges steadily. The comparison with the true model is shown in Figure 7.10a, while Figure 7.10b shows the comparison of the predicted traveltimes (solid line) with the input data (dotted lines). In this case the solution accurately matches (overlies) the synthetic model.

Although the traveltime inversion has given an accurate solution in this case, the constant velocity layers are unrealistic. Velocity variations in both horizontal and vertical directions must be taken into account in the traveltime variations, in which case the inversion will be more complicated. Since seismic amplitudes are significantly sensitive to the velocity variation in the vicinity of an interface, i.e. the impedance contrast, the assumption of constant velocity layers can be adopted in the amplitude inversion which however, may also allow elastic properties to vary along interfaces (Chapter 9). In practice, traveltime inversion with constant layer velocities should only be used to produce a preliminary estimate, which in turn is used for the estimation of the source amplitude A_0 and the absorption coefficients $\eta(z)$. These parameters are then assumed to be known *a priori* in the interface inversion.

Application examples of the multi-stage damped subspace method are shown in the following chapters.

Chapter 8

Practical approach to application

Abstract   For the application of amplitude inversion to real seismic amplitude data, we present a three-step practical approach. (1) Performing the estimation of recorded seismic amplitudes from reflection seismic gathers with the aid of prestack time migration, which enhances continuity and reflection strength and reduces reflection point dispersal and diffraction effects. Contraction of the Fresnel zone by migration also brings the amplitudes closer to the ray amplitudes assumed in the inversion. (2) Demigration of the amplitude data, so that a set of "true" observations is recovered for input to inversion. (3) To make the amplitude inversion robust, mitigating the effect of noise in the amplitude data by applying iteratively a locally reweighted regression, which can efficiently reject amplitude outliers. We apply this three-step approach to a reflection seismic profile from the North Sea to constrain the geometry of a stack of interfaces, by using both the amplitude inversion and a joint inversion with traveltime data. The application example will represent a valuable contribution to the discussion of how the combined effort of imaging and inversion of seismic data should be organized when dealing with field data.

8.1.   Introduction

In the previous chapters, we investigated the use of amplitude information in reflection seismic tomography using synthetic experimental data. For practical application, however, we always have a problem with the amplitude extraction from a specified seismic event. Raw seismic records are frequently distorted by wave propagation effects, such as diffractions, crossing arrivals and caustics. The distortion of seismic events makes the picking of traveltimes and amplitudes tedious. In some circumstances, we even have difficulty in identifying a reflection event. To overcome these difficulties, some researchers (e.g. Stork, 1992a; Grau and Lailly, 1993) suggested performing picking with the aid of migration. Migrating the data alleviates many of the problems in using unmigrated data. Identifying arrivals and estimating the amplitudes or other data parameters should be easier in migrated common-offset or common-reflection-point (CRP) gathers than on unmigrated gathers. Moreover contraction of the Fresnel zone by migration brings the amplitudes closer to the ray amplitudes assumed in the inversion. By taking advantage of migrated gathers, we try now to implement the amplitude inversion for interface geometry using real reflection seismic amplitudes.

Tomography using data from prestack migrated gathers was first proposed in the context of traveltime inversion. Stork (1992a) suggested determining a velocity model by optimizing the consistency of imaged reflection events on a CRP gather, i.e. adjusting the velocity by means of tomography to flatten the reflections on prestack depth-migrated CRP gathers. The input data for tomography are time deviations derived from the apparent depth deviations of reflection events on CRP gathers. Such deviations are presumably due to the inaccuracy of the velocity model adopted in migration. This method was also used by Kosloff *et al.* (1996). Grau and Lailly (1993) proposed a sequential migration aided reflection tomography (SMART) method, in which traveltimes picked from migrated gathers are demigrated kinematically and used as the input for tomography. In these methods, reflectors picked from depth-migrated gathers are used as a structural model for ray tracing in subsequent tomographic inversions to improve the velocity estimate. Bording *et al.* (1987) and Dyer and Worthington (1988), in their respective two-part schemes of migration plus tomography, also exploited the ability of depth-migration to determine suitable reflector locations for the velocity field obtained by tomography. However, the results of ray tracing are very sensitive to inaccuracies in the location of migrated reflections (Jannaud, 1995) and, unfortunately, reflector locations predicted by migration are intrinsically inaccurate, due to the inaccuracies in the velocity field (yet to be determined) used in the migration.

Ultimately, the estimation of useful subsurface models may benefit from the use of amplitude information. As we saw in Chapter 5, reflection amplitudes recorded at the surface are especially sensitive to interface geometry and relatively insensitive to velocity variations within layers bounded by interfaces. Even if the velocity distribution is not given accurately, we saw in Chapter 4 that the geometry of an interface can be determined satisfactorily by amplitude inversion. In this chapter therefore, we make the assumption that amplitudes are mainly sensitive to the geometry of reflecting horizons (i.e. focusing and defocusing effects of curved interfaces) and the velocity contrasts at the interfaces, and that the amplitude effects of continuously varying velocity anomalies within a layer are small. We explore models containing variable geometry reflectors separating constant velocity layers by means of a ray-based inversion, using amplitude data picked from prestack gathers after time migration.

We perform the time migration on the prestack seismic gathers in the common-offset domain, and estimate the amplitudes from these migrated gathers. We also apply demigration to the amplitude data so that they can be considered as diffraction-free amplitudes for the input to inversion. As these amplitudes still contain considerable noise, we shall winnow them iteratively using a locally reweighted regression (LOESS) method (Cleveland and Gross, 1991). After this preprocessing, we use the amplitude data, both separately and in a joint inversion with traveltime data, to constrain the geometry of a stack of interfaces. We demonstrate the method using seismic data from a North Sea reflection profile.

8.2. Amplitudes estimated from migrated gathers

8.2.1 Prestack time migration

The North Sea seismic data used in this chapter is shown in Figure 8.1. Each trace is the stack of a common-midpoint (CMP) gather after prestack time migration. Each CMP gather consists of 66 traces, with a minimum offset of 199 m and a group interval of 33.3 m. By migrating the gathers, the reflection events are more visible than on the original field data, which suffer from reflection point dispersal. Compared to depth-migration, which was used by Bording *et al.* (1987), Dyer and Worthington (1988), Stork (1992a) and Grau and Lailly (1993) to determine interfaces when a traveltime inversion was used to determine the velocity field, time migration has advantages: to collapse the diffractions in a time section successfully, a less accurate subsurface velocity model is required than for accurate depth-migration (Hatton *et al.*, 1986). If we perform common-offset time migration and then back out the implicit amplitude compensation in migration, it emerges that the subsequent interface inversion is not particularly sensitive to potential errors in the migration velocity.

The migration processing consists of a forward normal-moveout (NMO) correction step, followed by dip-moveout (DMO) correction and time migration on zero-offset traces (Deregowski, 1990). The process can remove reflection point dispersal in moderately dipping structures, and improve lateral resolution. The Fourier transform migration method (Stolt, 1978) is used for the zero-offset time migration. A laterally invariant velocity model is usually used and often results in better imaging than the use of an over-complex, under-constrained velocity model. Although such a 1-D velocity migration will not

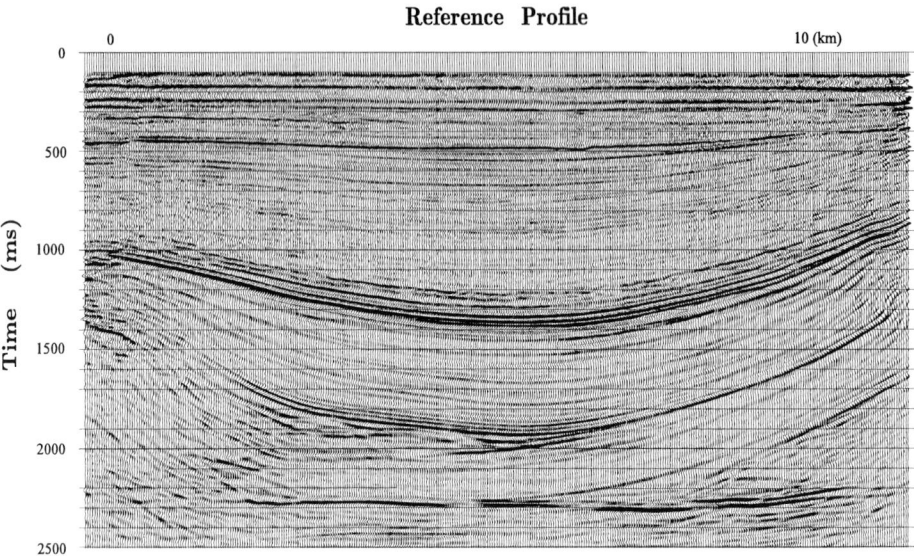

Figure 8.1. North Sea reflection seismic section used as the reference profile for the amplitude inversion test in which five reflections are considered.

make non-hyperbolic events hyperbolic, it will nevertheless significantly reduce the reflection point dispersal for shallow dipping events. In addition, due to the inherent simplicity of the velocity model and of the image ray, we can easily calculate the effect of amplitude compensation in the migration. This effect must be removed from the amplitudes picked from the migrated gathers.

8.2.2   Reference traveltimes

The essential properties of time-migrated common-offset data can be understood in terms of the CMP gather, which is the collection of traces with a common image ray location selected from the multiple images of the reflector. Since these traces are NMO corrected (because a forward NMO step is employed before the time migration process), they can be summed into one trace of the stacked section, as shown in Figure 8.1. From the stack profile of Figure 8.1 we pick the reflection time, which in turn is used as an initial reference time in a cross-correlation process for extracting the amplitudes. For the example dataset of Figure 8.1, five reflections (including the sea-bottom reflector) are picked; the reference traveltimes are shown in Figure 8.2, in which 601 CDPs are considered.

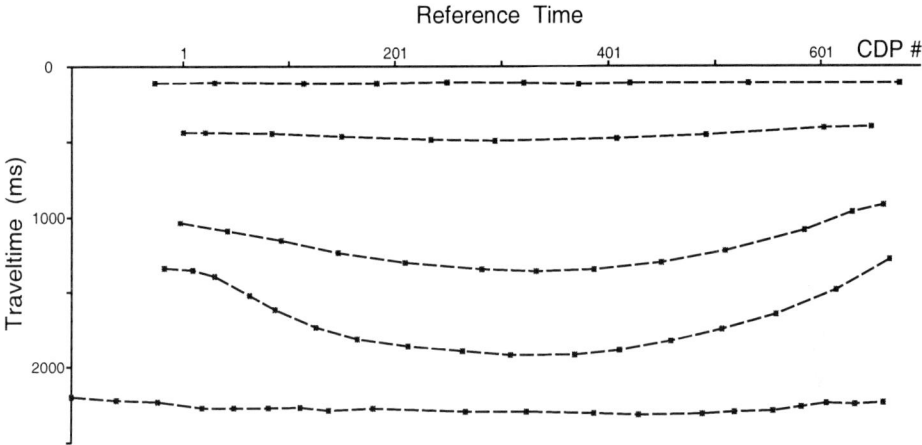

Figure 8.2. Reflection times picked from the reference profile in Figure 8.1. The reflection time is used as a reference time in a cross-correlation procedure for picking the relative reflection amplitudes on individual traces of CMP gathers.

Before we pick the amplitudes and consider the inversion procedure further, a pilot analysis of the reflections is needed. The link between a recorded primary reflection signal and a reliable seismic reflectivity model constructed from a well log is established and is shown in Figure 8.3. A broadband synthetic seismogram computed from well logs is matched with the prestack migrated CMP gather by using the coherence techniques developed by White (1980) and Walden and White (1984). In Figure 8.3, the traces from left to right are, respectively, the impedance log, the broadband synthetic, the estimated wavelet, the filtered synthetic, the data segment and the residuals. The normalized mean square error in the filtered synthetic is 0.03, signifying a very accurate match. The arrows indicate the points on the waveforms picked for reflection 4 and reflection 5.

8.2.3   Extraction of reflection amplitudes

The extraction of amplitudes for each trace of a CMP gather is based on a cross-correlation with the stack, where stability of the picking is achieved by cross-correlating waveforms.

There are hardly any offset amplitude data available for reflection 1, because of the muting of the shot gathers (to remove the direct and refracted wave trains). For the remaining four reflections, the signal-to-noise ratios of the amplitude data decrease markedly with increasing depth. As we know, for a multi-layered model with interfaces of a generally curved nature, a time migration process

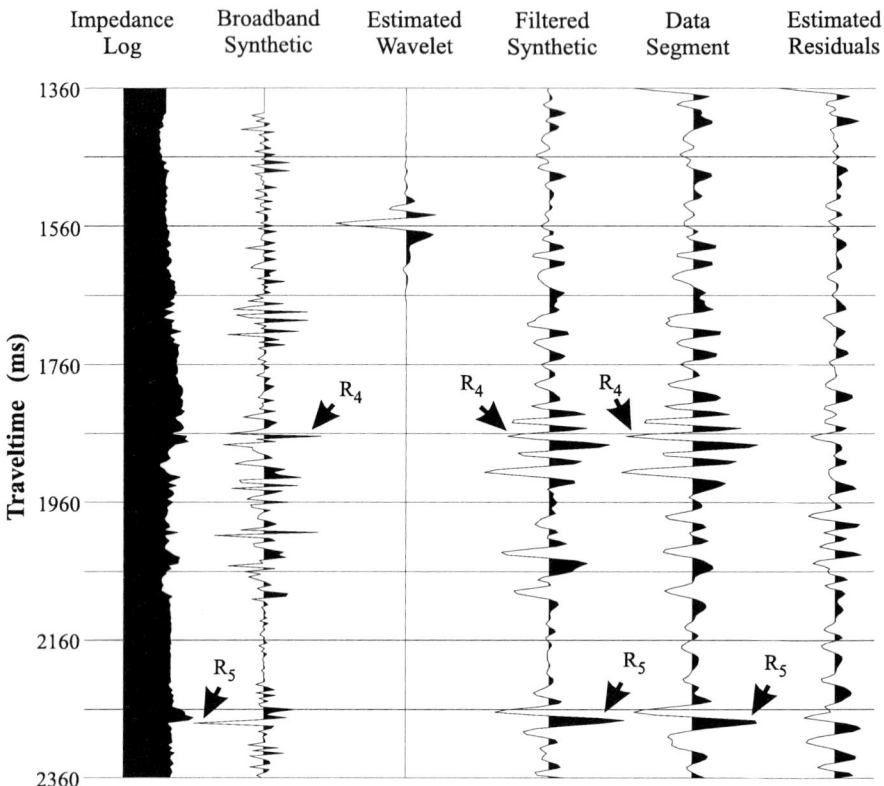

Figure 8.3.   Matching the broadband well-log synthetic seismogram with the migrated seismic trace at the well over a 1400-2300 ms time interval. The seismic wavelet in trace 3 is plotted with its time zero at 1560 ms. The filtered synthetic in trace 4 is the broadband synthetic filtered by the estimated seismic wavelet. The arrows indicate the troughs picked as reference times for reflections 4 and 5 in Figure 8.2.

can provide a complete migration only for the uppermost interface (Hubral, 1977). The stronger the velocity variations and the steeper the structure, the less complete is the migration. Thus the distortions of wave propagation effects in the original gathers are not completely removed, and the picking of amplitudes for deeper reflections is still problematic. Nevertheless, the time migration has made amplitude-picking much easier, by collapsing diffractions, unwrapping triplications etc., than picking on the original unmigrated gathers.

The common-offset stack technique, which is used in exploration seismic processing (Ostrander, 1984), could be adopted here to reduce further the effects of the remaining wave distortions after time migration. It sums traces from several CMP gathers that have the same offset, to produce a trace suitable

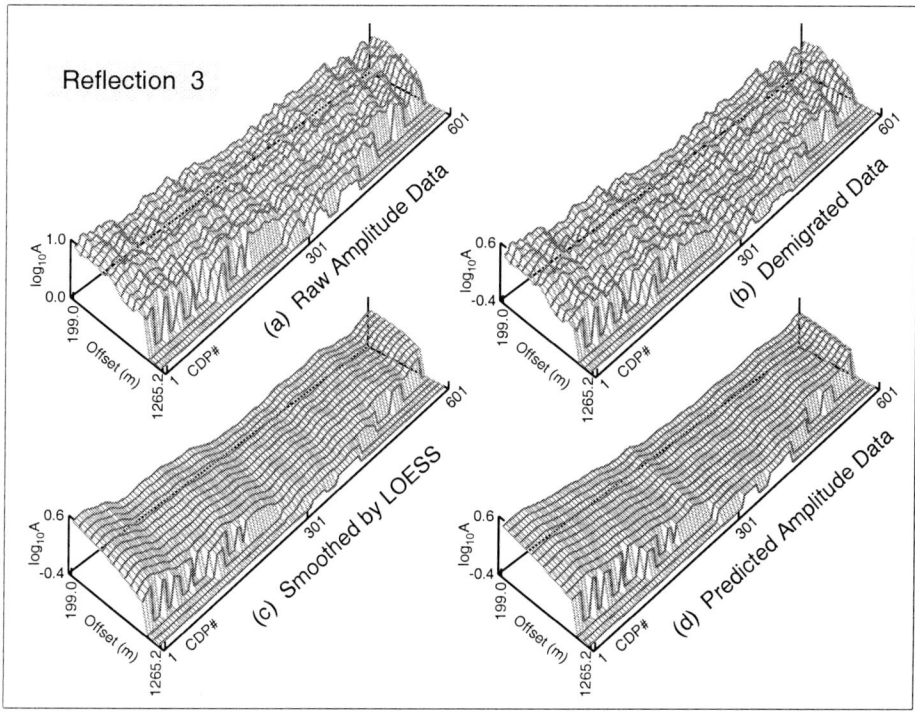

Figure 8.4. Amplitudes of reflection 3: (a) raw amplitudes picked from the prestack migrated gathers; (b) demigrated amplitudes; (c) demigrated amplitudes smoothed by the robust LOESS method; (d) the amplitudes predicted from the inversion model.

for amplitude-picking. In this chapter, however, we simply select the median amplitude at a given offset from three adjacent CDPs to form the input data set of the inversion. This process is similar to the common-offset stack but can partially avoid the smearing of amplitude information caused by trace summing. Consequently, amplitude data from 201 CDPs are used in the inversion. The horizontal distance between two selected CDPs is 50 m. The amplitude data from reflection 3, for example, are shown in Figure 8.4a.

8.3. Demigration of reflection amplitudes

Demigration processing consists of two steps. The first step is the demigration of position. Considering the 1-D model (i.e. velocities are depth controlled) and ignoring the effect of the overburden structure, we can easily determine the horizontal coordinate x_m of a reflection point, since the image ray of a zero-offset reflection is a vertical straight ray in this case. We also have the migrated time t_m (Figure 8.2) and the time slope β_m. The unmigrated time slope β_u, unmigrated time t_u and unmigrated position x_u may be obtained using the following three equations (Whitcombe, 1994; Whitcombe and Carroll, 1994):

$$\beta_u = \frac{2}{v}\sin\left[\tan^{-1}(\beta_m v/2)\right] , \tag{8.1}$$

$$t_u = \frac{t_m}{\sqrt{1 - \beta_u^2 v^2/4}} \tag{8.2}$$

and

$$x_u = x_m + t_u\,\beta_u v^2/4 , \tag{8.3}$$

where v denotes a constant velocity. For the 1-D case, the migration velocity is used. Each data point is then moved back to its unmigrated position with CDP coordinate x_u. Note that we assume a limited source-receiver offset and a small slope of reflection.

We now consider the second step, i.e. demigration of amplitudes. The migration process has partly, depending on its completeness and correctness, compensated the amplitudes of primary reflections due to spherical divergence along the ray path and the focusing/defocusing from reflector curvature. These amplitude compensations need to be removed so that "true" amplitudes are obtained for the tomographic inversion, although amplitude demigration would not compensate for defects in the preprocessing. The amplitude changes in a migration process are rather intricate, compared to the changes in the kinematic elements. However, with a 1-D (vertically varying) velocity model, approximate expressions for the changes in the reflection amplitudes can be used.

Denoting the effect due to spherical divergence along the ray by G_s, and the effect associated with the focusing and defocusing due to the curvature of the reflector by G_c, the demigration of an amplitude $\tilde{A}$ estimated from the migrated gathers can be explicitly written as

$$A = G_s\,G_c\,\tilde{A} , \tag{8.4}$$

where A is the "true" amplitude for the input to inversion. Appendix A.2 derives Taylor expansions in the offset coordinate y for G_s and G_c, assuming prestack time migration using a 1-D velocity model. The effect of spherical divergence can be approximated by

$$G_s \approx \nu_s\left[1 + \left(1 - \frac{v_{rms}^2}{v_0^2}\right)\nu_s^2 y^2\right]^{1/2} , \tag{8.5}$$

where ν_s is defined by

$$\nu_s = \frac{v_0}{v_{rms}^2 t(y)} \, , \tag{8.6}$$

v_0 is the velocity in the vicinity of the source point, v_{rms} is the root-mean-square velocity and $t(y)$ is the reflection traveltime. The focusing and defocusing arising from reflector curvature, which is compensated by migration, is predicted by the equation,

$$G_c \approx \left[1 - \Psi + \Psi^2 + \Psi(1 - 2\Psi)\nu_c^2 y^2 \right]^{1/2} , \tag{8.7}$$

expressed in terms of

$$\Psi = \frac{R}{R_c} \tag{8.8}$$

and

$$\nu_c = \frac{v_z}{v_{rms}^2 t(y)} \, , \tag{8.9}$$

where R is the length of the normal-incidence ray, R_c is the radius of curvature of the reflector and v_z is the local velocity in the vicinity of the reflector. Note that equation (8.7) is a small-dip approximation of expression (A.42) derived in Appendix A.2. A complete (and also correct) migration along the 2-D profile removes the effects of curvature in the direction of the profile, but the effect of curvature in the direction perpendicular to the line remains.

The velocities used in equations (8.5) and (8.7) are the migration velocities actually used in the prestack time migration. For the calculation of the ratio Ψ in equation (8.7), given the traveltimes at zero-offset $T_0(x)$ shown in Figure 8.2, we make the approximations

$$R = v_{rms} T_0 \quad \text{and} \quad \frac{1}{R_c} = \frac{1}{2} v_{rms} T_0'' \, ,$$

evaluated in terms of the traveltime at zero-offset and its second derivative with respect to the horizontal coordinate x. Linear or spline interpolations to the crudely picked reference times seem too coarse and do not provide stable estimates for T_0''. Instead we use a least-squares regression to invert for a smoothly varying function $T_0(x)$.

The reflection amplitudes of reflection 3, after the amplitude demigration, are shown in Figure 8.4b.

8.4.   Winnowing amplitudes by LOESS

The estimated demigrated amplitudes from the migrated CMP gathers contain considerable noise. We now winnow these noisy data, using a locally weighted regression (LOESS) which smooths the data by fitting a locally quadratic surface using weighted least-squares (Cleveland, 1979; Cleveland and Grosse, 1991).

The weight in the least-squares fitting is set initially to be a decreasing tri-cube function,

$$w(x_i) = \begin{cases} \left(1 - |u|^3\right)^3 , & |u| \leq 1 , \\ 0 , & |u| > 1 , \end{cases} \qquad (8.10)$$

where

$$u = \frac{x_i - x_{\text{tar}}}{r}$$

is a normalized distance from the target point x_{tar}, and r is a given radius from x_{tar} enclosing K neighbours. The weight $w(x_i)$ applied to the observation at point x_i is used in fitting the regression through the K neighbours of x_{tar}.

Once we have obtained a locally quadratic surface, we can also down-weight the contribution of observations identified by their large residuals from the fitted values as likely outliers. The weight is given by a bi-square function,

$$a(x_i) = \begin{cases} \left(1 - u^2\right)^2 , & |u| \leq 1 , \\ 0 , & |u| > 1 , \end{cases} \qquad (8.11)$$

where

$$u = \frac{1}{6} \frac{e_i}{e_{\text{m}}} ,$$

e_i is the residual at point x_i, and e_{m} is the median of the absolute residuals $|e_i|$ over the K neighbours. The procedure for down-weighting the large residuals is iterated several times, with the product weight $a(x_i)w(x_i)$.

Figure 8.4c shows the winnowed, or smoothed, amplitudes corresponding to the demigrated data of Figure 8.4b. The number of neighbourhoods (parameter K) used in the 2-D LOESS smoothing is ten per cent of the total number of data points. Four robust iterations were performed to prevent the effect of outliers. Since the amplitude inversion is significantly sensitive to the noise in the data, the reweighted least-squares smoothing essentially down-weights the large picking errors or outliers, thus making the procedure robust in finding the optimum solution. Although the smoothing is done arithmetically, logarithmic

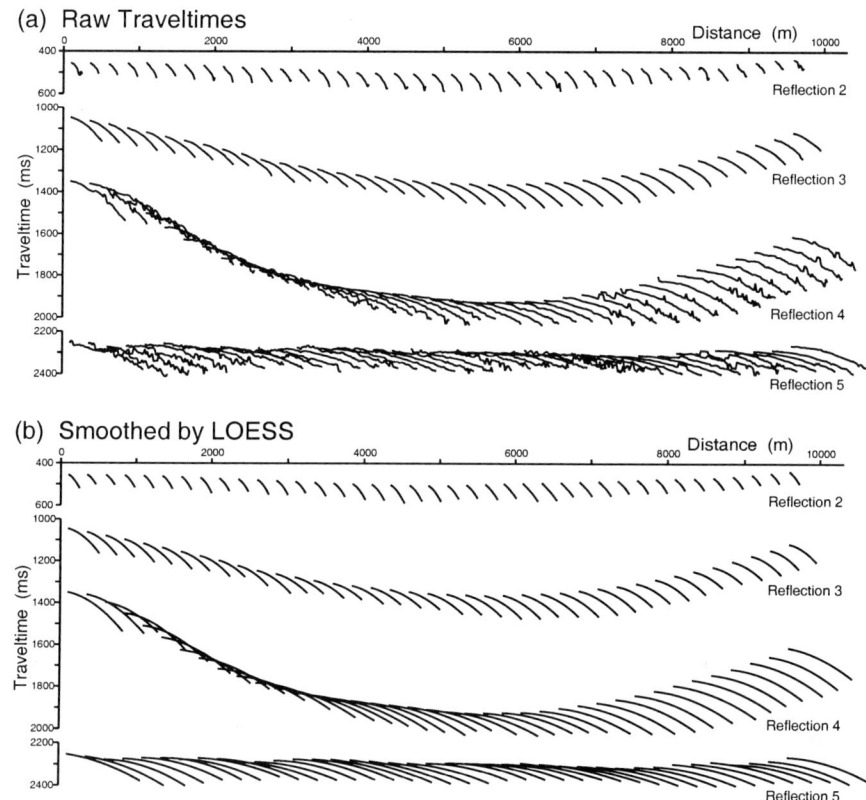

Figure 8.5. Reflection traveltimes and smoothed data sets after using the robust LOESS method. The robust LOESS used here eliminates only the outliers, and the winnowed output is a true regression of the data, with noise much reduced.

data are used as the input to inversion. Working in the logarithmic domain reduces the non-linearity of the problem and makes it more stable than inversion from arithmetic data.

For comparison, Figure 8.4d shows the amplitude data predicted from the inversion model described in the following section.

There is a risk that LOESS, or any other pre-inversion processing, could remove useful information which, once lost, would never be recovered by inversion. To verify the efficacy of the robust LOESS, we also apply the same winnowing process to the traveltime data. Figure 8.5 shows the raw and winnowed traveltime data. We can clearly see that the robust LOESS used here eliminates only the outliers, and the winnowed output is a true regression of the data, with noise much reduced. It is therefore reasonable to assume that the distributions of the data errors are now homogeneous, when we use the smoothed data as the input to the following inversion.

8.5.   Inversion procedure

8.5.1   Recapitulation of the inversion method

The amplitudes extracted from prestack time-migrated gathers (followed by demigration and smoothing as described in the previous section) are used, separately or jointly with traveltime data, to invert for the geometry of reflection interfaces. The subsurface model is assumed to be a multi-layered structure with curved interfaces separating constant velocity layers. The interface is parameterized by a truncated Fourier series (equation 6.25). In the inversion example, each interface is defined by $N = 20$ harmonic components with the fundamental wavenumber $\Delta k = 0.1$ km^{-1}. Thus there are 41 interface parameters, plus one velocity parameter, for each reflector in the inversion.

In the inversion, we use the multi-stage damped subspace method developed in the previous chapter. The 42 model parameters are divided into six groups: group 1 is the velocity and groups 2-6 are five groups of interface parameters. Partitioning of the five subsets of the 41 interface parameters is based on their sensitivities in the reflection seismic inversion, in which both amplitudes and traveltimes have different sensitivities to different wavenumber components of the reflection interface. The subspace inversion is performed in several stages. At each stage the inversion constrains a group of parameters by damping the remaining subsets.

In the inversion for reflections 2-5, respectively, 1699, 5326, 8037 and 8040 amplitude (and traveltime) data are used. The loss of data in shallow reflections is due to the muting process. For this real data example, the implementation of the inversion process consists of the following three steps:

1) A preliminary estimate of the model structure using the traveltime inversion.

2) Estimates of the source amplitude and the absorption coefficients.

3) The interface inversion using the amplitude data alone or jointly with traveltime data.

Before we describe the inversion results, we first describe the first two steps.

8.5.2   Preliminary estimate of model structure from traveltime information

A preliminary estimate of the subsurface model to be used as a known structure in estimating the source amplitude and the absorption coefficients, is ob-

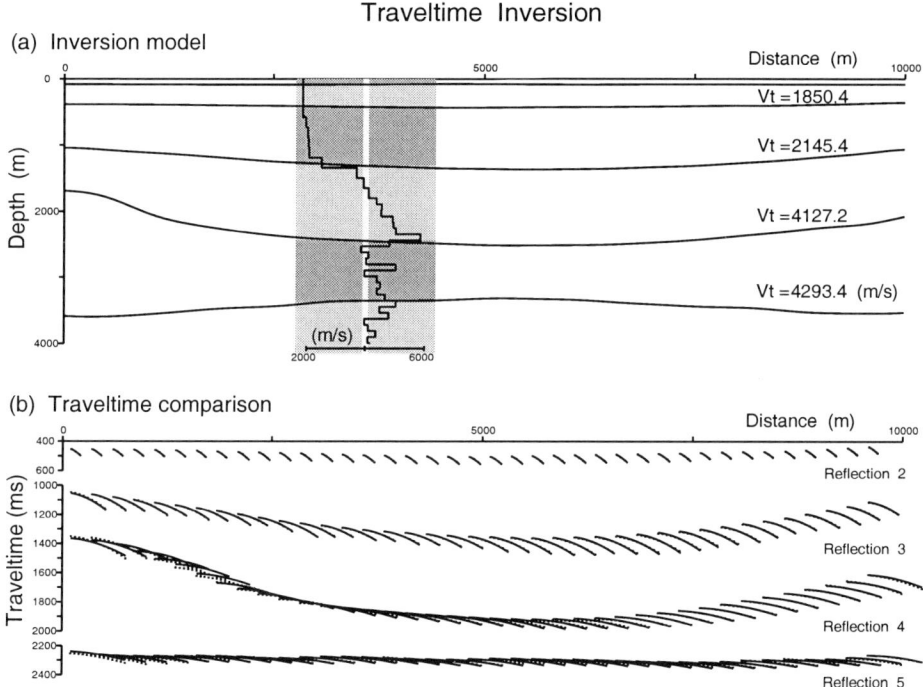

Figure 8.6. The result of traveltime inversion: (a) the inversion model, compared with the interval velocities from check shot survey; (b) comparison of predicted traveltimes (solid lines) and the observations (dotted lines). The vertical white bar in (a) indicates the location of the well. The model shown in (a) was used to estimate the source amplitude and the absorption coefficients.

tained by traveltime inversion. Figure 8.6 shows the result in which the first reflector (the sea-bottom) is converted from the reference time using a velocity equal to 1500 m/s in the water layer. The model is compared with the interval velocity curve from the check shot survey (the interval velocity curve is only available from a depth of 580 m). The vertical white bar indicates the location of the well. The check shot recording interval for the evaluation of velocities between 580 m and 1800 m is 152.4 m, and the recording interval for the re-mainder of the survey is 91.44 m. The background grey stripe indicates the division of the subsurface medium into layers.

The depths of reflectors 4 and 5 may be unambiguously identified on the impedance log in Figure 8.3. The check shot interval velocities shown in Figure 8.6 provide an overall idea of reflection 3. Reflector 3 is the interface between a near constant velocity layer and a layer with a linear increase in velocity. Reflector 4 is also the interface between two distinct formations, the layer with a relatively large velocity gradient and a layer below it with a much smaller velocity gradient. Reflection 5 represents a reflection from the top of a high velocity interval, or a high impedance interval as seen in Figure 8.3.

In Figure 8.6, "Vt" indicates the velocity obtained from traveltime inversion. With the parameterization of the stratified model consisting of a stack of constant velocity layers, the velocity values inverted from traveltime data are consistent with the layer velocities evaluated from the check shot information. For example, the time average interval velocity and the rms velocity value within layer 4 from the check shot survey are 4030.0 and 4132.0 m/s, respectively, whereas the interval velocity obtained from traveltime inversion is 4127.2 m/s. The interval velocity and the rms velocity value within layer 5 from the check shot survey are 4312.2 and 4327.6 m/s, respectively, whereas the interval velocity obtained from traveltime inversion is 4293.4 m/s.

8.5.3   Estimation of the source amplitude and the absorption coefficients

If we assume the model structure (obtained from the traveltime inversion) is known *a priori*, then we may estimate the source amplitude and the absorption coefficients.

With the high frequency assumption, the ray-amplitude is determined by

$$A = A_0 C_Q C / L , \tag{8.12}$$

where A_0 is the amplitude at the source point, C_Q is the inelastic attenuation factor, C is the product of all the relevant reflection and transmission coefficients, and L is the geometrical spreading function (Chapter 1). The inelastic attenuation is given by

$$C_Q = \exp[-\eta(z)\,\ell(z)] , \tag{8.13}$$

where $\eta(z)$ is the average absorption coefficient averaged from the sea-bottom to depth z, and $\ell(z)$ is the ray length of the ray reflected from the target interface at depth z. While both C and L are estimated based on the known model obtained from the previous traveltime inversion, we invert simultaneously for

$$A_0 , \quad \text{and} \quad \eta_2 , \quad \eta_3 , \quad \eta_4 , \quad \eta_5 ,$$

where η_j refer to the average absorption coefficients averaged from the sea-bottom to the interfaces 2-5, respectively.

These five parameters have different physical units, orders of magnitude apart. For this real data example, the source amplitude A_0 has a magnitude of order 10^5, and the absorption coefficients η_j have magnitudes of the order 10^{-3} or 10^{-4}. The Fréchet matrix $\mathbf{F}$ is very ill-conditioned. We now rescale the Fréchet matrix as

$$\mathbf{F}' = \mathbf{F}\mathbf{W} , \tag{8.14}$$

using a diagonal weighting matrix

$$\mathbf{W} = \text{diag}\{w_j\} \,. \tag{8.15}$$

That is, each column vector of the matrix $\mathbf{F}$ is multiplied by a scalar w_j. Weight w_1 (corresponding to the source amplitude parameter) is set equal to the initial estimate of the source amplitude A_0, and the remaining weights w_j (corresponding to the absorption coefficients η_j) are set as 10^{-3}. The inverse solution $\delta\mathbf{m}'$ from the scaled operator $\mathbf{F}'$ must then be multiplied by the diagonal matrix $\mathbf{W}$ to achieve the true model update,

$$\delta\mathbf{m} = \mathbf{W}\delta\mathbf{m}' \,. \tag{8.16}$$

We have only five parameters in this inversion, but the number of data is more than 10^4. For such a highly over-determined system the inversion converges to the solution after only one or two iterations. We have carried out a synthetic test in which the relative errors in the initial estimate were 20.0, 6.25, 1.64, 9.09 and 21.57 per cent, corresponding to A_0 and the four absorption parameters, respectively. After one iteration, the relative errors were reduced to 2.55, 0.35, 0.31, 0.23 and 0.18 per cent. The relative errors clearly show that the average absorptions to deeper layers are more accurately determined, since the average ray length increases relative to the shallower layers. In contrast, the estimation of an interval absorption coefficient must surely be more noise prone for deeper layers. Tests have also shown that the inversion solution does not depend strongly on the initial estimates.

For these seismic data, the estimated source amplitude A_0 is 3×10^5 and the four average absorption coefficients are 1.65, 1.32, 0.24, and 0.19 ($\times 10^{-3}$), respectively. As the depth increases, the average absorption generally decreases. Note that the source amplitude here is a relative value when relative amplitudes along the profile are used. The absorption coefficients are empirical measurements. Whether they precisely reflect the physical properties of the earth is not known. With the source amplitude and the absorption coefficients known, we are ready to use geometrical modelling to invert amplitude data alone or both traveltime and amplitude data simultaneously.

8.6. Inversion results

8.6.1 Amplitude inversion

We first examine the inversion result using amplitude data alone. In order to gain some insight into how well the amplitude data can constrain curved interfaces in a multi-layered structure, flat straight interfaces are used as the

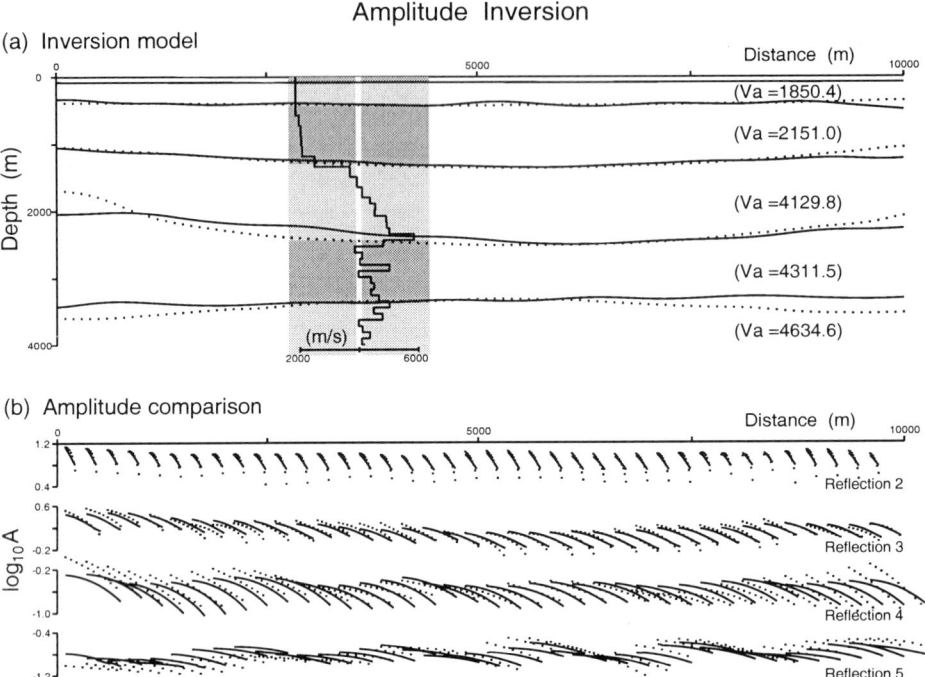

Figure 8.7. Inversion model obtained from amplitude inversion, showing the five reflectors and the interval velocities. The first interface is the sea-bottom. For comparison, the model obtained from traveltime inversion is shown by dotted lines. The velocity, Va, is the interval velocity obtained from amplitude inversion.

starting models for the inversion. There are no amplitudes available for reflection 1. For the remaining four reflections, a layer-stripping approach is used in the inversion.

The inversion result (solid lines) is shown in Figure 8.7, in which the dotted lines represent the result of traveltime inversion (Figure 8.6). The general consistency between the models from the amplitude inversion and from the traveltime inversion suggests that the amplitude inversion has recovered the basic features of the subsurface model. However, it should be recalled that in layer-stripping the properties of each layer depend on the inversion of the previous layers and error propagation depends on the number of interfaces.

We now compare the amplitudes predicted by inversion with those input for each reflection. The tapering used in muting the direct and refracted wave trains from the shot gathers has reduced the amplitudes of reflection 2 on the far offsets. Except for this tapering, the predicted amplitudes match the input data quite well. The predicted amplitudes of reflection 3 satisfactorily reproduce the offset and CDP variations of the input data. The data fits for reflections 4 and 5 might be also acceptable, but considerable coherent residuals exist.

There is an apparent trade-off between the goodness of data fit and the confidence in the model structure. The aberrant behaviour of the amplitude data of reflections 4 and 5 suggests that a better data fit might provide a geologically unreasonable model. The LOESS smoothing removes the random noise, but cannot remove locally coherent noise in the data. However, traveltime data are usually of good quality compared with amplitude data, especially for deep reflections. Further confidence in the structural model can be promoted by the joint use of amplitude and traveltime data in an inversion. Also, according to the conclusions drawn in Chapter 6, amplitudes are sensitive to high wavenumber components, whereas traveltimes can only constrain the long wavenumber components well. Thus both types of data should complement each other in such a joint inversion.

8.6.2 *Joint inversion of traveltime and amplitude data*

In order to constrain the interface structure effectively, both traveltime and amplitude data, d_{ampl} and d_{time}, are used simultaneously in the definition of the objective function. The data covariance matrix C_D^{-1} is set to be a data weighting matrix (equation 6.29) with a balancing factor κ defined by the ratio of traces of matrices F_{time} to F_{ampl} (equation 6.30). The ratio of matrix traces is an empirical quantity indicating the relative sensitivities of the traveltimes compared to the amplitudes. Numerical experiments in Chapter 6 showed that the ratio of matrix traces is much less than 1.0, because the amplitudes are more sensitive to variations in the interface model than are the traveltimes, and that the optimal value of g is in the range 0.25 to 0.75 for the interface inversion. To increase the influence of traveltimes on the interface inversion, we set the weighting factor $g = 0.75$ in this inversion, where we recalculate κ at each iteration.

A starting model consisting of flat straight interfaces is again used here. Since traveltimes can constrain the velocity v_q above the interface and amplitudes can constrain the velocity contrast at the interface, both v_q and v_{q+1} are included in the inversion for the qth interface. There are therefore 43 model parameters to be inverted in each interface inversion. The final model, obtained by the joint inversion using both types of data, is depicted in Figure 8.8a (again compared with that from the traveltime inversion). The interface geometry from the two inversions is generally consistent, with minor differences. These geometrical differences arise because traveltime data cannot constrain interface components with high wavenumbers. Interfaces 2 and 3 are modified more than interfaces 4 and 5, relative to the model from traveltime inversion.

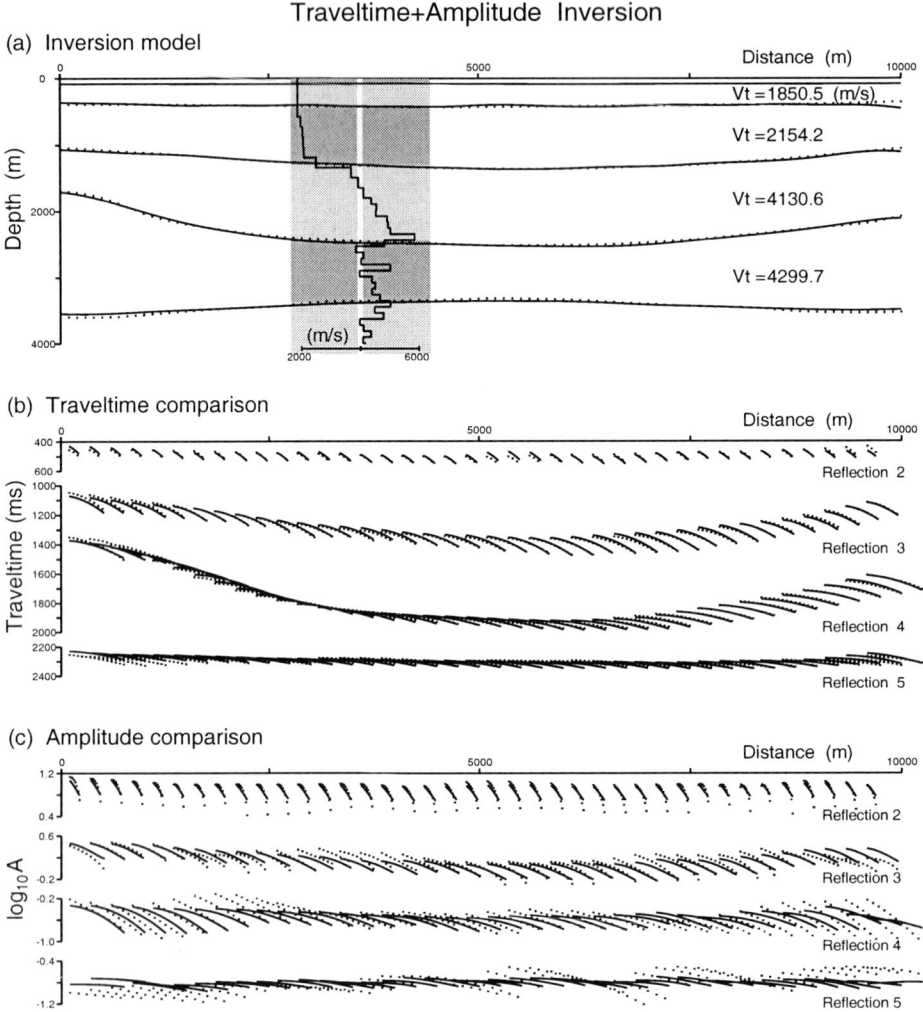

Figure 8.8. Joint inversion using both traveltime and amplitude data. (a) Inversion model (solid lines); for comparison, the model from traveltime inversion is also drawn (dotted lines). (b) Comparison of traveltimes. (c) Comparison of amplitudes. In (b) and (c), the solid lines are predictions from the inversion, and dotted lines are the data input to the inversion.

The comparison of traveltimes and amplitudes is shown in Figures 8.8b and 8.8c, in which solid lines represent predictions from the inversion model, and dotted lines represent the input to inversion. From these comparisons, we can see that traveltimes fit better than amplitudes, although the data fittings of traveltimes and amplitudes slightly compromise each other. With the traveltime constraint in the inversion, reflections 4 and 5 have considerable amplitude residuals, which are at least partly coherent errors.

The idea that traveltimes can constrain the velocities and amplitudes can constrain the velocity jumps is compromised by allowing only constant velocities in each layers. When, as in the real data case presented here, there is a significant vertical gradient, the traveltime inversion can give only an average velocity for the layer and the jumps between these will not correspond to the true interface contrasts. It seems quite likely that this is implicated in the systematic amplitude errors seen in the final inversion of the real data. To overcome this problem, we may use a modified model parameterization in which we de-couple the background constant-velocity layers and the variation of velocity contrasts at the interfaces. The inversion result is presented in the next chapter.

Figures 8.6, 8.7 and 8.8 also include the velocities obtained from the check shot survey at the location of the well. The velocities from all inversions agree reasonably well with the check shot velocities, in spite of our assumption that each layer consists of a homogeneous velocity. This gives us some confidence in the validity of our approach.

Chapter 9

Simultaneous inversion for model geometry and elastic parameters

Abstract   We study how to use both reflection traveltime and amplitude data in a joint
inversion, to invert simultaneously for the interface geometry and the elastic
parameters at the reflectors. Such an inverse problem has different physical
dimensions in both data and model spaces. We propose some practical approaches
to tackle the dimensional difficulties. Using the joint inversion that may properly
take care of the structural effect, we can improve the estimate of the subsurface
elastic parameters. We combine the three estimated elastic parameters into one,
$\gamma(x)$, the ratio of the S-wave velocity contrast to the P-wave velocity contrast,
and analyse the deviation of this parameter from a normal background trend. The
latter, $\Delta\gamma(x)$, referred to as an *abnormality factor*, promises to have potential
application in lithological interpretation. We demonstrate the inversion method
by applying it to real seismic data from the North Sea.

9.1.   Introduction

In the previous chapters, we saw that in the reflection seismic experiment, the
information contained in amplitudes and in traveltimes is complementary, being
sensitive to different features of the subsurface model, and that a joint inversion
of both types of data can potentially reconstruct the model geometry of a multi-
layered structure. In this chapter, we also see that by properly correcting for
the structural effect, we can improve the estimation of the elastic parameters
of subsurface targets in the traditional inversion of amplitude variation versus
offset (AVO).

AVO inversion has been used extensively for lithology and fluid prediction in many regions (e.g. Ostrander, 1984; Hilterman, 1990; Hampson, 1991; Castagna *et al.*, 1993; Spratt *et al.*, 1993; Mallick, 1995; Buland *et al.*, 1996; Simmons and Backus, 1996; Ramos and Davis, 1997). In conventional AVO inversion we commonly assume that all offset-dependent amplitude effects, other than the reflection coefficient, are corrected and then the contrasts in elastic parameters can be estimated from amplitude data (Ursin and Dahl, 1992; Ursin and Ekren, 1995; Ursin *et al.*, 1996). However, wave propagation effects can significantly affect AVO measurements (Swan, 1991; Martinez, 1993). These effects include inelastic attenuation, interbed multiples due to fine layering, surface multiple reflections, transmission losses and geometrical spreading.

The geometrical spreading may also be regarded as a geological effect, which is mainly influenced by the geometry of the target reflector and interfaces within the overburden media. We see that the essential part of this effect is the focusing and defocusing due to the curvature of the target reflector, and that it is necessary to remove this structural effect to derive the AVO attributes correctly. Therefore, in the present chapter, we attempt to invert simultaneously for the interface geometry and the elastic parameters at reflectors.

The inverse problem that we are dealing with has different physical dimensions in both data and model spaces. In the data space, we use traveltimes and amplitudes jointly. In the model space, we consider different classes of physical parameters, including the velocities, the interface depths and the different elastic parameters. We present practical approaches to tackle these dimensional difficulties, and also modify the multi-stage damped, subspace method, developed in the previous two chapters, for use in this multi-dimensional case.

Following the development in Chapter 3 and using an approximation of the reflection coefficient, which is accurate to second order in the slowness p, we can constrain the following three elastic parameters along an interface: the relative P-wave velocity contrast, $\Delta\alpha/\alpha$, the relative S-wave velocity contrast, $\Delta\beta/\beta$, and the ratio of average S-wave to average P-wave velocities, β/α. We parameterize the spatial variation of each of these elastic parameters with respect to the horizontal location, x, by truncated Fourier series. Finally, we present an application of the inversion method to real data from the North Sea.

9.2.   Ray-amplitude and its approximation

Within the ray assumption, a reflection amplitude is given by

$$A = A_0 C_Q C / L(\ell) , \qquad (9.1)$$

where A_0 is the source amplitude, C_Q is the inelastic attenuation along the ray path, C is the product of reflection and transmission coefficients calculated with reference to the Zöppritz equations, and $L(\ell)$ is the geometrical spreading. A practical approach to the estimate of A_0 and C_Q is given in the previous chapter. The geometrical spreading, $L(\ell)$, can be expressed explicitly in terms of the curvature of interfaces encountered along the ray path (Chapter 1 and Appendix A.1). In order to be able to invert for elastic parameters efficiently, we use approximations to the Zöppritz equations for the calculation of reflection and transmission coefficients (Chapter 3). When using the approximations rather than the exact Zöppritz equations, the Fréchet derivatives of amplitudes with respect to elastic parameters can be calculated analytically.

In Chapter 3, we derived quadratic approximations to the *P-P* wave reflection and transmission coefficients, R_{PP} and T_{PP}, with respect to the ray parameter or the elastic contrasts at an interface (equations 3.26 and 3.27). As Ursin and Tjåland (1996) showed, up to three parameters can be estimated from precritical *P-P* reflection coefficients when the full Zöppritz equations are used. The quadratic approximations are expressed finally in terms of three elastic parameters: the relative *P*-wave velocity contrast, $\Delta\alpha/\alpha$, the relative *S*-wave velocity contrast, $\Delta\beta/\beta$, and the ratio of the average *S*-wave to the average *P*-wave velocities, β/α. They are key parameters in the description of rock properties at the reflector. An empirical systematic relation between the *P*-wave velocity and the bulk density (Gardner *et al.*, 1974) was used to eliminate density from these expressions.

In conventional AVO inversion, a linearized approximation to the *P-P* reflection coefficient with respect to the elastic contrasts (Bortfeld, 1961; Chapman, 1976; Richards and Frasier, 1976; Aki and Richards, 1980; Shuey, 1985) is used. Previous studies using this approximation found that it was difficult to estimate more than two of these key elastic parameters in AVO inversion (de Nicolao *et al.*, 1993; Ursin and Tjåland, 1993). In those circumstances, the usual assumption was that the *S*-wave to *P*-wave velocity ratio β/α was known *a priori*. With such a constraint the inversion could produce a biased solution. Chapter 3 has shown, however, that when using the quadratic equation (3.26) in amplitude inversion, the condition number of the matrix of Fréchet derivatives of amplitudes with respect to these three elastic parameters, is reduced from that which using the linearized formula. Therefore, we can, in principle, invert for three parameters, subject to the signal-to-noise ratio in the amplitude data. Readers may refer to Chapter 8 for techniques used to improve the signal-to-noise ratio in real amplitude data.

9.3.   Inversion method

9.3.1   *Parameterization in inverse modelling*

In the approach to the inversion problem, let us consider a 2-D stratified structure model, as depicted in Figure 9.1, where smooth and curved interfaces separate homogeneous layers. In each layer the P-wave velocity is assumed to be constant. However, the inversion allows the elastic properties to vary along the horizontal interfaces. This model parameterization is based on the following assumptions:

1) Amplitude data do not constrain absolute interval velocities, which are essentially constrained by reflection traveltimes.

2) The stack of constant velocity layers is a good macro-model in a specific area as far as the reflection traveltimes are concerned.

3) Amplitudes are significantly sensitive to velocity perturbations in the vicinity of a reflection point, when compared with the velocity variation along the whole ray path from source to receiver.

The elastic property of a reflector is measured in terms of three elastic parameters, which are laterally variable along the interface and denoted by

$$\Delta\alpha/\alpha = P(x) , \qquad \Delta\beta/\beta = S(x) , \quad \text{and} \quad \beta/\alpha = G(x) , \qquad (9.2)$$

where these parameters are functions of the spatial coordinate x. The process of extracting the variation of the parameters, $P(x)$, $S(x)$ and $G(x)$, along a target reflector is often referred to as reflector attribute extraction in the context of AVO inversion. In AVO inversion the geometrical effect due to the curvature of interfaces is not uauslly considered. In this chapter, we advocate a simultaneous inversion for both the interface geometry and the elastic parameters. Both the geometrical spreading for a multi-layered structure and the reflection/transmission coefficients at interfaces are represented analytically as function of the interface geometry and the elastic parameters.

The depth, $z = Z(x)$, of an interface and the variation of the elastic parameters, $P(x)$, $S(x)$, and $G(x)$, along the horizon are parameterized by a truncated Fourier series,

$$f(x) = a_0 + \sum_{n=1}^{N} [a_n \cos(n\pi k_0 x) + b_n \sin(n\pi k_0 x)] , \qquad (9.3)$$

where k_0 is the basis wavenumber and $\{a_0, a_n, b_n, (n = 1, N)\}$ are amplitude coefficients of the harmonic terms. The model vector, $\mathbf{m}$, that we invert for in a non-linear inversion thus consists of four sets of amplitude coefficients

$$\{a_0^{(J)}, a_n^{(J)}, b_n^{(J)}, \quad \text{for} \quad n = 1, \cdots, N\} , \qquad (9.4)$$

where $J = 1, \cdots, 4$, representing Z, P, S and G, respectively. In the following inversion tests, we set the number of harmonic terms to $N = 20$. Including

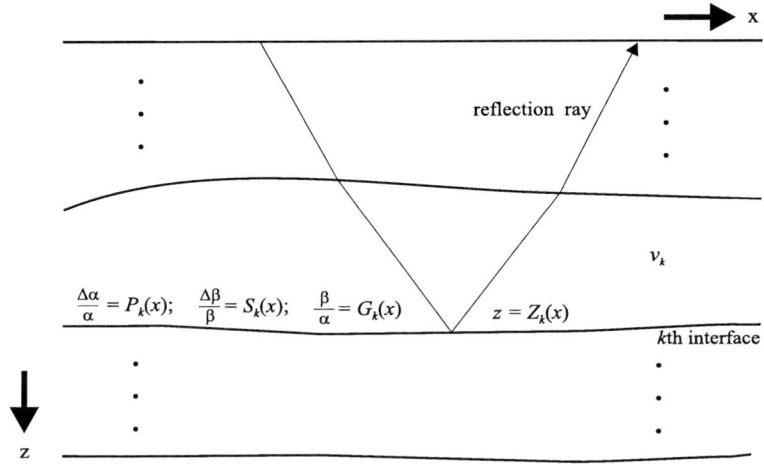

Figure 9.1. A 2-D stratified structure model considered in the joint inversion of traveltime and amplitude data, inverting simultaneously for interface geometry and elastic parameters along the interfaces.

the interval velocity above each reflector, there are $1+4(2N+1)$ parameters in total in the inversion. The data vector, $\mathbf{d}$, in the inversion consists of reflection traveltimes, $\mathbf{d}_{\text{time}}$, and reflection amplitudes, $\mathbf{d}_{\text{ampl}}$.

9.3.2   The Fréchet matrix and the inversion method

The Fréchet derivatives of traveltimes and amplitudes with respect to interface perturbations are calculated by using a first-order approximation, as described in Chapter 2. The Fréchet derivatives of amplitudes A with respect to the coefficients, $\{\xi_i \equiv a_n \text{ or } b_n, (n = 0, \cdots, N)\}$ of the three elastic parameters, $P(x)$, $S(x)$ and $G(x)$, along a specified reflector are

$$\frac{\partial(\log_{10}A)}{\partial\xi_i} = (\log_{10}e)\frac{1}{R_k}\frac{\partial R_k}{\partial\xi_i} , \qquad (9.5)$$

where R_k is the reflection coefficient of the kth (target) reflector, and $\partial R_k/\partial\xi_i$ is calculated analytically from equation (3.26). Finally, we have the full Fréchet matrix,

$$\mathbf{F} = \begin{bmatrix} \mathbf{F}_A^{\text{time}} & \mathbf{F}_B^{\text{time}} & \mathbf{0} & \mathbf{0} & \mathbf{0} \\ \mathbf{0} & \mathbf{F}_B^{\text{ampl}} & \mathbf{F}_C^{\text{ampl}} & \mathbf{F}_D^{\text{ampl}} & \mathbf{F}_E^{\text{ampl}} \end{bmatrix} , \qquad (9.6)$$

where $\mathbf{0}$ is the null matrix, and $\mathbf{F}_J^L$ represents a submatrix of the Fréchet derivatives of data of type L (traveltime or amplitude) with respect to model parameters of class J. In the example considered here, model parameters are divided into five classes: A for interval velocity, B for interface geometry, and C, D and E for the three elastic parameters respectively.

For the different classes of physical parameters considered simultaneously in an inversion procedure, the Fréchet matrix is rescaled by

$$\tilde{\mathbf{F}} = \mathbf{FW} , \tag{9.7}$$

where $\mathbf{W}$ is a weighting matrix used to balance the sensitivities of the different classes of physical parameters. Denoting the data residual vector by $\delta\mathbf{d}$ and the data covariance matrix by $\mathbf{C}_{\mathrm{D}}^{-1}$, with a rescaled Fréchet matrix defined by equation (9.7), the gradient vector,

$$\mathbf{g} = \mathbf{F}^{\mathrm{T}}\mathbf{C}_{\mathrm{D}}^{-1}\delta\mathbf{d} ,$$

is then modified according to

$$\tilde{\mathbf{g}} = \tilde{\mathbf{F}}^{\mathrm{T}}\mathbf{C}_{\mathrm{D}}^{-1}\delta\mathbf{d} \equiv \mathbf{W}^{\mathrm{T}}\mathbf{g} , \tag{9.8}$$

and the Hessian matrix, $\mathbf{H} = \mathbf{F}^{\mathrm{T}}\mathbf{C}_{\mathrm{D}}^{-1}\mathbf{F}$, is modified according to

$$\tilde{\mathbf{H}} = \tilde{\mathbf{F}}^{\mathrm{T}}\mathbf{C}_{\mathrm{D}}^{-1}\tilde{\mathbf{F}} \equiv \mathbf{W}^{\mathrm{T}}\mathbf{HW} . \tag{9.9}$$

The scaled model update can be obtained using a matrix-based method such as the subspace algorithm (Kennett *et al.*, 1988),

$$\delta\tilde{\mathbf{m}} = -\mathbf{A}(\mathbf{A}^{\mathrm{T}}\tilde{\mathbf{H}}\mathbf{A})^{-1}\mathbf{A}^{\mathrm{T}}\tilde{\mathbf{g}} , \tag{9.10}$$

where $\mathbf{A}$ is the projection matrix composed of the basis vectors. The model vector is defined as a linear combination of these basis vectors. Finally, the model update is given by

$$\delta\mathbf{m} = \mathbf{W}\delta\tilde{\mathbf{m}} . \tag{9.11}$$

The problem remaining is how to define in practice the weighting matrix $\mathbf{W}$ and the data covariance matrix $\mathbf{C}_{\mathrm{D}}^{-1}$.

9.3.3   *Working definitions of matrices $\mathbf{W}$ and $\mathbf{C}_{\mathrm{D}}^{-1}$*

We now describe a practical approach to the definition of the weighting matrix $\mathbf{W}$ and the data covariance matrix $\mathbf{C}_{\mathrm{D}}^{-1}$. The definitions are designed to balance the contributions, in terms of sensitivities, of the different model parameters and the different data types in the inversion.

The weighting matrix is defined as a diagonal matrix,

$$
\mathbf{W} = \begin{bmatrix} w_A\mathbf{I} & \cdots & \cdots & \cdots & \mathbf{0} \\ \vdots & w_B\mathbf{I} & & & \vdots \\ \vdots & & w_C\mathbf{I} & & \vdots \\ \vdots & & & w_D\mathbf{I} & \vdots \\ \mathbf{0} & \cdots & \cdots & \cdots & w_E\mathbf{I} \end{bmatrix} ,
\tag{9.12}
$$

where the weighting factors w_J ($J = A, \cdots, E$) for each submatrix are given by

$$
w_A = \frac{\text{trace}(\mathbf{F}_B^{\text{time}})}{\text{trace}(\mathbf{F}_A^{\text{time}})} , \qquad w_B = 1 , \qquad w_C = \frac{\text{trace}(\mathbf{F}_B^{\text{ampl}})}{\text{trace}(\mathbf{F}_C^{\text{ampl}})} ,
$$

$$
w_D = \frac{\text{trace}(\mathbf{F}_B^{\text{ampl}})}{\text{trace}(\mathbf{F}_D^{\text{ampl}})} , \qquad w_E = \frac{\text{trace}(\mathbf{F}_B^{\text{ampl}})}{\text{trace}(\mathbf{F}_E^{\text{ampl}})} ,
$$

and the trace of a matrix is defined as the sum of its eigenvalues. The data covariance matrix is then defined by

$$
\mathbf{C}_D^{-1} = \begin{bmatrix} \mathbf{I} & \mathbf{0} \\ \mathbf{0} & \kappa^2\mathbf{I} \end{bmatrix} ,
\tag{9.13}
$$

where the dimensionless balancing factor, κ, is given by

$$
\kappa = \left(\frac{\text{trace}(\mathbf{F}_B^{\text{time}})}{\text{trace}(\mathbf{F}_B^{\text{ampl}})} \right)^g ,
\tag{9.14}
$$

and the secondary free parameter g is set equal to 0.75, following Chapter 6.

Even within the same class of physical parameters (such as the interface components) the traveltimes and amplitudes have different sensitivities to the model components. In Chapter 7 we described a multi-stage damped subspace algorithm used in the joint inversion. When we also consider the different classes of physical parameters, the formula for the damped subspace algorithm becomes

$$
\delta\mathbf{m} = -\mathbf{W}\mathbf{A}[\mathbf{A}^T(\mathbf{W}^T\mathbf{H}\mathbf{W} + \mathbf{D})\mathbf{A}]^{-1}\mathbf{A}^T\mathbf{W}^T\mathbf{g} ,
\tag{9.15}
$$

where $\mathbf{D} = \text{diag}\{\mu_i\}$ and μ_i is the damping factor corresponding to the ith group of model components, with unit of (*model parameter*)$^{-2}$.

A multi-stage damped subspace scheme is depicted schematically in Figure 9.2. Five different classes of physical parameters (i.e. interval velocity v_k, interface z, elastic parameters $\Delta\alpha/\alpha$, $\Delta\beta/\beta$ and β/α) are considered simultaneously

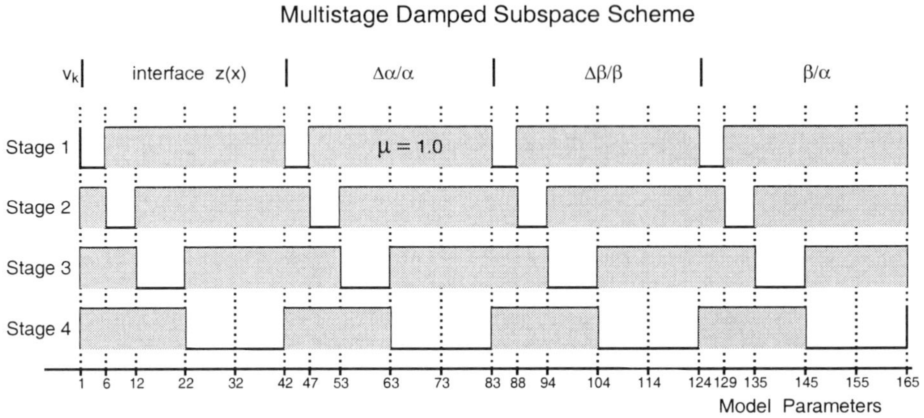

Multistage Damped Subspace Scheme

Figure 9.2.   Multi-stage damped subspace scheme. Five different classes of physical parameters (interval velocity v_k, interface z, $\Delta\alpha/\alpha$, $\Delta\beta/\beta$ and β/α) are considered simultaneously in the inversion. Model components of each class of physical parameter are divided into several groups, corresponding to different wavenumber ranges. By this mean, the model space is divided into a set of subspaces, or subsets. The inversion procedure, at each stage, constrains a group of subsets by damping the remaining subsets ($\mu = 1.0$ within grey areas).

in the inversion. The model components of each class of physical parameter are divided into several groups, corresponding to different wavenumber ranges. In this manner, the model space is divided into a set of subspaces, or subsets. The inversion procedure constrains a group of subsets at each stage, by damping the remaining subsets ($\mu = 1.0$, grey areas in Figure 9.2).

9.4.   Inversion example

9.4.1   Example dataset

The simultaneous inversion for model geometry and elastic parameters is demonstrated using real seismic reflection data from the North Sea. Figure 9.3 shows a stacked example profile, after prestack time migration, and indicates five reflections R_i ($i = 1, \cdots, 5$) which are considered in the following inversion. Close examination of Figure 9.3 clearly shows that there are gradual phase changes at points A and B, on reflections 3 and 4, respectively, and phase changes at C and D in reflection 5, where the reflection event varies from single wavelet to multi-wavelet. These phase changes could correspond to changes in the elastic properties along the reflections.

We now pick traveltimes from the five reflections R_i ($i = 1, \cdots, 5$), and amplitudes from the four reflections R_i ($i = 2, \cdots, 5$). These reflection traveltimes and amplitudes are extracted from prestack time-migrated CRP gathers.

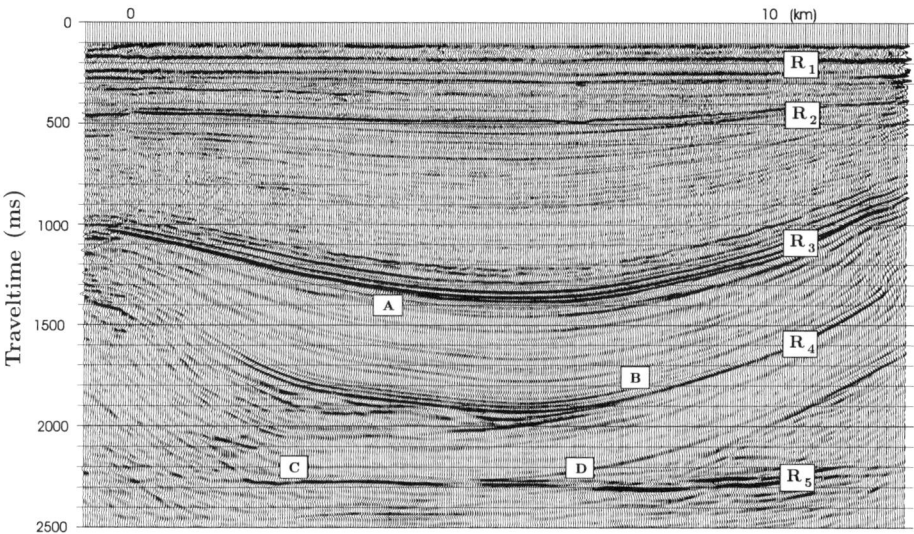

Figure 9.3. Reflection seismic profile from the North Sea. Amplitudes and traveltimes extracted from prestack migrated CDP gathers are used in the inversion for interface geometry and elastic parameters along reflectors. Four reflections ($R_2 - R_5$) are considered in the inversion. Labels A, B, C and D indicate phase changes of reflection events.

A demigration process is carried out so that we have a set of "true" observations, used as the input of inversion. The actual input data have been winnowed to remove the outliers. This data set has also been used in Chapter 8, where the demigration processing and the data winnowing are described in detail.

9.4.2 Inversion results

A layer-stripping approach is adopted, i.e. the model geometry and the elastic parameters used in the overburden are those obtained from the preceding inversion steps. The initial model for each interface is an interface curve obtained from traveltime inversion. The initial estimates for the three elastic parameters along each interface are constants [i.e. all coefficients in the parameterization equation (9.4) are zero-valued, except that of a_0], where $\Delta\alpha/\alpha$ is calculated based on the constant velocities from the previous traveltime inversion, $\Delta\beta/\beta$ is simply set equal to $\Delta\alpha/\alpha$, and β/α is initially set equal to 0.45. Except for the sea-bottom which is reconstructed from traveltime information, four individual simultaneous inversions for reflections 2-5 are performed.

Inversion results for reflectors 2, 3 and 4, 5 are shown in Figures 9.4 and 9.5, respectively. In each of these figures, (a) depicts the inversion model and (b)

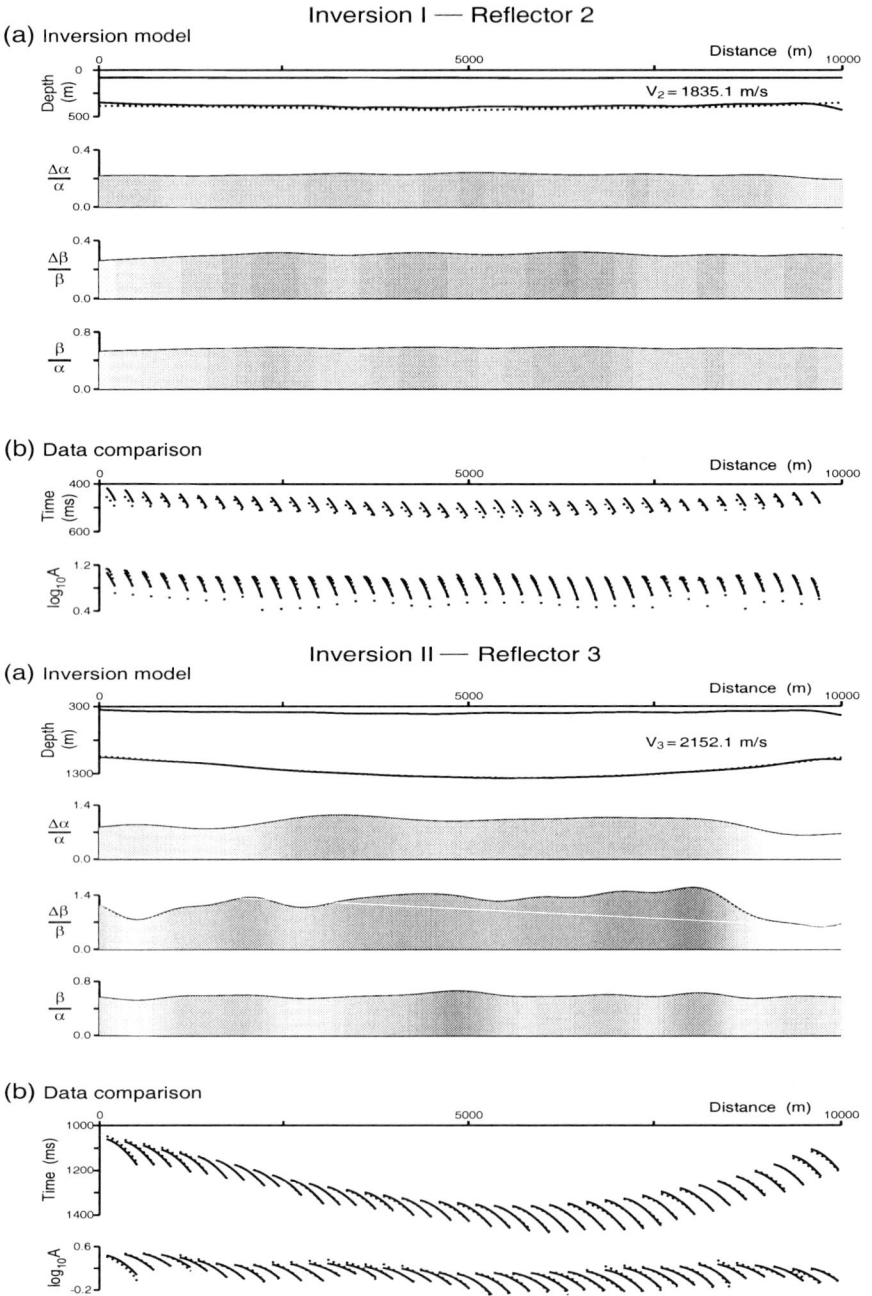

Figure 9.4. Simultaneous inversion — Reflectors 2 and 3. (a) Inversion model: interface geometry (solid line) compared with inversion result from traveltime inversion (dotted line), and three elastic parameters ($\Delta\alpha/\alpha$, $\Delta\beta/\beta$ and β/α). (b) Data comparison: predicted traveltimes and amplitudes (solid lines) compared with input data (dotted lines).

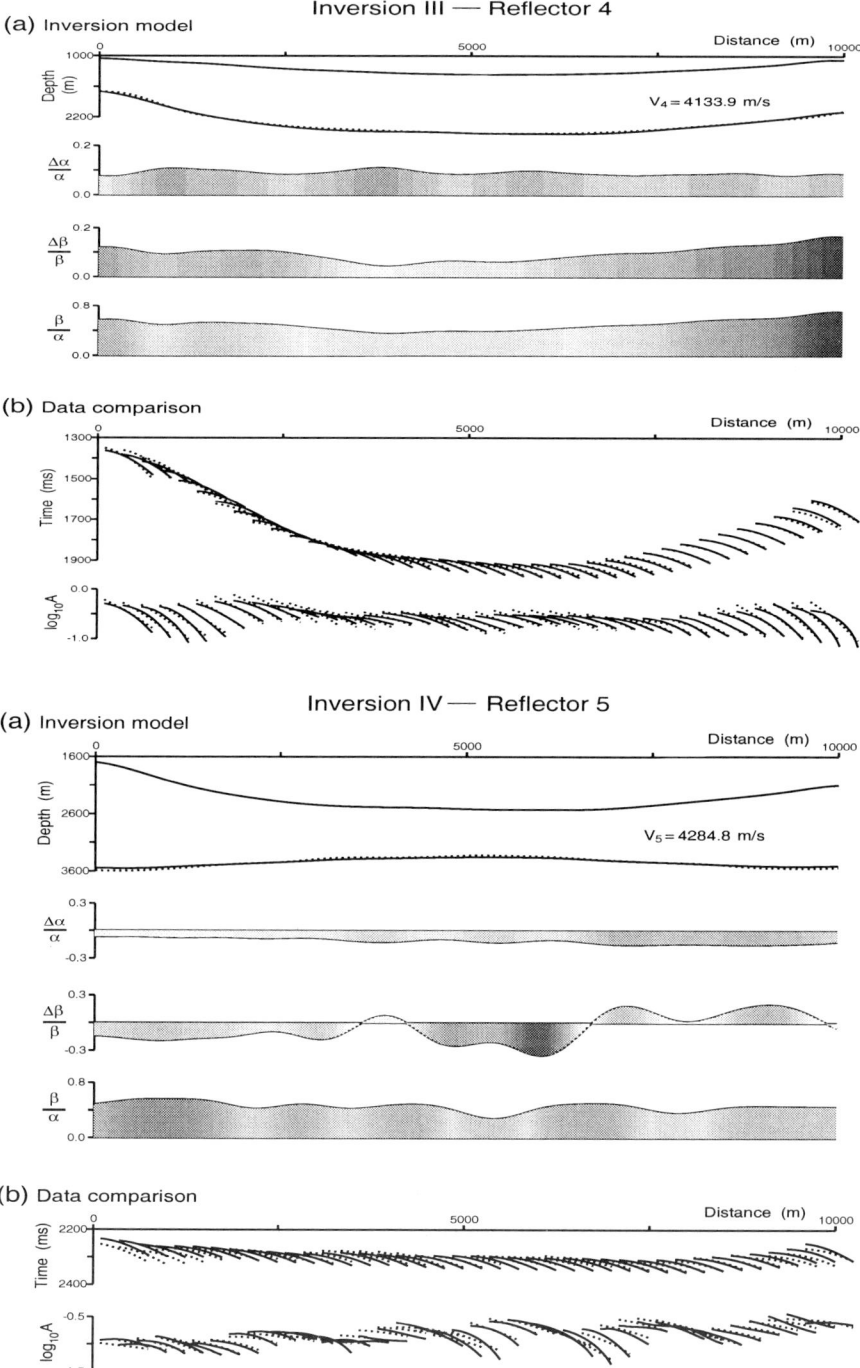

Figure 9.5. Simultaneous inversion — Reflectors 4 and 5.

shows the data comparison. An inversion model consists of the interface geometry and three elastic parameters $\Delta\alpha/\alpha$, $\Delta\beta/\beta$, and β/α. The inverted interface geometry (solid line) is compared with the initial model from the traveltime inversion result (dotted line). The constant interval-velocity from the inversion is also shown in the figure. The horizontal variations in the elastic parameters are shown by both curves and the grey scale, which facilitates comparison of their spatial variations.

As mentioned previously, a problem in AVO inversion with a linearized reflection coefficient is that the background S-wave to P-wave velocity ratio trend has to be known *a priori*. This ratio is often assumed to be a constant (Castagna and Smith, 1994; Ursin *et al.*, 1996). From the experiments carried out with the real data presented in this chapter, it was found that by using a constant β/α and the linearized formula, the inversion tended to produce physically unreasonable solutions for $\Delta\alpha/\alpha$ and $\Delta\beta/\beta$, such as $\Delta\beta/\beta \geq 2$ (which implies a negative velocity). Using the non-linear approximation (3.26), we can now invert for $\Delta\alpha/\alpha = P(x)$, $\Delta\beta/\beta = S(x)$ and $\beta/\alpha = G(x)$, simultaneously. In order to see the inter-dependence of solutions for these three parameters, we estimate the correlation coefficient between any two parameters, using

$$\varepsilon_{\eta_i,\eta_j} = \frac{\text{cov}\{\eta_i(x),\eta_j(x)\}}{\sqrt{\text{var}\{\eta_i^2(x)\}\text{var}\{\eta_j^2(x)\}}}, \tag{9.16}$$

where $\text{var}\{\eta_i^2(x)\}$ is the sample variance of model parameter $\eta_i(x)$ and $\text{cov}\{\eta_i(x), \eta_j(x)\}$ is the covariance between parameters $\eta_i(x)$ and $\eta_j(x)$. Numerical results for the solution models shown in Figures 9.4 and 9.5 are listed in the following table,

	$\varepsilon_{\Delta\alpha/\alpha,\Delta\beta/\beta}$	$\varepsilon_{\Delta\alpha/\alpha,\beta/\alpha}$	$\varepsilon_{\Delta\beta/\beta,\beta/\alpha}$	
R_2	0.179	0.197	0.992	
R_3	0.794	0.210	0.613	(9.17)
R_4	-0.470	-0.421	0.995	
R_5	-0.764	0.845	-0.603	

which shows that $\Delta\beta/\beta$ and β/α are correlated at reflectors R_2 and R_4 but not at reflectors R_3 and R_5. The correlations could be partly due to intrinsic variation of elastic properties, although solution artifacts are possible.

Amplitudes have different sensitivities with respect to the three elastic parameters. An empirical measure of the sensitivity can be obtained from the trace of the Fréchet matrix. In order to gain some insight into the differences in the parameter sensitivities, sample numerical results are shown as follows:

$$\text{trace}(\mathbf{F}_{\Delta\alpha/\alpha}) : \text{trace}(\mathbf{F}_{\Delta\beta/\beta}) : \text{trace}(\mathbf{F}_{\beta/\alpha})$$

$$= \begin{cases} 6.0 : 1.0 : 1.0 & \text{for } R_2, \\ 11.5 : 1.0 : 3.5 & \text{for } R_3, \\ 12.5 : 2.5 : 1.0 & \text{for } R_4, \\ 20.0 : 1.5 : 1.0 & \text{for } R_5, \end{cases} \tag{9.18}$$

though these sample values may depend, to a certain extent, on the structural geometry. The observation that the first kernel trace($\mathbf{F}_{\Delta\alpha/\alpha}$) increases with depth relative to the other two kernels suggests that the degree of difficulty in solving $\Delta\beta/\beta$ and β/α increases with depth. If a simple matrix inversion method were used, not all parameters would be equally well resolved. Including poorly resolved parameters in the inversion would cause the solution to be unstable and inefficient. However, the weighting matrix proposed in equation (9.12) can efficiently balance the sensitivities of different physical parameters and thus effectively stabilize the inverse procedure.

Lithology log data (*cf.* Figure 8.3) revealed that reflector R_4 is an interface between muddy chalk and chalk. Normally, $\Delta\beta > 0$, and β/α is expected to be around 0.5. Reflector R_5 is the interface between evaporites (dolomite/salt) and shale. Thus, $\Delta\beta < 0$ and β/α would be about 0.55-0.60. The results shown in Figure 9.5 appear quite close to these expectations.

9.5. Measurements for lithological interpretation

Once a set of elastic parameters is obtained, the variation of alternative elastic parameters such as Poisson's ratio can be easily evaluated. For the purpose of lithological interpretation, an attempt may be made to combine the variations of all three elastic parameters, $\Delta\alpha/\alpha$, $\Delta\beta/\beta$ and β/α, in one diagnostic function. For instance, the contrast of Poisson's ratios at the interface, $\Delta\sigma(x)$, is given by

$$\Delta\sigma = \frac{(\beta/\alpha)^2}{[1 - (\beta/\alpha)^2]^2} \left(\frac{\Delta\alpha}{\alpha} - \frac{\Delta\beta}{\beta} \right). \tag{9.19}$$

Note that the relative contrast $\Delta\sigma/\sigma$ should not be used because of the singularity [$\sigma \to 0$ when $(\beta/\alpha)^2 \to 0.5$].

Figure 9.6 shows the relative variation in Poisson's ratio obtained from the example data sets, for reflections 2-5. The magnitude of $\Delta\sigma(x)$ is close to zero for all reflections except reflection 3. In reflection 3, however, the function $\Delta\sigma(x)$ along the spatial coordinate x oscillates strongly and seems too erratic. In addition, the physical meaning of the contrast of Poisson's ratio is not easily interpreted.

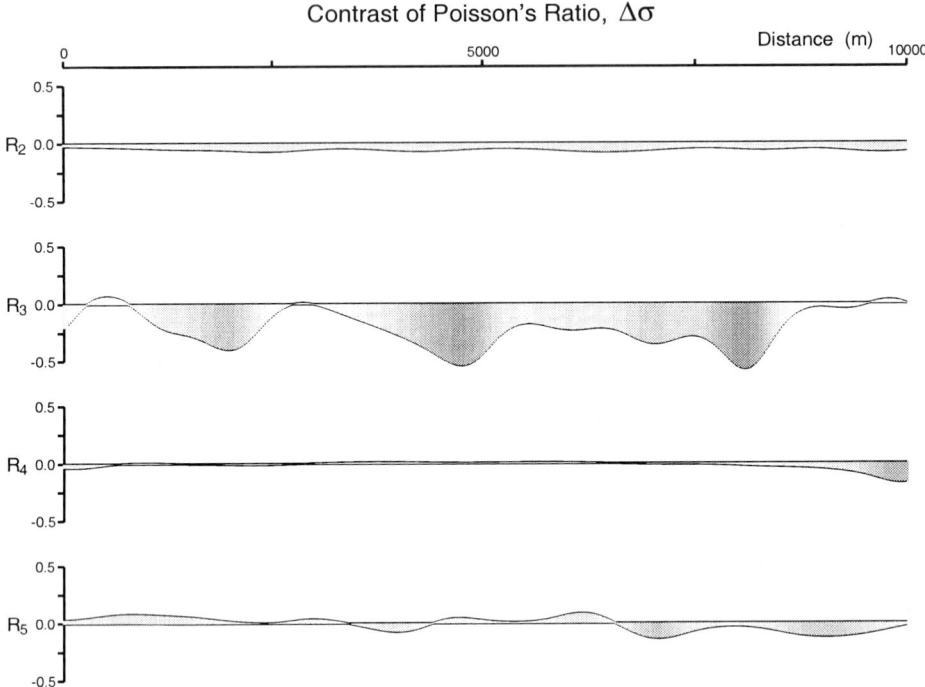

Figure 9.6. Contrasts of Poisson's ratios $\Delta\sigma(x)$, at reflectors 2-5 (depicted as R_2-R_5, respectively).

As an alternative to $\Delta\sigma$, we examine the spatial variation in the ratio between S-wave and P-wave velocity contrasts,

$$\gamma(x) = \frac{\Delta\beta(x)}{\Delta\alpha(x)} \, , \qquad (9.20)$$

where the dimensionless parameter γ measures the degree of correlation or anti-correlation between changes in the P-wave and S-wave velocity contrasts and thus indicates a lithological property of the reflector. This measurement may also be estimated from the three elastic parameters,

$$\gamma = \frac{\beta}{\alpha} \frac{\Delta\beta/\beta}{\Delta\alpha/\alpha} \, . \qquad (9.21)$$

For a normal, uniform reflection, $\gamma(x)$ might be approximated by a straight line, say γ_0. The physical meaning of such an approximation is that a linear relationship between the P-wave velocity α and the S-wave velocity β exists (e.g. Castagna *et al.*, 1985), i.e.

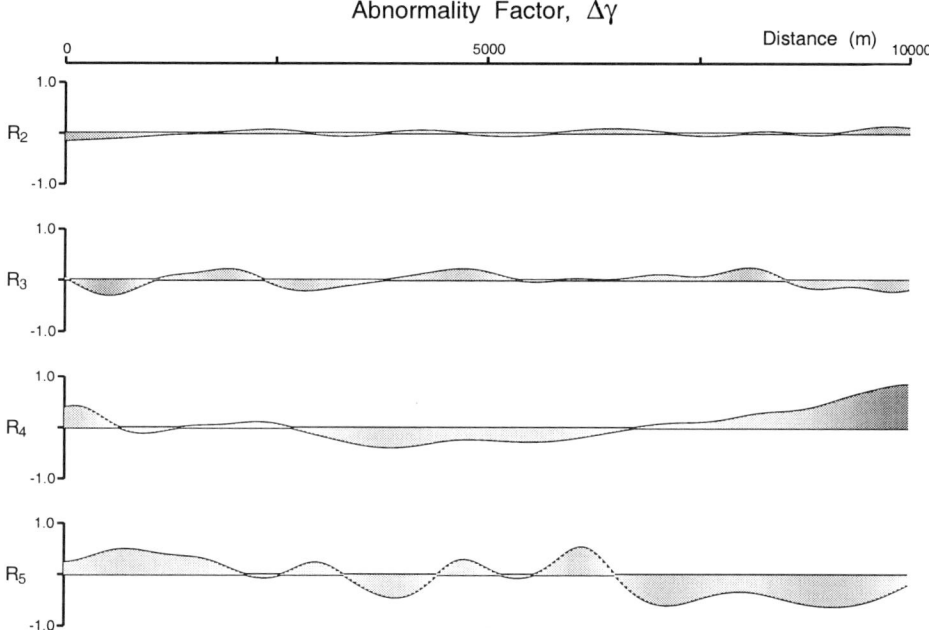

Abnormality Factor, $\Delta\gamma$

Figure 9.7. The abnormality factor, $\Delta\gamma(x)$, the deviation of $\gamma(x)$ from its linear trend. The parameter $\gamma(x)$ along the horizon x is the ratio of the S-wave to the P-wave velocity contrasts at the interface, $\gamma = \Delta\beta/\Delta\alpha$.

$$\beta \propto \gamma_0\alpha. \tag{9.22}$$

In the sample profile shown in this chapter, γ_0 values estimated for reflectors 2-5 are 0.76, 0.74, 0.52 and 0.58, respectively. Any inhomogeneous variation can be calculated using $\Delta\gamma(x)$, the deviation of $\gamma(x)$ from the linear trend γ_0,

$$\Delta\gamma(x) = \gamma(x) - \gamma_0 , \tag{9.23}$$

which shows quantitative change of the rock property along a specified horizon. Therefore, we refer to $\Delta\gamma(x)$ as an "abnormality factor". This abnormality factor can be understood as an alternative to the "fluid factor" proposed by Smith and Gidlow (1987), in which they assumed that deviations from the "mudrock line" for water-saturated clastic silicate rocks (Castagna *et al.*, 1985) can be used as a gas indicator in some regions.

Figure 9.7 shows the estimates of $\Delta\gamma(x)$ obtained from the inversion. The polarity changes from positive to negative or from negative to positive may be associated with the lithological changes. It can be observed that some of these changes also correspond to the phase changes in seismic events, shown in Figure 9.3.

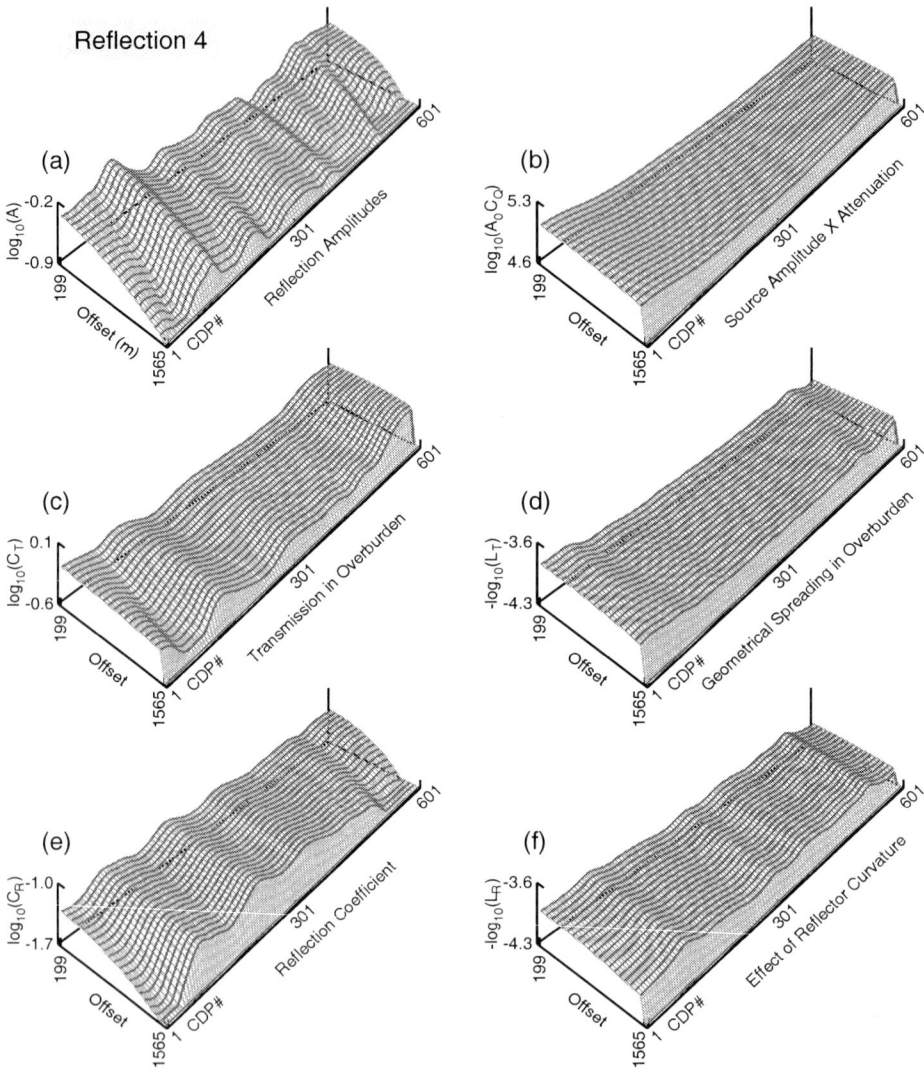

Figure 9.8.   Reflection amplitudes (reflection 4) and various effects on seismic amplitudes: (a) total amplitudes; (b) the effect of inelastic attenuation; (c) transmission coefficients at interfaces within overburden media; (d) the overburden effect in geometrical spreading; (e) the reflection coefficient at the target interface; (f) the effect of reflector curvature in geometrical spreading. Calculations are based on the inversion model (Figure 9.5).

9.6. Structural effects on amplitude variation

AVO inversion is traditionally used to obtain information on the shear modulus contrast. In the current study, we estimate the elastic parameters and the model geometry simultaneously. This approach can be used to improve traditional AVO inversion by accounting for the structural effects properly.

Figure 9.8 shows various effects on the seismic amplitudes in reflection 4. Since a fairly satisfactory data fit has been reached in the inversion, the amplitude prediction from the inversion, shown in (a), is used in the following analysis. In Figure 9.8, (b) shows the effect of the inelastic attenuation, C_Q, (c) shows the cumulative effect of the transmission coefficients, C_T, at interfaces 1-3 within the overburden media, (d) shows the overburden effect, L_T, on the geometrical spreading, (e) shows the reflection coefficient, C_R, at the target interface and, finally, (f) shows the effect of reflector curvature on the geometrical spreading, L_R. The calculations are based on the inversion model for the reflector shown in Figure 9.5.

Attenuation along the ray propagating through the medium is ray-length dependent (Figure 9.8b). For a fixed offset, horizontal variation is dependent on the depth of the target reflector. As the number of layers increases, transmission loss is a significant problem encountered in AVO inversion, as can be seen from Figure 9.8c. The overburden effect (Figure 9.8d), due to the focusing and defocusing associated with the curvature of interfaces within overburden media, seems relatively small compared to the transmission coefficients in overburden. With a limited offset, the focusing effect and the defocusing effect on the downgoing and up-going transmission paths within the overburden largely cancel each other out.

The spatial variation of the reflection coefficient in Figure 9.8e is, of course, remarkable. However, it is not a simple copy of the total amplitude variation. When the depth of the reflector increases, the difference between these increases. This is partly due to the transmission loss within the overburden, and partly due to the effect of reflector curvature.

From Figure 9.8f it can be seen that the curvature of a reflector plays an important role in the inverse modelling of reflection amplitudes. In a conventional AVO inversion, which considers only the angle-dependent coefficients in reflection amplitudes, the removal of this effect is necessary to derive the AVO attributes correctly. A prestack migration is often used to remove the structural effect. Since many currently available migration algorithms are not amplitude preserving, the amplitude inversion is best performed on the raw seismic data. In the simultaneous inversion for interface geometry and elastic parameters presented in this chapter, reflection amplitudes (and traveltimes) extracted from prestack time-migrated gathers are demigrated and can be considered as "true" observations for the inversion processing.

Lateral variation in velocities in overburden layers is not considered here. In general, amplitudes are sensitive to the transverse derivatives of slowness within the medium, not to the absolute velocity values. However, we have also found that a reflection amplitude is most sensitive to the velocity in the vicinity of the reflection point. This is because the amplitude is sensitive to the contrast in impedance at the reflector.

Chapter 10

Decomposition of structural effect and AVO attributes

Abstract   The structural amplitude effect, associated with focusing and defocusing due to the reflector curvature, makes an important contribution to reflection seismic amplitudes. In this chapter we develop an approach which reconciles the estimation of the structural amplitude effect with that of the attributes of amplitude variation versus offset (AVO). We take account the structural amplitude effect in raw amplitudes explicitly based on a structural model reconstructed from traveltime inversion, and extract the AVO attributes not just locally but also horizontally to see the global variation along the reflection. We decompose such lateral variations of AVO attributes by the Chebyshev expansion, and demonstrate the methodology with an example of weak shallow gas-water contact appearing on a 2-D seismic profile of a site survey in the North Sea.

10.1.   Introduction

In the context of *AVO inversion*, as in the previous chapter, we invert for three elastic parameters from the reflection amplitude data. In the conventional *AVO analysis*, we attempt to extract the so-called AVO attributes, such as the intercept and the slope etc. In this chapter, we apply conventional AVO analysis to raw amplitude data. Our goal is to perform the AVO analysis not just locally (by CDP) but also horizontally to explore the global variation along a reflection. As in the AVO inversion in the previous chapter, we also take into account the structural effect when we extract the AVO attributes. We estimate the structural amplitude effect explicitly including the focusing and defocusing of curved reflectors, using a structural model from traveltime inversion.

Following the work in Chapter 8, we can estimate seismic reflection amplitude from migrated gathers. Migrating the data reduces many of the problems in using unmigrated data. Mosher *et al.* (1996) showed that migration could improve both the signal-to-noise ratio and the spatial resolution of AVO analysis. In the context of the ray-path-based inversion, contraction of the Fresnel zone by migration brings the amplitudes closer to the ray-amplitudes. The amplitudes estimated from migrated gathers are demigrated so that they can be considered as diffraction-free, raw amplitudes.

These simulated raw amplitude data are decomposed explicitly into the structural amplitude component and the AVO attributes, based on a simplified theoretical ray assumption. In this chapter, we parameterize the *lateral variations* of AVO attributes by Chebyshev polynomials of the first kind, which facilitates the interpretation of the AVO attributes.

We conduct a sample analysis on a weak gas-water contact appearing on the 2-D reflection seismic profile of a site survey in the North sea. We see that the Chebyshev coefficients (typically c_3) indicate clearly the lateral variation of the AVO attributes on the gas-water contact.

10.2.   Decomposition of ray-amplitude

10.2.1   Separation of structural effect and AVO effect

To separate the structural effect from the AVO effect, i.e. the angle dependence of the reflection coefficient, the ray-amplitude is given explicitly by

$$A(\ell) = A_0 C_Q R / L(\ell) , \tag{10.1}$$

where A_0 is the amplitude at the source point, C_Q is the inelastic attenuation factor, $L(\ell)$ is the geometrical spreading, evaluated along the reflection ray path ℓ, and R is the reflection coefficient. The source amplitude A_0 and the attenuation C_Q along the ray path are assumed to vary only slowly laterally and can be included in R as a constant weighting factor. Other amplitude factors such as fine layering (the layer thickness being less than the seismic wavelength) may be assumed to be multiplicative scaling factors applied to the observed amplitude data. The logarithmic (absolute) amplitude is used in the inversion. Using log variables converts the multiplicative structure to an additive data structure so as to reduce effect of multiplicative scaling in the least-squares inversion.

The structural amplitude effect includes the effect of focusing and defocusing due to the curvature of the target reflector and the curvature of interfaces within the overburden media. The overburden effects are negligible for a limited offset, since the focusing and the defocusing effects on the down-going and up-going transmission paths largely cancel each other out. The principal structural effect is then the one from the target reflector. It can be evaluated by

$$L(\ell) = \sqrt{\ell \, |\ell + \Delta \ell|} \,, \tag{10.2}$$

where

$$\Delta \ell = \frac{2\ell_1 \ell_2}{\cos \theta} \left[1 + \left(\frac{dz}{dx} \right)^2 \right]^{-3/2} \frac{d^2 z}{dx^2} \,, \tag{10.3}$$

and where ℓ_1 and ℓ_2 are the lengths of the incident ray and the reflected ray, respectively, $\ell = \ell_1 + \ell_2$, θ is the incident angle and $z = Z(x)$ is the reflector geometry. These formulae are deduced from the geometrical spreading function for multi-layered structures derived in Appendix A.1. Equation (10.3) shows that the geometrical spreading depends on the slope and the curvature of the reflector, and the angle of the ray incident at the reflector.

The AVO effect can be characterized by the parameter R. In traditional AVO analysis, it is often modelled by a linear approximation (Shuey, 1985),

$$R = R_0 + B \sin^2 \theta, \tag{10.4}$$

where R_0 is the intercept amplitude at zero-offset and B is the slope of the amplitude along the offset. A pair of AVO attributes, the intercept R_0 and the slope B, can be produced at each CDP point for a given time sample. Thus, both attributes are a function of the horizontal location x.

10.2.2 Decomposition of AVO attributes

The lateral variation of an AVO attribute, $R_0(x)$ or $B(x)$, is parameterized by a Chebyshev expansion,

$$Y(x) = \sum_{n=1}^{N} c_n F_{n-1}(x) \,, \tag{10.5}$$

where N is the number of Chebyshev terms, each of which is given by

$$F_n(x) = \cos \left[n \cos^{-1} \left(\frac{x - x_a}{x_b - x_a} - \frac{x_b - x}{x_b - x_a} \right) \right] \,, \tag{10.6}$$

defined over the range of interest $[x_a, x_b]$, and c_n are the coefficients that form the "model" vector we invert for. The Chebyshev approximation in equations (10.5) and (10.6) is evaluated using Clenshaw's (1962) recurrence formula, affecting the two equations simultaneously (Press *et al.*, 1989). The Chebyshev polynomials satisfy a discrete orthogonality relation. A general form of orthogonal polynomials is used by Johansen *et al.* (1995) to track the amplitude variation of a CDP gather.

The extraction of the polynomial coefficients can be considered as a decomposition of the lateral variation of AVO attributes. One of the useful properties of the Chebyshev polynomials is that the approximation can be truncated to a polynomial of lower degree $m \ll N$ in an elegant way, yielding the "most accurate" approximation of degree m. In other words, the accuracy of those low-order coefficients used in the AVO interpretation is not influenced by the maximum degree N that is chosen. If we wish, during the course of later interpretation, we can easily add higher degree terms which have more extrema.

10.3.   The inverse problem

10.3.1   Model parameterization and ray tracing

The AVO analysis discussed here is in fact a ray-path-based inverse problem. The ray path between a source and a receiver at the surface is traced through a current estimate of the subsurface model, which is parameterized as a "smooth velocity model" in contrast to a blocky model representation (Lailly and Sinoquet, 1996; Pereyra, 1996). In the smooth model representation, a reflector is embedded in the globally defined velocity distribution. In this chapter, the smooth velocity distribution is assumed to be the 1-D variable, $v(z)$, where z is the depth.

The reflector is also described by a Chebyshev expansion (equation 10.5). Once the Chebyshev series for the reflector $z = Z(x)$ is determined, the Chebyshev coefficients of the first derivative of $Z(x)$ can be obtained using

$$c'_{n-1} = c'_{n+1} + 2(n-1)c_n , \qquad (10.7)$$

which is a recurrence formula, starting with the values $c'_{N+1} = c'_N = 0$. Similarly the Chebyshev coefficients, c''_n, of the second derivative of $Z(x)$ can be computed from c'_n. The first and the second spatial derivatives of $Z(x)$ are required in equation (10.3) to calculate the geometrical spreading.

Ray tracing is implemented by Fermat's principle,

$$\nabla_{\mathbf{x}} T(\gamma) = 0 , \qquad (10.8)$$

where γ is the ray path and T is the traveltime $T = \sum d\ell_k/v_k$, expressed in terms of $d\ell_k$ the length of the kth ray segment between knots $\mathbf{x}_{k-1}$ and $\mathbf{x}_k$, and v_k the corresponding velocity. Denoting the target reflector as the Kth interface and assuming $K-1$ interpolation levels between the surface and the target reflector, we have a total of $2K-1$ intersection points to be determined for a ray travelling from a source point $\mathbf{x}_0$ to a receiver point located at $\mathbf{x}_{2K}$. The ray tracing process is then characterized by the non-linear system,

$$\frac{\partial T}{\partial x_k} = 0 , \qquad \text{for} \quad k = 1, \cdots, 2K - 1 . \qquad (10.9)$$

This ray tracing method is much faster than the ray-shooting method, but it requires a starting solution close to the true answer which can be easily obtained for this 1-D smooth model.

10.3.2   Inversion method

The inversion process is performed iteratively. At each iteration the model update, $\delta\mathbf{m}$, is given by

$$\delta\mathbf{m} = -\mathbf{A}[\mathbf{A}^{\mathrm{T}}(\mathbf{F}^{\mathrm{T}}\mathbf{F} + \mu\,\mathbf{C}_{\mathrm{M}}^{-1})\mathbf{A}]^{-1}\mathbf{A}^{\mathrm{T}}\mathbf{F}^{\mathrm{T}}\delta\mathbf{d} , \qquad (10.10)$$

where $\mathbf{A}$ is a projection matrix which puts the model parameters into several subspaces, $\mathbf{F}$ is the Fréchet matrix, $\mathbf{C}_{\mathrm{M}}^{-1}$ is the model covariance matrix, μ is a trade-off parameter and $\delta\mathbf{d}$ is the data residual.

The model covariance matrix that we used is defined by

$$\mathbf{C}_{\mathrm{M}}^{-1} = \frac{1}{2}[\mathbf{I} + \mathbf{B}] , \qquad (10.11)$$

where $\mathbf{I}$ is the identity matrix, and $\mathbf{B}$ is a matrix operator to penalize the first spatial derivatives of the model parameters. Following the discussion in Chapter 6, we define the following scalar product,

$$\left[\hat{\mathbf{B}}\right]_{mn} = \int_X \frac{dF_{m-1}(x)}{dx}\,\frac{dF_{n-1}(x)}{dx}\,dx , \qquad (10.12)$$

where X is the model range from which rays are reflected and F_n is the Chebyshev basis function. The normalized matrix $\mathbf{B}$ in equation (10.11) is given by

$$\mathbf{B} = \frac{N}{\text{trace}(\hat{\mathbf{B}})}\,\hat{\mathbf{B}} . \qquad (10.13)$$

The trade-off parameter μ in equation (10.10) is determined using the relation

$$\mu = \rho\,\frac{\text{trace}(\mathbf{F}^{\mathrm{T}}\mathbf{F})}{\text{trace}(\mathbf{C}_{\mathrm{M}}^{-1})} , \qquad (10.14)$$

where the ratio of traces is an empirical quantity used to normalize the matrices, and ρ then easily controls the total amount of regularization in the inversion. In the inversion example, following Chapter 6, $\rho = 0.01$ is used.

10.4.   Sample dataset of gas-water contact

10.4.1   Reflection seismic profile

The AVO analysis is carried out on the seismic profile of a site survey in the North Sea. The site survey is conducted before drilling to prevent accidental drilling into shallow gas, which can often cause a blowout and lead to loss of the drilling rig and loss of life. It involves the use of a high frequency seismic source to produce a high resolution image of the shallow subsurface. If a shallow gas body is found, an alternative drilling location is sought.

The presence of shallow gas would normally create a bright reflection, caused by a large (negative) reflection coefficient at the top of the gas body. A traditional "bright spot" occurs when the negative reflection coefficient of the contact between an overlying seal and a water-bearing reservoir increases in magnitude over the gas-bearing area. A dim spot occurs when a positive coefficient in the water leg turns into a weak negative (or zero) coefficient over the gas-bearing leg. This latter situation often occurs in carbonates, where the detection of a gas-water contact may also help in identifying a shallow gas body.

In November 1990, an oil company moved an exploration rig into the North Sea. The aim was to drill an exploration well to ~3000 m. A site survey previously conducted over the area had shown some high amplitude reflections, although none directly under the drilling location. Operations went as planned until the drill reached 478.5 m. At this depth, gas bubbles became visible at the well head and near the rig. The rig was moved quickly off the location without incident. Large quantities of gas were then observed escaping at the sea surface over the location. An investigation was launched to ascertain why this gas body was not detected by the site survey. It was found that the gas-water contact was invisible in the profile of the site survey conducted.

A possible cause might be the manner in which the data were originally processed. The key step in the processing sequence is the velocity analysis. If no picks were taken within the gas-bearing sequence then the reflection from the gas-water contact would not stack well, and the stacked amplitude would be weak as a result. Using a careful velocity analysis, we recompute a stack profile (Figure 10.1), on which the gas-water contact (appearing at about 560 ms of two-way time between CDP locations 200 and 400) is now visible. Note the polarity in Figure 10.1 where a black peak corresponds to a negative reflection coefficient: the top of the gas sand reflector is picked as the peak, the gas-water contact is picked as the trough, and they have opposite polarities.

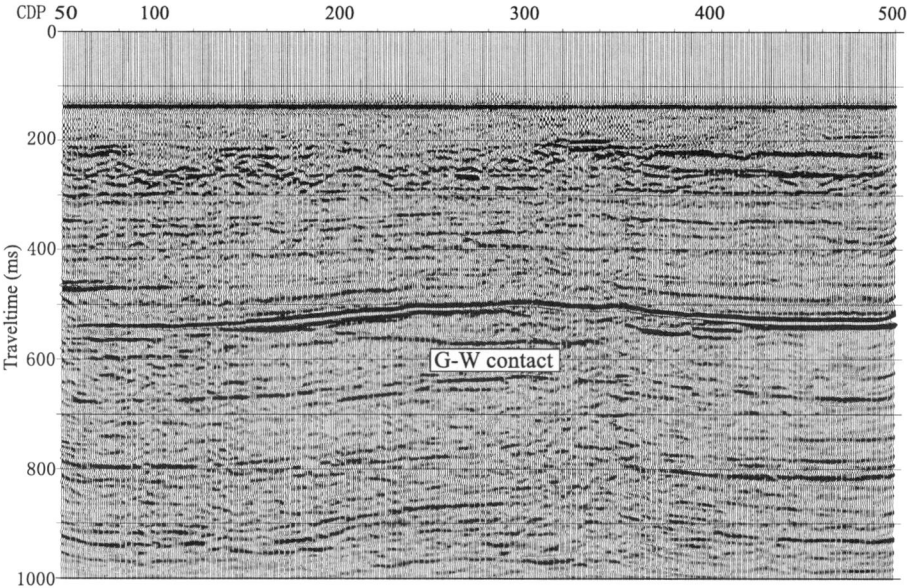

Figure 10.1. Reflection seismic profile from a site survey of the North Sea, showing a weak reflection from a gas-water contact at 560 ms (between CDP 200 and CDP 400). Note the polarity where a black peak corresponds to a negative reflection coefficient: the top of the gas sand reflector is picked as the peak, and the gas-water contact is picked as the trough.

The common-midpoint (CMP) gathers are computed from the common offset sections after prestack time migration, but without prestack predictive deconvolution and without any bandpass filter, both of which removed some of the seismic reflection energy from the gas-water contact in the original processing. Those gathers are used to pick amplitudes, as described in the following section.

10.4.2   Amplitudes extracted from prestack gathers

Our initial objective is to perform amplitude inversion on raw amplitude data. In practice, however, the extraction of amplitudes from raw gathers is always problematic because of the distortion of the waveform caused by wave propagation effects. Therefore, a minimum amount of preprocessing is applied to facilitate amplitude picking from reflection events. Firstly, we use a very simple prestack time migration to enhance continuity and reflection strength and, more importantly, to reduce diffraction effects (for the ray-amplitude assumption). Secondly, we use an adaptive beamforming multiple suppression technique (Hu and White, 1998) on the migrated gathers to separate primaries

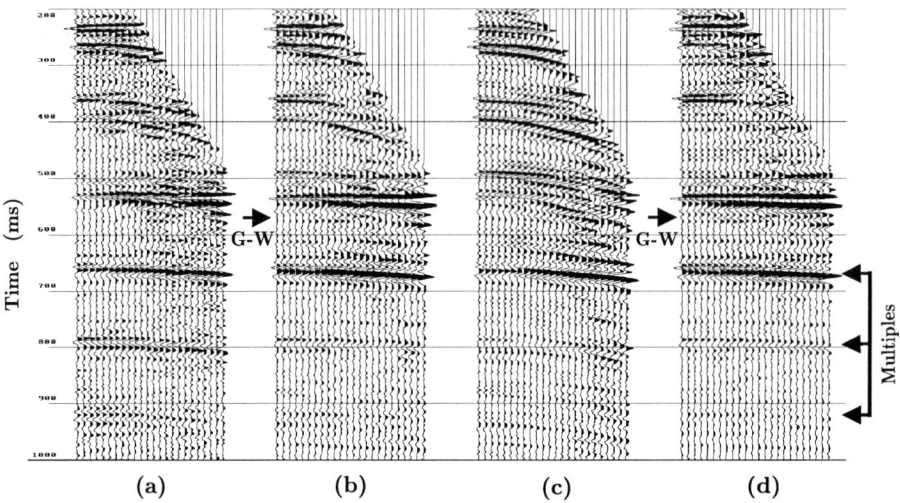

Figure 10.2. A sample gather (CDP 200) of prestack migration, followed by beamforming for the separation of primaries and multiples: (a) NMO-corrected gather; (b) the result of time migration; (c) multiples estimated by adaptive beamforming method; (d) primaries and incoherent noise.

from multiples. A sample gather (CDP 200) is shown in Figure 10.2, which depicts (a) the normal-moveout (NMO) corrected gather, (b) the result of prestack time migration, (c) the multiples estimated by the beamforming method, and (d) the primaries with non-coherent noise. After the migration and demultiple processing, the weak gas-water contact reflection (at about 560 ms in Figure 10.2d) becomes visible. However, the simple multiples (at 670, 890 and 910 ms) are only partly suppressed. These multiples can, in general, be attenuated using deconvolution, which is not applied here.

The picking of reflection amplitudes is based on a cross-correlation between the stack (as shown in Figure 10.1) and the individual traces within the gather. When we extract the amplitudes from the gathers, traveltimes are also obtained concomitantly. De-migration on reflection amplitudes (and traveltimes) follows, as described in Chapter 8. The result is then considered as diffraction-free, raw amplitudes (and traveltimes) for the input to inversion. These amplitudes (and traveltimes) still contain considerable noise, which would cause ambiguities in the solution of AVO inversion (Drufuca and Mazzotti, 1995). We apply the LOESS method to winnow the datasets, rejecting the outliers from the data.

Figure 10.3 depicts the datasets of logarithmic (absolute) amplitude variation versus offset of the reflections from the top of the gas body and from the gas-water contact. Both plots have the same vertical scale of the logarithmic amplitudes. From the plots (the range between CDP 200 and CDP 400), we can

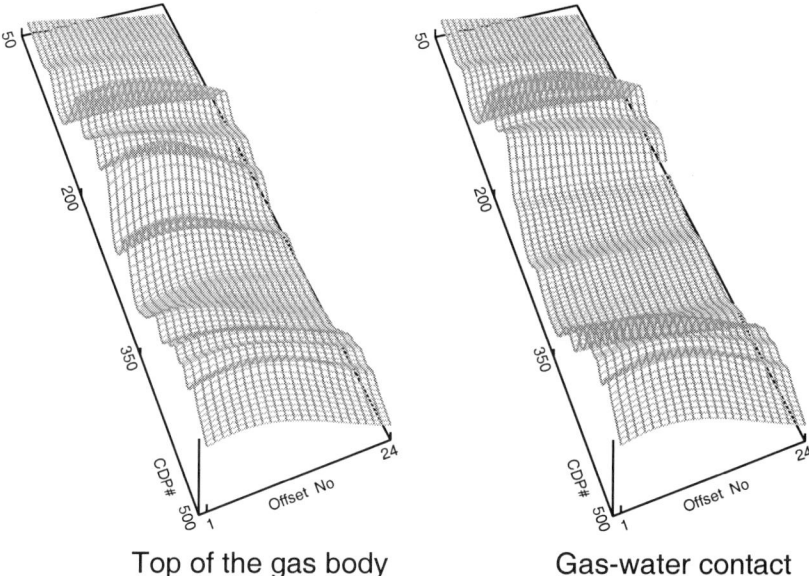

Top of the gas body   Gas-water contact

Figure 10.3. Amplitude variation versus offset datasets of the reflections from the top of the gas body and from the gas-water contact. Both plots have the same vertical scale of the logarithmic (absolute) amplitudes. Offset 1 is the near offset.

see clearly that the reflection from the gas-water contact is much weaker than the reflection from the top of the gas body. Every 3rd CDP is shown, because a 3-point median process was used to select the raw data, and each gather has 24 traces. These demigrated datasets are used as the input to the following inversion process.

10.5.   Inversion results

10.5.1   *Traveltime inversion for interface geometry*

The inversion process is divided into two steps: traveltime inversion to constrain the geometry of reflectors and amplitude inversion, based on the structural model from traveltime inversion, to recover the AVO attributes. For the traveltime inversion in this example, a 1-D vertically variable velocity model is assumed. This 1-D velocity function is an approximation to the 2-D velocity model obtained from standard velocity analysis. During the preprocessing, two velocity analyses were carried out. The first is the analysis on the raw data, before prestack time migration. The second is performed on the gathers after time migration and a subsequent demigration process.

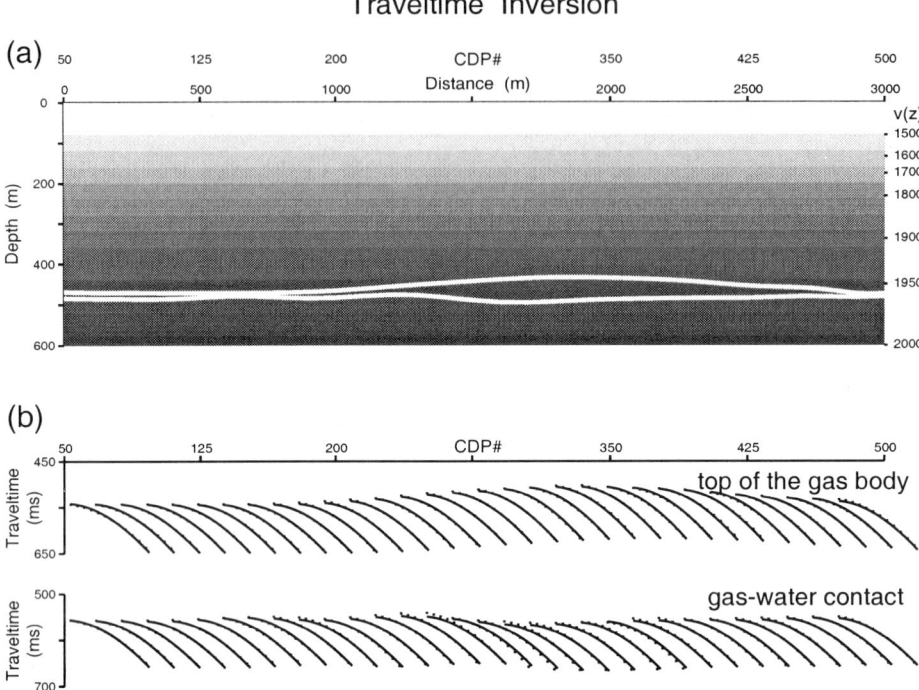

Figure 10.4. Traveltime inversion for interface geometry: (a) The geometry of reflectors, obtained from traveltime inversion, embedded in the 1-D velocity model; (b) comparison of prediction (solid lines) and the input (dotted lines) of the inversion. The solid lines here plot on top of the dotted lines and are hardly visible.

Traveltime inversions for two reflectors, the top of the gas body and the gas-water contact, are performed separately. The resultant geometry of the reflectors, embedded in the 1-D velocity model, is shown in Figure 10.4a. The velocity distribution of the model is depicted by a grey-scale. Reference velocity values (in m/s) are shown at the right-hand margin. The inversion models the reflectors within the horizontal range $x = [0, 3000]$ m, where the coordinate of CDP 50 is $x = 0$ m and that of CDP 500 is $x = 3000$ m. A comparison of the prediction (solid lines) and the input (dotted lines) of the inversion is shown in Figure 10.4b. The dotted lines are overlain by the solid lines and are hardly visible. Note that in practice an option for building these reflectors exists. For instance, one might pick the events in the migration section and then stretch vertically from time to depth.

The resultant reflector geometry in Figure 10.4 is consistent with the reflections shown in the stack profile of Figure 10.1. The reconstructed reflector of

Amplitude Inversion

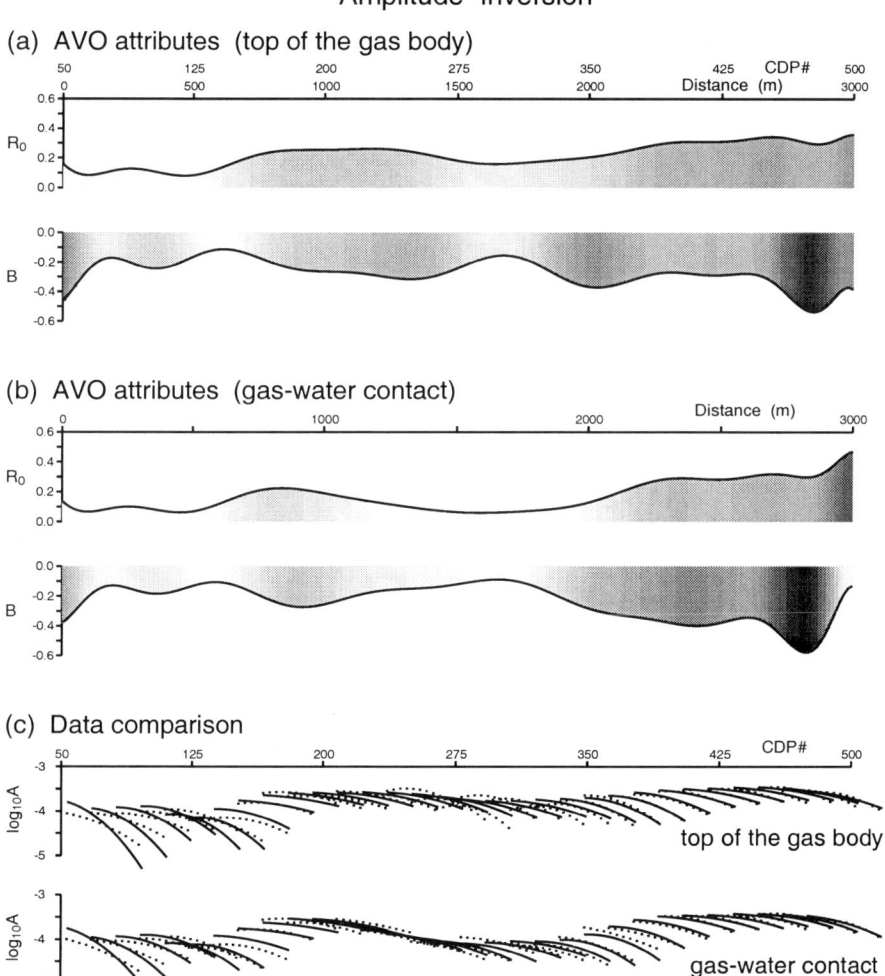

Figure 10.5. Amplitude inversion for AVO attributes: (a) AVO attributes from the top of the gas body; (b) AVO attributes from the gas-water contact; (c) comparison of prediction (solid lines) and the input (dotted lines) of the inversion.

the top of the gas body is a simple anticline structure. The gas-water contact shows clear push-down consistent with the thickness of the gas. This oscillation, reflecting the local velocity variation within the gas body, is caused by the simplified 1-D velocity assumption. Otherwise, it would be completely flat. In Figure 10.1, the gas seems to "pinch out" round about CDP 160 where there is a phase change on the top of the sand reflector.

10.5.2   Amplitude inversion for AVO attributes

Using the structural model obtained from the traveltime inversion, a ray-based amplitude inversion was then carried out. The geometrical spreading, evaluated along the ray-path, takes account of the effect of focusing and defocusing of the ray tube, reflected from a curved interface. If we do not consider the lateral variation of attenuation and the variation of source amplitudes, the remaining component of amplitude variation is attributed to an empirical reflection coefficient, which is modelled by the AVO attributes R_0 and B.

Results of the amplitude inversion are shown in Figure 10.5, in which (a) and (b) are AVO attributes of the reflections from, respectively, the top of the gas body and the gas-water contact, and (c) is the data comparison between the input amplitudes (dotted lines) and the predicted amplitudes (solid lines) from the inversion. The horizontal variations in the AVO attributes are shown by both curves and the grey scale, which facilitates comparison of their spatial variations. Within the area between 1000 m and 2300 m, the magnitude of the zero-offset amplitude, $R_0(x)$, of the reflection from the gas-water contact is smaller than that from the top of the gas body. The slope variation $B(x)$ also differs slightly between these two limits. Outside the area between 1000 m and 2300 m, the AVO variations for the two reflections are almost identical. From Figure 10.5c we see that there is a considerable data misfit for the first few CDPs. This misfit is probably caused by the edge effect of the interface parameterization.

Figure 10.6 shows the product of R_0 and B, the AVO product, which is often used to verify classical bright spots (Class III, Rutherford and Williams, 1989). This is based on the observation that a low-impedance gas sand encased in shale will have a larger (absolute) AVO intercept and a larger (absolute) AVO slope compared to reflectors not associated with gas.

10.6.   The Chebyshev spectra of the AVO attributes

Displaying the polynomial coefficients c_n may facilitate the interpretation of *lateral variations* of the AVO attributes. Explicitly, the coefficients $\{c_n\}$ have the following physical meanings: c_1 is related to the mean value; c_2 relates to the mean gradient, describing whether an AVO attribute is rising laterally or falling; c_3 indicates whether the AVO is changing from falling to rising, or vice versa; and c_4 or higher-order coefficients indicate strong noise (possibly due to fine-layering effects in the data).

Figure 10.7 shows the Chebyshev spectra of the AVO attributes, estimated from both the top of the gas body and the gas-water contact over the range from CDP 50 to CDP 500. We can make the following observations.

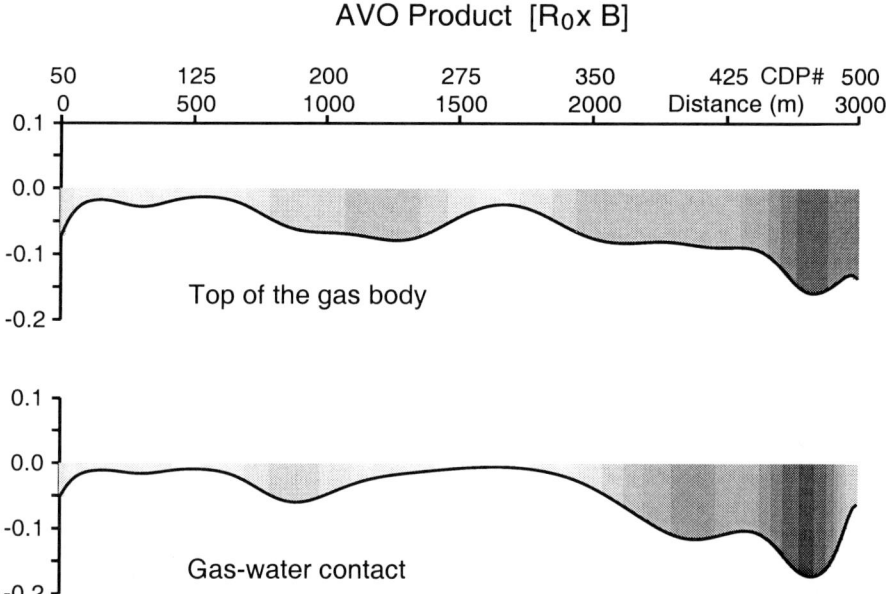

Figure 10.6.   AVO product $R_0 \times B$, estimated from the top of the gas body and the gas-water contact.

1) The coefficient c_3 of $R_0(x)$ in the reflection of the gas-water contact clearly shows that the zero-offset amplitudes for the gas-water reflection change from falling laterally, to rising.

2) The spectra of the $B(x)$ coefficients suggest that for the top of the gas body the AVO slope changes from rising (i.e. becoming less negative) to falling, indicated by coefficient c_3, whereas that of the gas-water contact decreases gradually according to coefficient c_2.

3) In addition, the mean value (c_1) of the AVO slope of the latter is about one-half of the value of the former.

Note that the individual coefficients of the Chebyshev spectrum might be dependent on the horizontal range of the model parameterization, and that the interpretation of AVO attributes should, in general, be examined in conjunction with well data. Unfortunately well data are not available here, since the drilling stopped when gas blew out.

As the AVO attributes are now decomposed into Chebyshev spectra, we can easily adopt an alternative indicator $5R_0+B$ and represent it also by Chebyshev series. This indicator, proposed by Dong (1996), is ideally for highlighting hydrocarbon presence.

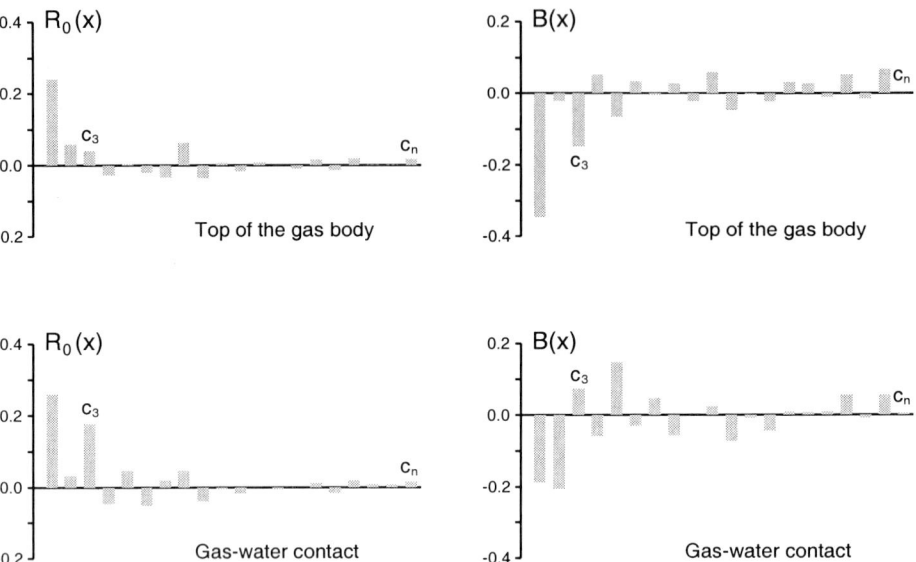

Figure 10.7. Chebyshev spectra of the AVO attributes, estimated from the top of the gas body and from the gas-water contact. The horizontal axis is the index of Chebyshev coefficients, $\{c_n, n = 1, 20\}$.

When pore fluid changes from brine to hydrocarbons, both the bulk modulus (κ) and the density (ρ) decrease but the shear modulus (μ) remains unchanged because of the invariant rock skeleton. However, since the AVO intercept, slope and their product (Figures 10.5 and 10.6) depend on the shear modulus, they are not the most sensitive parameters to fluid content change in porous rocks. The variation of bulk modulus, shear modulus and density may be expressed in terms of the relative changes in *P*-wave and *S*-wave velocities in a matrix form,

$$\begin{pmatrix} \Delta\alpha/\alpha \\ \Delta\beta/\beta \\ \Delta\rho/\rho \end{pmatrix} = \frac{1}{\rho} \begin{pmatrix} 1/(2\alpha^2) & 2/(3\alpha^2) & -1/2 \\ 0 & 1/(2\alpha^2) & -1/2 \\ 0 & 0 & 1 \end{pmatrix} \begin{pmatrix} \Delta\kappa \\ \Delta\mu \\ \Delta\rho \end{pmatrix}. \qquad (10.15)$$

Using this equation, the intercept R_0, slope B and curvature C can be expressed as

$$\begin{pmatrix} R_0 \\ B \\ C \end{pmatrix} = \frac{1}{\rho} \begin{pmatrix} 1/(4\alpha^2) & 1/(3\alpha^2) & 1/4 \\ 1/(4\alpha^2) & -5/(3\alpha^2) & -1/4 \\ 1/(4\alpha^2) & 1/(3\alpha^2) & -1/4 \end{pmatrix} \begin{pmatrix} \Delta\kappa \\ \Delta\mu \\ \Delta\rho \end{pmatrix}. \qquad (10.16)$$

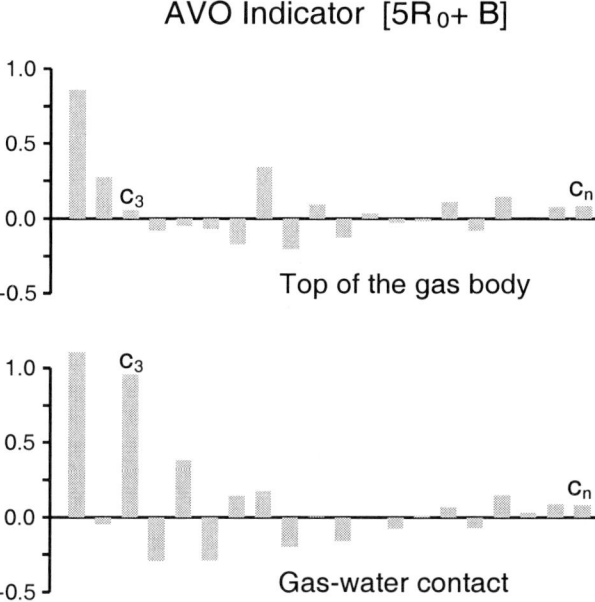

Figure 10.8. Chebyshev spectra of the AVO indicator $5R_0+B$, resulting from the top of the gas body and from the gas-water contact, respectively. Horizontal axis is the index of Chebyshev coefficients, $\{c_n, n = 1, 20\}$.

It is easy to see that the influence of shear modulus change $\Delta\mu$ on slope B is negatively five times the influence on R_0 (and C).

The $5R_0+B$ indicator, where

$$5R_0 + B = \frac{3}{2}\frac{\Delta\kappa}{\rho\alpha^2} + \frac{\Delta\rho}{\rho}, \qquad (10.17)$$

has the property that bulk modulus and density changes reinforce each other, and is highly desirable for porous rock with hydrocarbons. One might argue that, as R_0 is weighted by a factor of 5 and the effect of slope B is hardly reflected in the indicator, this indicator mainly reflects the effect of the intercept R_0. But the adjusted Smith-Gidlow fluid factor (Smith and Gidlow, 1987; Smith and Sutherland, 1996) suggested coincidentally that the main contribution of the fluid effect stems from the intercept, rather than the AVO slope.

The Chebyshev expansion of the $5R_0+B$ indicator is simply the sum of weighted coefficients of R_0 and B series. Figure 10.8 depicts the resultant spectra of the indicator for reflections of both the top of the gas body and the gas-water contact. Note that the coefficient c_3 in the reflection of the gas-water contact represents the $5R_0+B$ indicator changing from falling to rising. In contrast, there is no such indication at the top of the gas body (c_3 is rather small). Therefore, the Chebyshev spectra (typically coefficient c_3) may be used to indicate clearly the lateral variation of the AVO attributes on the gas-water contact.

To further enhance sensitivity, we may propose a hydrocarbon indicator which emphasizes only the change in the most fluid-sensitive elastic parameter $\Delta\kappa$. If the Chebyshev spectrum of curvature C is also obtained, an alternative indicator,

$$3R_0 + B + 2C = \frac{3}{2}\frac{\Delta\kappa}{\rho\alpha^2} , \qquad (10.18)$$

would be the most sensitive hydrocarbon indicator, as the change in density ρ during fluid substitution is a small change. The extension of amplitude inversion from the Chebyshev spectra of two parameters (R_0 and B) to the Chebyshev spectra of three parameters (R_0, B and C) is straightforward.

Chapter 11

Amplitude tomography in practice

Abstract   In the previous chapters, we studied the use of ray-amplitudes, separately and jointly with traveltimes in reflection seismic inversion. In this chapter we point out the basic improvement we needed, rather than the road ahead, in order to bring the seismic amplitude inversion into a technology that is applicable in practice; in other words, to gradually relax the assumptions involved in amplitude inversion and even in the traveltime inversion. We see that the work includes better estimation of amplitude and traveltime, incorporating more information in the inversion, improvement on forward calculation, consideration of many other factors influencing the seismic reflection amplitude, and establishing the near-surface velocity model. Finally, we present a proposal of seismic trace inversion for ray impedance in the prestack domain.

11.1.   Introduction

In the previous chapters, we studied the use of ray-amplitudes, separately and jointly with traveltimes in reflection seismic inversion. We may draw the following major conclusions in order of significance or priority:

1) Traveltime and amplitude in reflection seismic show different dependencies on the geometry of reflection interfaces, and different dependencies on the variation in interval velocities. Both for reflector geometry and for interval

velocity variations, the reflection traveltime is sensitive to the model components of small wavenumber, whereas the reflection amplitude is more sensitive to the high-wavenumber components. Therefore, traveltime and amplitude data do contain complementary information in the context of tomographic inversion.

2) Joint inversion of traveltime and amplitude data significantly improves the inversion solutions over the use of one type of data alone. Since reflection traveltime and amplitude are sensitive to different features of the earth model, including amplitude data constraints in reflection traveltime tomography therefore may better resolve the known ambiguity between reflector depth uncertainty and interval velocity uncertainty in traveltime inversion.

3) Ray amplitude generally decreases with increasing length of the ray path, and a major part of the amplitude decay is attributable to the geometrical spreading. The analytical expression for the geometrical spreading of a 2-D stratified velocity structure reveals that it depends on the velocity contrasts, the angle of incidence, and the local slope and curvature of the interfaces.

4) Practical approaches to dealing with different data types and different classes of model parameters with different physical dimensions in a simultaneous inversion have proven to be robust in producing reliable inversion results. The multi-stage damped, subspace method is effective in obtaining a geologically realistic subsurface model, and can overcome specifically the two main difficulties appearing in the amplitude inversion for multi-layered structures: (a) a high wavenumber oscillation on shallow interfaces and (b) a failure to recover the long wavelength components of deep, curved interfaces.

5) Success in the joint inversion of traveltime and amplitude, demonstrated using real seismic data, is partially due to the practical usage of the data covariance matrix that balances the relative contribution, not directly by the data magnitude but by the relative sensitivity, of different data types. To invert simultaneously for different classes of model parameters, a working definition of the model covariance matrix, expressed in terms of the weighted sum of the correlation between different model parameters and the correlation between their first derivatives, has proven to be effective.

6) Seismic amplitudes can be extracted from real reflection gathers, using the aid of prestack time migration. The purpose of the migration is to enhance continuity and reflection strength. Moreover, contraction of the Fresnel zone by migration brings the amplitude closer to the ray amplitude assumed in the inversion.

7) Using a non-linear approximation for amplitudes, it is possible to invert for not only two elastic parameters (as in traditional AVO inversion) but for three parameters, simultaneously, i.e. the *S*-wave to *P*-wave velocity ratio, the contrast in *P*-wave velocity and the contrast in *S*-wave velocity, which are all essential for the complete description of the elastic property of a reflector. The "abnormality factor", combining these three elastic parameters into one, can be used as an effective fluid factor indicating lithological change along the interface.

8) The effect of interface geometry is an important component in reflection amplitudes. By accounting for the structural effects properly, it is possible to improve the estimate of elastic parameters of the subsurface targets in traditional AVO analysis. Spatial variations (with respect to the CDP coordinate x) of the gradient, which measures how a reflection event changes amplitude as a function of offset, as well as the intercept term, i.e. the amplitude at zero-offset, are modelled by the Chebyshev expansion, which facilitates the interpretation of the AVO attributes.

9) The computational time for amplitude inversion is comparable with traveltime inversion. This is due to (a) the use of a quadratic approximation of the reflection and transmission coefficients, and (b) the use of a first-order approximation to the perturbation of ray-amplitude due to the model perturbation, which speed up the calculation of the Fréchet matrix in the procedure.

Further research will be mainly to transform the methodology into a practical applicable technique, in other words, to relax the assumptions involved in the amplitude inversion and even in the traveltime inversion. The work includes better estimation of amplitude and traveltime, incorporating more information in the inversion, improvement on forward calculation, consideration of many other factors influencing the seismic reflection amplitude, and establishing the near-surface velocity model. Finally, we present a proposal of seismic trace inversion for ray impedance in the prestack domain.

11.2. Estimation of amplitudes, traveltimes and data uncertainties

The picking of traveltimes and amplitudes can be improved by using a beam-forming algorithm. The basic idea is to create beams with maximum amplitudes by using an inversion technique which determines delay times, τ_j. This algorithm also provides the measurement of uncertainty in the extracted amplitudes (at delay time τ_j). The latter can be used to build a data covariance matrix for the amplitude and/or traveltime inversions.

This method can be performed on a CDP gather. Assume we have N recorded seismic traces $\{y_j(t); \, j = 1, \cdots, N\}$ which are normalized to unit maximum amplitude. Normalization ensures that all N traces contribute similarly to the beam. The corresponding N linearly independent beams are generated by the weighted sum of delayed waveforms,

$$y_i^{\mathrm{B}}(t, \mathbf{x}) = \sum_{j=1, j \neq i}^{N} w_j y_j(t + \tau_j), \qquad \text{for} \quad i = 1, \cdots, N, \qquad (11.1)$$

where $\mathbf{x}$ is an unknown vector consisting of the weights w_j and the delays τ_j. A least-squares solution $\mathbf{x}$ can be obtained by minimizing the objective function,

$$J(\mathbf{x}) = \|e_i(\mathbf{x})\|^2, \qquad (11.2)$$

where $e_i(\mathbf{x})$ are the rms residuals between the beam traces and the record traces, given by

$$e_i(\mathbf{x}) = \left\{ \frac{1}{M} \sum_{k=1}^{M} \left[y_i(t_k) - y_i^{\mathrm{B}}(t_k, \mathbf{x}) \right]^2 \right\}^{1/2}, \qquad \text{for} \quad i = 1, \cdots, N, \qquad (11.3)$$

within a chosen time window of the length M, starting at t_1. The inversion can be implemented iteratively. At each iteration, a locally quadratic approximation of $J(\mathbf{x})$ can be made around the current estimate, so that

$$J(\mathbf{x} + \delta\mathbf{x}) \approx J(\mathbf{x}) + \delta\mathbf{x}^{\mathrm{T}} \hat{\mathbf{g}} + \frac{1}{2} \delta\mathbf{x}^{\mathrm{T}} \mathbf{H} \delta\mathbf{x}, \qquad (11.4)$$

where $\hat{\mathbf{g}} = \nabla J$ is the gradient and $\mathbf{H} = \nabla\nabla J$ is the Hessian matrix.

The "model" vector $\mathbf{x}$, is composed of N weighting factors $\{w_j; \, j = 1, \cdots, N\}$, and N delay times $\{\tau_j; \, j = 1, \cdots, N\}$. To avoid problems associated with differences in the physical dimensions of the two parameter classes, a two-dimensional subspace can be used. During each iteration, the two-dimensional subspace is produced by first calculating the updated gradient vector $\hat{\mathbf{g}}$ from the objective function $J(\mathbf{x})$, and then partitioning this vector into two vectors, $\mathbf{a}_1$ and $\mathbf{a}_2$, corresponding to the two parameter classes,

$$\mathbf{a}_1 = [\hat{g}_1^w, \, \hat{g}_2^w, \, \cdots, \, \hat{g}_N^w \mid 0, \, 0, \, \cdots, \, 0]^{\mathrm{T}},$$
$$\mathbf{a}_2 = [\, 0, \, 0, \, \cdots, \, 0 \mid \hat{g}_1^\tau, \, \hat{g}_2^\tau, \, \cdots, \, \hat{g}_N^\tau]^{\mathrm{T}}. \qquad (11.5)$$

The form of the model update can then be expressed as

$$\delta\mathbf{x} = \alpha_1 \mathbf{a}_1 + \alpha_2 \mathbf{a}_2, \qquad (11.6)$$

where α_1 and α_2 are estimated by minimizing the quadratic approximation $J(\mathbf{x})$.

The beamforming method described above can used to extract traveltimes and amplitudes of reflection seismic gathers. The algorithm differs from conventional cross-correlation methods in its robustness in the presence of high random noise levels and local geological scattering, and in its ability to obtain time delays and weights simultaneously by minimizing residuals of all possible waveform fittings. The weights $\{w_j, j = 1, \cdots, N\}$ give the measurements of uncertainty of the extraction, which are important quantities required in the traveltime and amplitude inversions.

11.3. Tomographic inversion incorporating more information and using an improved forward calculation

Nowack and Matheney (1997) developed an inversion formula for velocity and attenuation structure using the envelope amplitude, the instantaneous frequency and the traveltime of selected seismic phases, namely seismic AFT (Amplitude, Frequency and Time) attributes. The extraction of these attributes is performed by complex trace analysis. The instantaneous frequency data are then converted to the so-called attenuation factor,

$$t^* = \int u(\mathbf{x})Q^{-1}(\mathbf{x})ds \,, \tag{11.7}$$

where $u(\mathbf{x})$ is the slowness (inverse velocity) and $Q^{-1}(\mathbf{x})$ is the inverse of the quality factor. In order to obtain values of the attenuation factor t^*, a matching procedure for the instantaneous frequencies between the observed data and a reference pulse is performed. This removes approximately the effects of the source spectra (Matheney and Nowack, 1995). The amplitude and traveltime, along with the attenuation factor t^*, are used in the inversion to reconstruct the variations in slowness and attenuation.

When three-component data are available, the polarization angle for the P-wave can be used as an alternative seismic attribute in the inversion. This method was used by Hu and Menke (1992) and Farra and Le Bégat (1995). Any method for including more information in the traveltime and amplitude inversions would represent an important extension of the present research.

In seismic amplitude inversion, the reflection amplitude picked from a seismic profile is not actually the amplitude of an infinite frequency component. In complex structures where the wave at any point is the sum over many slightly different diffracted paths, the amplitude picked is the maximum amplitude, or the combination of the diffractions. Nichols (1996) suggested computing the wavefield at a few discrete frequencies in the seismic band and picking the strongest peak in the Fourier sum as the traveltime. This is the so-called travel-time of the maximum energy arrival. In the context of reflection traveltime and amplitude inversion, this method may be used to estimate the traveltime and amplitude that are valid in the seismic frequency band, not the usual high-frequency approximation. Instead of solving the eikonal equation for the traveltime, the Helmholtz equation is used to estimate the wavefield for a few frequencies. The estimates of traveltime, amplitude and even phase are obtained by parametrical fitting to the wavefield.

Asymptotic ray theory, used throughout this book, is valid for laterally hetero-geneous media, except at the location of caustics. In the vicinity of a caustic, the wavefield can be calculated by considering the contributions from neighbour-ing, non-Fermat rays (Červený, 2001). For a 1-D earth model this approach is known as the WKBJ seismogram method. It was developed by Chapman (1976, 1978, 1985) and is equivalent to disc-ray theory (Wiggins and Madrid, 1974; Wiggins, 1976). A thorough review of WKBJ theory and other methods can be found in Chapman and Orcutt (1985), and numerical algorithms are provided in Chapman *et al.* (1988). The generalization of the WKBJ seismogram method to 2-D and 3-D problems is known as the Maslov theory (Maslov, 1972; Maslov and Fedoriuk, 1981). It was developed for seismological problems by Frazer and Phinney (1980) and Chapman and Drummond (1982), and was reviewed by Thomson and Chapman (1985). Kendall and Thomson (1993) and Liu and Tromp (1996) implemented the Maslov theory for body-wave propagation in global earth models.

11.4. Consideration of factors influencing amplitudes

In further investigation of seismic amplitude inversion in reflection tomo-graphy, the range of factors that influence the amplitude of a reflection recorded at the surface should be considered.

Many factors affect seismic amplitude. O'Doherty and Anstey (1971) stated that the relevant factors defining the amplitude of a reflection signal are: spher-ical divergence, absorption, the reflection coefficient of the reflecting interface, the cumulative transmission loss at all interfaces above this, and the effect of multiple reflections. Besides the wave interaction with the subsurface, it is also necessary to account for the generation of elastic waves by the source, the re-ception at the surface and subsequent processing. Each of these stages imposes its characteristic influence on the appearance of a reflection event.

Array directivity

The character of seismic radiation is controlled by the nature of the source and its configuration relative to the surface. The effective radiation pattern changes with source depth and is also modified when multiple sources are deployed in an array. Where the individual sources do not interact, the resulting radiation can be estimated by linear superposition. However, for marine arrays a more complex nonlinear representation is required.

The most faithful reproduction of the seismic wavefield is achieved by taking the seismic records from a single receiver but this has the disadvantage that a large amount of amplitude energy travelling near horizontally is retained on the seismic traces. Such energy is normally suppressed by groups of receivers whose outputs are combined in a single seismic trace, but this imposes a selective frequency and slowness filtering on the data, which influence components well away from those for which the suppression is designed.

The source and receiver array directivities affect the amplitude of a reflected seismic signal by the product of the source and receiver array directivities, and the ghost effects at the source array and at the receiver array, respectively (Amundsen and Ursin, 1991; Ursin and Ekren, 1995; Buland *et al.*, 1996). All individual terms in the array amplitude function depend on frequency ω and slowness p.

Transmission loss along the reflection path

When seismic energy penetrates the earth to reach a subsurface reflector, there must be transmission losses as the waves pass through the overburden both on the way to and the way back from the reflector. As transmission coefficients are angle dependent, the amplitude of the *P*-wave incident on the reflector of interest may also be angle dependent. This problem is most significant when the reflectivity above the target is very strong. A strong reflection zone will act as a barrier to transmission and leave a region of muted reflections beneath. Such features are often observed in the presence of localized regions of strong cyclic bedding.

Based on the relationship of an anti-correlation between the power spectrum of the reflection coefficient series and the amplitude spectrum of the pulse transmitted through it, O'Doherty and Anstey (1971) proposed an approximation for the transmission response of a set of thin layers. Shapiro *et al.* (1994) further generalized O'Doherty and Anstey's formula, and expressed the relative amplitude and phase distortions caused by the thin layers with respect to the reference medium as

$$T(\omega) \propto T_0 \exp\{-[\phi_r(\omega,\ \theta) + i\phi_i(\omega,\ \theta)]L\}\ , \tag{11.8}$$

where $|T(\omega)|$ is the amplitude spectrum of the transmitted pulse, T_0 denotes the transmissivity for a homogeneous isotropic reference medium (that describes a phase shift), L is the thickness of the thinly layered stack, and ϕ_r and ϕ_i are the scattering attenuation and the phase-shift coefficient, respectively. Both the functions ϕ_r and ϕ_i are dependent on the incidence angle θ and the frequency ω.

For a thinly layered and sufficiently thick constant-density medium, which exhibits an exponential correlation function $\sigma^2\exp(-\Delta z/\ell)$ of the 1-D random velocity distribution $v(z)$ as a function of depth z, the ϕ_r and ϕ_i functions are expressed by

$$\phi_r(\omega,\ \theta) = \frac{1}{\cos^2\theta}\frac{\omega^2\ell\sigma^2}{v_0^2 + 4\omega^2\ell^2\cos^2\theta} \qquad (11.9)$$

and

$$\phi_i(\omega,\ \theta) = \frac{\omega\sigma^2}{v_0\cos\theta}\left(\frac{1}{2} - \frac{\omega^2\ell^2}{v_0^2 + 4\omega^2\ell^2\cos^2\theta}\right), \qquad (11.10)$$

where σ is the relative standard derivation of the 1-D velocity distribution $v(z)$, ℓ is the correlation length, and v_0 is the velocity of the reference medium.

Equation (11.9) provides the P-wave attenuation coefficient for a stack of thin layers in the case where the P-wave velocity fluctuations and the angle of incidence θ are small ($\sigma^2/\cos^2\theta \ll 1$) and the thickness L is large compared to the correlation length ℓ ($L \gg \ell$). Shapiro and Hubral (1996) showed that fine elastic multi-layering is characterized by a frequency-dependent anisotropy.

The nature of reflection amplitudes

The simple interface reflection coefficient is only a starting point towards understanding offset-dependent reflectivity. The principal control on the amplitude of a reflected wave arises from the nature of the interface which reflects the incident energy. The material contrasts at individual interfaces are rarely large, and coherent reflections often arise from the combined effects of many thin beds. Such reflected energy is built up from a complex interference pattern of primary reflections and internal multiples in the fine bedding (or fine layering) which depend on the thickness and the arrangement of the bed. Horizontal changes in the character of the local bedding can have significant effects on the waveform interference pattern and so change the apparent reflector strength.

The term "fine bedding" is used here for the description of combined reflectors for the target reflection, while "fine layering" refers to a stack of thin layers in the overburden. In other words, fine bedding relates to reflection and fine layering relates to transmission loss. Both are related to the rapid spatial variations of the medium parameters at scales smaller than the seismic wavelength. In the inversion as well as in migration, it is common to ignore these effects. This is understandable, since the resolving power of any imaging technique is limited to details of the order of half the wavelength. However, the small scale variations manifest themselves as an apparent anisotropic dispersion of the seismic wavefield. The reflectivity method, as outlined by Müller (1985), can be used to generate the wave responses, including the interbed multiples, at an arbitrary angle of incidence (e.g. Martinez 1993). A correction term can also be explicitly defined by a multidimensional cross-correlation of the reflection measurements (e.g. Wapenaar and Herrmann, 1996).

Velocity anisotropy should also be taken into account when analysing the AVO response of shales for example, because shales, which make up 75% of the sedimentary cover of the hydrocarbon reserves, are almost invariably anisotropic (Kim *et al.*, 1993). Investigations by de Hoop *et al.* (1999), using the generalized Radon transform inversion (Beylkin, 1985; de Hoop and Bleistein, 1997), showed what information about anisotropy seismic amplitudes could reveal, and how to use this information to image rock properties.

Seismic data processing

Many of the commonly applied processes have a nonlinear component whose influence on the amplitudes of reflection is not easy to predict without detailed analysis. In the standard NMO procedure, a time-varying moveout stretch is applied to the traces. This has a particularly dramatic effect on shallow reflectors. The post-NMO result will therefore include transfer of energy between higher and lower frequency components. Another operation whose influence is pervasive is the "muting" of shot gathers, by removing the residual horizontally travelling energy with a pair of numerical scissors. Much of the shallow reflections is also lost, with the result that the relative amplitude of shallow and deep reflectors will be altered.

It should also be noted that most migration procedures are not amplitude preserving. Such techniques move energy to the appropriate spatial location, but the weighting procedures are usually based on simple approximations for acoustic wave propagation which achieve amplitude fidelity over a limited range of propagation angles (e.g. Tygel *et al.*, 1992; Bleistein *et al.*, 2001). The comparison of observed and calculated amplitudes is therefore best conducted on seismic sections before migration. When prestack migration is used to enhance a reflection event, the formulae for the effect of spherical divergence and the effect associated with the focusing and defocusing of the curved reflector mentioned above must be used for the demigration processing of the amplitudes extracted from the migrated gathers.

The amplitude factors reviewed above must be considered collectively in the course of the analysis and the inversion of reflection amplitude data. However, due to the trade-offs involved, including these amplitude factors in the amplitude inversion is not a trivial problem. Other factors, not listed above, also exist, for instance, the effect of near-surface velocity variations which can have a strong focusing or defocusing influence on the radiated seismic field. In the following section, we discuss the use of turning-ray tomography to investigate the immediate vicinity of the surface (or just below the seabed in marine work), which is one of the most complex zones in the earth.

11.5.   Turning-ray tomography for near-surface velocity structure and attenuation

Seismic reflection tomography assumes that for a given data set the near-surface model is already known, and attempts to derive the macro velocity model from reflection seismic data. To establish the near-surface model in order to be able to derive, e.g., reliable statics for processing, refraction or turning-ray tomography can be used (White, 1989; Zhu *et al.*, 1992; Stefani, 1995).

Turning-ray traveltime inversion complements surface reflection traveltime inversion. The relative merit of the surface reflection traveltime inversion is the greater lateral resolution at depth, since there are many rays travelling quasi-vertically. Because these vertical rays require a reflector, an ambiguity exists between the velocity and the reflector depth in the standard reflection traveltime inversion. To overcome this ambiguity, in this book, seismic amplitude data are also used in the reflection tomography. However, the velocity-depth ambiguity is absent in the turning-ray traveltime inversion (Stefani, 1995). We may consider traveltime inversion and amplitude inversion separately in turning-ray tomography.

Turning-ray traveltime inversion makes no assumptions regarding the existence or geometry of reflectors or refractors in the near surface. Consequently it performs about as well when the near-surface geometry/velocity is very complex as when it is characterized by simple layering. The associated cost is that the velocity inversion must smooth areas of abrupt velocity change, such as the transition from the weathering layer to the top of a fast refractor. However, using an overthrust example, Stefani (1995) showed that even in such a "hard rock" case the velocity smoothing is mild enough to allow accurate estimation of both statics times and migration velocities.

Turning-ray amplitudes represent energy that leaves a seismic source and, due to a positive velocity gradient, returns to receivers at the surface without being reflected in the subsurface. Because turning rays are independent of deeper reflectivities, and as they propagate in the shallowest formation, their amplitudes are affected primarily by transmission and attenuation in the near surface. This makes them ideal candidates to use in analysing the near-surface attenuation without the need to make assumptions about the deeper geology and reflectivity.

Turning-ray amplitude inversion is a deterministic seismic processing algorithm that uses first-arrival data to generate a near-surface attenuation model, which is used to compensate prestack and poststack reflection data for amplitude loss caused by shallow anomalies such as gas zones. Shallow gas zones are common near-surface features in areas of hydrocarbon exploration and production. Frequently the source of the gas can be traced back to a reservoir at depth, while in other cases the shallow gas is biogenic in origin. In either case, gas trapped in the overburden will strongly attenuate seismic *P*-waves and degrade the amplitude and frequency content at depth, thus making identification and interpretation of deeper reflectors difficult. Near-surface attenuation models obtained from turning-ray amplitude inversion may be used to compensate deeper reflection amplitudes for attenuation caused by shallow anomalies.

Brzostowski and McMechan (1992) discussed a way of estimating near-surface attenuation that could be used to mitigate amplitude variations caused by near-surface heterogeneities. Their method requires ray tracing through a near-surface model. Deal *et al.* (2002) presented a simple and robust method that simplifies the complexities and ray tracing involved in attenuation tomography, and inverts the first-arrival amplitudes from long offsets (approximating vertical rays) to generate surface-consistent scale factors in the near-surface beneath the source and receiver locations. Use of these scale factors to correct seismic data, instead of the conventional automatic gain control, may improve the amplitude fidelity of deeper reflections.

11.6. Prestack seismic trace inversion for ray elastic impedance

Finally, we propose a prestack seismic trace inversion method that, using a recursive formula for the reflection coefficient, transforms a seismic trace into a ray-impedance trace. Ray impedance is a lithological parameter that measures the acoustic or elastic impedance along a seismic ray with a specific ray parameter, p. This so-called ray-impedance parameter is an extension of the concept of acoustic and elastic impedance. Using ray impedance, the reflection coefficient given in Chapter 3 can be expressed as a recursive formula, so that the ray-impedance inversion can be implemented on a prestack seismic trace with constant p, in the same way as conventional acoustic impedance inversion on a post-stack trace with the assumption of zero-offset.

Connolly (1999) proposed a pragmatic seismic trace inversion technology, called elastic-impedance inversion, which would allow the first-order AVO effects to be incorporated routinely into seismic and rock-property analysis and interpretation, without using specialist software and with a minimal increase in effect. VerWest *et al.* (2000), Ma (2002) and Santos and Tygel (2002) described alternative approaches to the problem, using different approximations for elastic impedance to give highly accurate reflection coefficients, particularly at high angles of incidence where Aki and Richards' (1980) three-term linearization begins to break down. Whitcombe *et al.* (2002) modified the definition of elastic impedance so that it could be correlated directly to bulk modulus, Lamé's parameter, the α/β ratio, shear impedance and shear modulus, and could be optimized as a fluid or lithology discriminator. However, the elastic impedance inversion can only be applied to an angle-stack trace, i.e. a partial stack of seismic segments with constant incident angle.

In this section, we consider the concept and implementation of seismic trace inversion in the prestack domain, using the ray-impedance parameter. We emphasize the concept of the acoustic impedance or the elastic impedance along a seismic ray with constant ray parameter p, in contrast to the constant incidence angle required by the elastic-impedance inversion. Because the ray parameter $p = k/\omega$, where k and ω are the wavenumber and the angular frequency, respectively, ray impedance may also be called plane-wave impedance.

When the reflectivity is defined in terms of ray impedance for a specific ray parameter p, we can use a convolution model to calculate a synthetic seismogram in the τ-p domain, where τ is the intercept time, i.e. the reflection time at $p = 0$. For the purpose of lithological interpretation, we may interpolate the $RI(\tau, p)$ image into an $RI(\tau, \theta)$ curve, where θ is the P-wave incident angle. The latter is much more accurate than the $EI(\theta)$ estimate, where θ must be a constant at all the time samples τ, and therefore it can provide a more reliable pore-fluid interpretation result.

Definition of ray impedance

Using the acoustic impedance as the lithological parameter, the reflectivity at an interface may be presented as

$$R_i = \frac{AI_{i+1} - AI_i}{AI_{i+1} + AI_i} = \frac{1}{2}\ln\frac{AI_{i+1}}{AI_i} \, , \qquad (11.11)$$

where AI_i and AI_{i+1} are the acoustic impedances above and below the ith interface, respectively, and R_i is expressed in the recursive form. We now define the ray impedance so that the reflectivity of a plane wave with a specified ray parameter p can be expressed as

$$R_i(p) = \frac{RI_{i+1} - RI_i}{RI_{i+1} + RI_i} = \frac{1}{2}\ln\frac{RI_{i+1}}{RI_i} \, , \qquad (11.12)$$

where RI_i and RI_{i+1} are the ray impedances above and below the ith interface. Once equation (11.12) is defined in the recursive form, the ray impedance for each individual layer can be obtained recursively from

$$RI_{i+1} = \prod_{\ell=1}^{i} \frac{1 + R_\ell}{1 - R_\ell} RI_1 \,, \tag{11.13}$$

where RI_1 is the ray impedance within the first layer. This recursive formula can be used to convert a seismic trace with constant p into a ray-impedance trace. We call this seismic trace inversion procedure *RI*-logging.

The normal acoustic impedance in equation (11.11) is defined as

$$AI_i = \rho_i \alpha_i \,, \tag{11.14}$$

where ρ_i is the density and α_i is the *P*-wave velocity. In equations (11.12) and (11.13), the ray acoustic impedance is defined as

$$RI_i(p(\theta_i)) = \frac{\rho_i \alpha_i}{\sqrt{1 - \alpha_i^2 p^2}} = \frac{\rho_i \alpha_i}{\cos\theta_i}, \tag{11.15}$$

and the ray elastic impedance is defined as

$$RI_i(p(\theta_i)) = \frac{\rho_i \alpha_i}{\sqrt{1 - \alpha_i^2 p^2}} \left(1 - \beta_i^2 p^2\right)^{2(r+2)} \tag{11.16a}$$

$$= \frac{\rho_i \alpha_i}{\cos\theta_i} \left(1 - \frac{\beta_i^2}{\alpha_i^2} \sin^2\theta_i\right)^{2(r+2)}, \tag{11.16b}$$

where θ_i is the *P*-wave incident angle, φ_i is the *P-SV* wave reflection angle, β_i is the *S*-wave velocity of the ith layer, and $r = (\Delta\rho/\rho)/(\Delta\beta/\beta)$.

The ray impedance can be derived from the linear expression of the *P*-wave reflection coefficient given in Chapter 3, that is

$$R(p) \approx R_f(p) - \frac{2\Delta\mu}{\rho} p^2 \,, \tag{11.17}$$

where R_f is the fluid-fluid reflectivity when a plane wave strikes the fluid interface, and μ is the shear modulus. The first and second terms are referred to as the fluid-fluid term and the rigidity term. Expression (11.17) is fairly instructive for pore-fluid substitution modelling (Hilterman, 2001): in the fluid term, only the *P*-wave velocity can change the way the amplitude increases or decreases with the incident angle during fluid substitution; in the rigidity term, $\Delta\mu$ is the same for the hydrocarbon- and water-saturated states.

The fluid-fluid item R_f may be expressed in the recursive form:

$$R_f(p(\theta_i)) = \frac{\alpha_{i+1}\rho_{i+1}\cos\theta_i - \alpha_i\rho_i\cos\theta_{i+1}}{\alpha_{i+1}\rho_{i+1}\cos\theta_i + \alpha_i\rho_i\cos\theta_{i+1}}$$

$$\approx \frac{1}{2}\ln\left(\frac{\alpha_{i+1}\rho_{i+1}}{\cos\theta_{i+1}} \Big/ \frac{\alpha_i\rho_i}{\cos\theta_i}\right), \qquad (11.18)$$

which provides the definition of ray acoustic impedance in equation (11.15). Since

$$\Delta\mu \approx \beta^2\Delta\rho + 2\rho\beta\Delta\beta \qquad \text{and} \qquad p = \frac{\sin\varphi}{\beta},$$

the rigidity term can be expressed as

$$R_g(p(\varphi)) \approx -2\left(\frac{\Delta\rho}{\rho} + 2\frac{\Delta\beta}{\beta}\right)\sin^2\varphi \approx -2(r+2)\frac{\Delta\beta}{\beta}\tan^2\varphi, \qquad (11.19)$$

where here we assume that the *P-SV* wave reflection angle φ is less than $30°$ and that $\Delta\rho/\rho \approx r\Delta\beta/\beta$, analogous to equation (3.25). Following the derivation of equation (3.15), we have the approximation,

$$\Delta\varphi \approx \frac{\Delta\beta}{\beta}\tan\varphi, \qquad (11.20)$$

derived from Snell's law. We then obtain the expressions

$$R_g(p(\varphi)) \approx -2(r+2)\tan\varphi\,\Delta\varphi \approx 2(r+2)\frac{\Delta\cos\varphi}{\cos\varphi}$$

$$\approx \frac{1}{2}\ln\frac{\cos^{4(r+2)}\varphi_{i+1}}{\cos^{4(r+2)}\varphi_i}. \qquad (11.21)$$

Combining expressions (11.18) and (11.21), we obtain the reflectivity, expressed as

$$R(p(\theta_i, \varphi_i)) = \frac{1}{2}\ln\left[\left(\alpha_{i+1}\rho_{i+1}\frac{\cos^{4(r+2)}\varphi_{i+1}}{\cos\theta_{i+1}}\right) \Big/ \left(\alpha_i\rho_i\frac{\cos^{4(r+2)}\varphi_i}{\cos\theta_i}\right)\right], \qquad (11.22)$$

which satisfies the recursive requirement in expression (11.12). Thus, we may define the ray elastic impedance as

$$RI(\theta_i, \varphi_i) = \alpha_i\rho_i\frac{\cos^{4(r+2)}\varphi_i}{\cos\theta_i},$$

which is reformulated in equation (11.16) in terms of the incidence angle θ_i or the ray parameter p.

Comparison with conventional elastic impedance

Conventional elastic impedance is defined by Connolly (1999) in the expression for the reflectivity of a plane wave at the ith interface:

$$R_i(\theta) = \frac{EI_{i+1} - EI_i}{EI_{i+1} + EI_i} = \frac{1}{2}\ln\frac{EI_{i+1}}{EI_i} \,, \tag{11.23}$$

where EI_i is the elastic impedance defined by

$$EI_i(\theta) = \rho_i\alpha_i(\rho_i^{-4K\sin^2\theta}\alpha_i^{\tan^2\theta}\beta_i^{-8K\sin^2\theta}) \tag{11.24}$$

and K is the square of the ratio of the average S- to P-wave velocities.

Note that there are two restrictions to the EI_i definition, i.e. both the incident angle θ and K must be constant for all the subsurface interfaces. We can verify that if θ and K are not constant, the reflectivity R in (11.23) cannot be expressed as a recursive formula. Aki and Richards' (1980) linear approximation of the reflection coefficient can be expressed as (equation 3.21)

$$R(\theta) = \frac{1}{2}\left[(1 - 4\gamma^2\sin^2\theta)\frac{\Delta\rho}{\rho} + (1 + \tan^2\theta)\frac{\Delta\alpha}{\alpha} - 8\gamma^2\sin^2\theta\frac{\Delta\beta}{\beta}\right] \,, \tag{11.25}$$

where $\gamma = \beta/\alpha$. For the ith interface, with

$$\frac{\Delta\rho}{\rho} \approx \ln\frac{\rho_{i+1}}{\rho_i} \,, \quad \frac{\Delta\alpha}{\alpha} \approx \ln\frac{\alpha_{i+1}}{\alpha_i} \quad \text{and} \quad \frac{\Delta\beta}{\beta} \approx \ln\frac{\beta_{i+1}}{\beta_i} \,,$$

we have the following expression

$$R_i(\theta) \approx \frac{1}{2}\ln\frac{\rho_{i+1}^{1-4\gamma_i^2\sin^2\theta_i}\alpha_{i+1}^{1+\tan^2\theta_i}\beta_{i+1}^{-8\gamma_i^2\sin^2\theta_i}}{\rho_i^{1-4\gamma_i^2\sin^2\theta_i}\alpha_i^{1+\tan^2\theta_i}\beta_i^{-8\gamma_i^2\sin^2\theta_i}} \,. \tag{11.26}$$

This expression is not in the recursive form. Only when $\gamma_i^2 = K$ and $\theta_i = \theta$, both being constant, will expression (11.26) become the recursive formula (11.23). Thus, in the definition of elastic impedance, Snell's law is not followed. To use it as a pragmatic parameter, we need to construct a seismic trace along which the incident angle is constant. In addition, K is assumed to be constant for all the layers from the recording surface down to deep layers of the earth, which is very different from reality.

However, in the ray impedance defined by equation (11.16), both the incident angle θ_i and the squares of the S- to P-wave velocity ratio vary with depth. Therefore, inversion using the reflectivity expression defined by ray impedance has the following two advantages:

1) Snell's law holds along a ray path with ray parameter p. The incident angle θ_i varies with depth. The refraction angle θ_{i+1}, which can be obtained from θ_i following Snell's law, is the incident angle at the $(i+1)$th interface if we consider a stratified earth model, as we generally do in the acoustic impedance inversion.

2) It is more accurate. By removing the restriction of constant K required by the elastic impedance definition, the reflectivity expression presented in terms of ray impedance is much more accurate. In addition, the ray impedance is derived from the linear expression (11.17) which is valid when $\Delta\mu/\rho$ or $\Delta\mu$ is small; the elastic impedance is derived from the linear expression (11.25) which assumes that $\Delta\alpha/\alpha$, $\Delta\beta/\beta$ and $\Delta\rho/\rho$ are all small (Aki and Richards, 1980). In most cases, expression (11.17) is more accurate than expression (11.25) and therefore ray impedance is more accurate than elastic impedance in describing earth properties.

Seismic trace inversion in the prestack domain

Seismic trace inversion in the prestack domain consists of the following three basic steps:

1) Point-source τ-p transform;

2) RI-logging, producing $RI(\tau)$ from each p trace;

3) Interpolating $RI(\tau, p)$ into $RI(\tau, \sin\theta)$.

Prestack seismic trace inversion is implemented in the τ-p domain. The τ-p transform used is the so-called cylindrical τ-p transform, which is the wave-equation solution of a wave propagating through a stratified earth and takes into account the point-source effect (Chapman, 1978; Wang and Houseman, 1997). The problem with an exact implementation of the cylindrical τ-p transform is that it is too slow to apply routinely on complete datasets. However, it is possible to implement it as a two-step procedure (Wapenaar et al., 1992). First, a rescaling of point source amplitudes is applied. This transforms a point-source response into a line-source response using a lateral integration. Then a conventional Cartesian τ-p transform is applied. Both the rescaling and the Cartesian τ-p transform are extremely fast. Although the method suffers from a slightly lower S/N ratio it is nearly 2 orders of magnitude faster than the exact cylindrical τ-p transform.

Using the recursive formula (11.13), we can convert each seismic trace in the τ-p domain into an RI-logging trace. The record $RI(\tau, p)$ in the τ-p domain may be interpolated into $RI(\tau, \sin\theta)$ using Snell's law, $\sin\theta = \alpha p$, based on the P-wave velocity function $\alpha(\tau)$.

The $RI(\tau)$ trace for different incident angles, θ, can be used routinely, in the same way as an EI curve, for the lithological analysis and interpretation. As reported by Connolly (1999), the oil sands correlated closely with the areas of low elastic impedance, and the relationship was used to estimate the in-situ volumes for the field from the inverted 30° seismic volume. Because RI is more accurate than EI, the interpretation using RI will be much more reliable in practice.

An assumption made in any seismic trace inversion method is that the earth is locally stratified. We expect, however, that the prestack seismic trace inversion presented here, used together with the tomographic inversion, may produce a more detailed picture of the subsurface structure from reflection seismic data.

Appendices

A.1. Derivation of the geometrical spreading function

This appendix gives the derivation of the geometrical spreading function presented in Chapter 1. A two-dimensional model, which requires that both the ray path and the normal to the interface lie in the same vertical plane, is considered. Although the structure is considered to be two-dimensional, geometrical spreading of the rays in 3-D is assumed.

The ray geometrical spreading function, $L(\ell)$, in which ℓ is distance from a source point, N_0, measured along the ray path, describes the amplitude variation due to geometrical spreading. Considering now the amplitude variation due to geometrical spreading and assuming the product of reflection-transmission coefficients to be 1, we have

$$A(\ell) = A_0 \frac{\ell_0}{L(\ell)} \,, \tag{A.1}$$

where A_0 is taken here to be the amplitude of the wave at some distance ℓ_0 sufficiently close to the source for there to be no intervening interfaces, but sufficiently far for near-source effects to be neglected. The wave front is spherical.

As we follow the ray out from the source in a homogeneous layer, the geometrical spreading function is simply

$$L(\ell) = \ell \,, \qquad \text{for } \ell \le \ell_1 \,, \tag{A.2}$$

until we reach distance ℓ_1 at which the ray intersects an interface at point N_1 and refraction or reflection occurs. Beyond distance ℓ_1 the geometrical spreading function is modified by refraction or reflection from a curved interface. Changes in amplitude due to acoustic impedance contrast across the interface are included separately using the Zöppritz relations.

Assume the interface is represented by a single-valued function $z = Z(x)$ in which x is the horizontal coordinate and z is the depth below some reference level. The dip of the interface (Figure A.1) is given locally by

$$\theta = \tan^{-1} \left(\frac{dz}{dx} \right) \,. \tag{A.3}$$

To calculate the effect of the curvature of the interface on the geometrical spreading function, we consider the change in divergence of two adjacent rays with angular separation $\Delta\psi$ at the source point N_0. These two rays impinge on the interface with angles of incidence φ and φ_m (as shown in Figure 1.1). The corresponding angles of reflection (or refraction) φ' and φ'_m are determined by Snell's law

$$\frac{\sin\varphi}{\sin\varphi'} = \frac{\sin\varphi_m}{\sin\varphi'_m} = \frac{v_1}{v_2} \,, \tag{A.4}$$

where v_1 and v_2 are the relevant velocities of the incoming and outgoing rays (reflected or refracted, converted or unconverted). We denote the apparent angular separation of the two

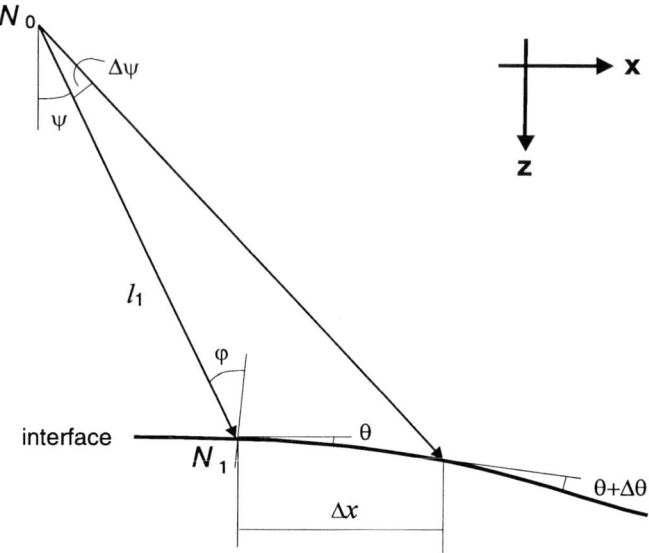

Figure A.1.   Geometry of interface. The ray takes off at point N_0 with take-off angle ψ. Given a small perturbation of the take-off angle $\Delta\psi$, the increment of horizontal distance of the incident point is Δx, the increment of slope angle of the interface is $\Delta\theta$, where θ is the slope angle of the interface locally.

outgoing rays from a virtual source N_0' as $\Delta\psi'$ and we now proceed to derive the dependence of $\Delta\psi'$ on $\Delta\psi$.

The difference in x-coordinate of the two incidence points (see Figure 1.1) is

$$\Delta x = \pm\ell_1 \frac{\cos\theta}{\cos\varphi}\,\Delta\psi\,, \tag{A.5}$$

where the "+" and "−" signs refer to the incident ray with an acute or an obtuse angle with the z-axis (or to the down- and up-going rays, respectively). Finding the change in slope angle of the interface between the two incidence points, we obtain

$$\Delta\theta = \left[1 + \left(\frac{dz}{dx}\right)^2\right]^{-1}\frac{d^2z}{dx^2}\,\Delta x\,, \tag{A.6}$$

or

$$\Delta\theta = \frac{\ell_1}{\cos\varphi}\Theta\,\Delta\psi\,, \tag{A.7}$$

where

$$\Theta(x) = \pm\left[1 + \left(\frac{dz}{dx}\right)^2\right]^{-3/2}\frac{d^2z}{dx^2}\,. \tag{A.8}$$

From Figure 1.1 we see that

$$\varphi_m = \varphi + \Delta\psi + \Delta\theta , \tag{A.9}$$

and we define

$$\Delta\varphi' = \varphi'_m - \varphi' . \tag{A.10}$$

From equations (A.9) and (A.10) (in the limit of small angular increments),

$$\sin\varphi_m = \sin\varphi + (\Delta\psi + \Delta\theta)\cos\varphi \tag{A.11}$$

and

$$\sin\varphi'_m = \sin\varphi' + \Delta\varphi'\cos\varphi' , \tag{A.12}$$

and substituting from equations (A.4) and (A.7), we have

$$\Delta\varphi' = \frac{v_2}{v_1}\frac{\cos\varphi}{\cos\varphi'}\left(1 + \frac{\ell_1}{\cos\varphi}\Theta\right)\Delta\psi . \tag{A.13}$$

The two outgoing rays appear to emanate from a virtual image of the source N'_0 (as shown in Figure 1.1). The apparent divergence of the outgoing rays $\Delta\psi'$ is obtained from $\Delta\varphi'$ by

$$\Delta\psi' = \Delta\varphi' \pm \Delta\theta , \tag{A.14}$$

where the "+" sign refers to the reflection case and the "−" sign refers to the refraction case. Thus

$$\Delta\psi' = \left[\frac{v_2}{v_1}\frac{\cos\varphi}{\cos\varphi'} + \left(\frac{v_2}{v_1}\frac{\cos\varphi}{\cos\varphi'} \pm 1\right)\frac{\ell_1}{\cos\varphi}\Theta\right]\Delta\psi . \tag{A.15}$$

The virtual image of the source appears to be at distance ℓ'_1 from the incidence point. The distance ℓ'_1 is related to ℓ_1 and to the change in divergence of the beam, by evaluating the area of the interface segment illuminated by the beam and equating the incident and emergent beam intensities:

$$\frac{\ell'_1\Delta\psi'}{\cos\varphi'} = \frac{\ell_1\Delta\psi}{\cos\varphi} . \tag{A.16}$$

Combining equations (A.15) and (A.16) we thus obtain

$$\frac{1}{\ell'_1} = \frac{1}{\ell_1}\frac{v_2}{v_1}\frac{\cos^2\varphi}{\cos^2\varphi'} + \frac{1}{\cos\varphi'}\left(\frac{v_2}{v_1}\frac{\cos\varphi}{\cos\varphi'} \pm 1\right)\Theta , \tag{A.17}$$

where ℓ' is not necessarily positive.

Now we consider the divergence of the rays in the perpendicular direction. As the structure is assumed to be 2-D, the dip of the interface in this direction is by definition zero, and equation (A.17) with $\Theta = 0$ and $\varphi = 0$ can be used to define the apparent distance ℓ''_1 (in the perpendicular direction) to the virtual image of the source:

$$\frac{1}{\ell''_1} = \frac{v_2}{\ell_1 v_1} . \tag{A.18}$$

To evaluate the geometrical spreading function at ray distance $\ell > \ell_1$ (beyond the reflection/refraction point), we now relate the spherical divergence of the beam to the virtual image. From equations (A.1) and (A.2), at N_1 the amplitude of the incident wave A_1 is

$$A_1 = A_0\ell_0/\ell_1 . \tag{A.19}$$

If we do not consider changes in amplitude due to the acoustic impedance contrast across the interface, we have

$$A_1^2 |\ell_1'| \, \ell_1'' = A_2^2 |\ell_1' + \ell_2|(\ell_1'' + \ell_2) \,,$$

and therefore

$$A_2 = A_0 \frac{\ell_0}{\ell_1 \left[\left| 1 + \dfrac{\ell_2}{\ell_1'} \right| \left(1 + \dfrac{\ell_2}{\ell_1''} \right) \right]^{1/2}} \,. \tag{A.20}$$

We thus have

$$L(\ell) = \ell_1 \left[\left| 1 + \frac{\ell_2}{\ell_1'} \right| \left(1 + \frac{\ell_2}{\ell_1''} \right) \right]^{1/2}, \qquad \text{for} \quad \ell_2 = \ell - \ell_1 \,. \tag{A.21}$$

Similarly, the wave emerges from N_2 and arrives at N_3, from where the virtual image of the source appears to be at distances ℓ_2' and ℓ_2''. These distances ℓ_2' and ℓ_2'' are related to $|\ell_1' + \ell_2|$ and $(\ell_1'' + \ell_2)$, respectively, as follows:

$$\frac{1}{\ell_2'} = \frac{1}{|\ell_1' + \ell_2|} \frac{v_3}{v_2} \frac{\cos^2\varphi_2}{\cos^2\varphi_2'} + \frac{1}{\cos\varphi_2'} \left(\frac{v_3}{v_2} \frac{\cos\varphi_2}{\cos\varphi_2'} \pm 1 \right) \Theta_2 \tag{A.22}$$

and

$$\frac{1}{\ell_2''} = \frac{v_3}{(\ell_1 v_1 + \ell_2 v_2)} \,, \tag{A.23}$$

where v_3 is the local velocity along ray segment ℓ_3 between N_2 and N_3, φ_2 and φ_2' are the incident and reflection or refraction angles at the point N_2, and Θ_2 is the factor describing the effect of local curvature of the second interface (equation A.8). From the relationship,

$$A_2^2 |\ell_2'| \, \ell_2'' = A_3^2 |\ell_2' + \ell_3|(\ell_2'' + \ell_3) \,,$$

we obtain the ray geometrical spreading function analogous to equation (A.21),

$$L(\ell) = \ell_1 \left[\left| 1 + \frac{\ell_2}{\ell_1'} \right| \left(1 + \frac{\ell_2}{\ell_1''} \right) \right]^{\frac{1}{2}} \left[\left| 1 + \frac{\ell_3}{\ell_2'} \right| \left(1 + \frac{\ell_3}{\ell_2''} \right) \right]^{1/2}, \tag{A.24}$$

$$\text{for} \quad \ell_3 = \ell - (\ell_1 + \ell_2) \,.$$

Generalizing, we obtain the ray geometrical spreading function for a distance ℓ up to the $(K+1)$th intersection with an interface as

$$L(\ell) = \ell_1 \prod_{i=2}^{K+1} \left[\left| 1 + \frac{\ell_i}{\ell_{i-1}'} \right| \left(1 + \frac{\ell_i}{\ell_{i-1}''} \right) \right]^{1/2}, \qquad \text{for} \quad \ell_{K+1} = \ell - \sum_{i=1}^{K} \ell_i \,, \tag{A.25}$$

with

$$\frac{1}{\ell_i'} = \frac{1}{|\ell_{i-1}' + \ell_i|} \frac{v_{i+1}}{v_i} \frac{\cos^2\varphi_i}{\cos^2\varphi_i'} + \frac{1}{\cos\varphi_i'} \left(\frac{v_{i+1}}{v_i} \frac{\cos\varphi_i}{\cos\varphi_i'} \pm 1 \right) \Theta_i \tag{A.26}$$

and

$$\frac{1}{\ell_i''} = \frac{v_{i+1}^i}{\displaystyle\sum_{j=1}^{i} \ell_j v_j} \, , \tag{A.27}$$

The factor Θ defined by equation (A.8), describes the curvature of an interface $z = Z(x)$ at coordinate x. A similar expression for $L(\ell)$ has also been derived by Červený and Ravindra (1971).

A.2. Derivation of reflection amplitude demigration

The demigration process (Chapter 8) includes both the effect of spherical divergence and the effect associated with the focusing and defocusing due to reflector curvature.

A.2.1 The effect of spherical divergence

The effect of spherical divergence considered here is the effect of a ray tube propagating through a horizontally layered medium, in which the P-wave velocity varies vertically. In this 1-D case, the relative geometrical spreading is defined by (Červený, 1985; Hron *et al.*, 1986)

$$(\det \mathbf{Q})^{1/2} = \left(\frac{y}{p} \frac{dy}{dp} \cos\phi_0 \cos\phi_N \right)^{1/2} \, , \tag{A.28}$$

where p is the ray parameter, y is the source-receiver horizontal separation, ϕ_0 is the take-off angle of the ray, ϕ_N is the angle to the vertical axis of the ray at receiver point, and $\mathbf{Q}$ can be understood as a matrix transforming global, ray coordinates to local, ray-centred coordinates. The effect of spherical divergence in the demigration process can be expressed as the reciprocal of the normalized geometrical spreading (Ursin, 1990),

$$G_s = \frac{v_0}{(\det \mathbf{Q})^{1/2}} \, , \tag{A.29}$$

where v_0 is the P-wave velocity at the source point. When the source and the receiver are located at the same level, we have $\cos\phi_0 = \cos\phi_N$ for the 1-D velocity model. The effect of spherical divergence can then be rewritten as

$$G_s = \left(q_0^2 \frac{y}{p} \frac{dy}{dp} \right)^{-1/2} \, , \tag{A.30}$$

where $q_0 \equiv v_0^{-1} \cos\phi_0$ is the vertical slowness at the source point, given by

$$q_0 = \left(\frac{1}{v_0^2} - p^2 \right)^{1/2} . \tag{A.31}$$

In practice, the ray parameter p can be obtained from the slope of a CMP gather,

$$p = \frac{d}{dy}t(y) ,$$ (A.32)

where $t(y)$ is the traveltime. Considering the following hyperbolic approximation,

$$t(y) = \left[t^2(0) + \frac{y^2}{v_{\text{rms}}^2}\right]^{1/2} ,$$ (A.33)

we have

$$p = \frac{y}{v_{\text{rms}}^2 t(y)}$$ (A.34)

and

$$dp = \frac{1}{v_{\text{rms}}^2 t(y)}\left[1 - \frac{1}{v_{\text{rms}}^2 t^2(y)} y^2\right] dy .$$ (A.35)

Substituting equations (A.34) and (A.35) into equation (A.30), we get

$$G_s = \frac{v_0}{v_{\text{rms}}^2 t(y)}\left[\frac{v_{\text{rms}}^2 t^2(y) - y^2}{v_{\text{rms}}^2 t^2(y) - (v_0/v_{\text{rms}})^2 y^2}\right]^{1/2} ,$$ (A.36)

expressed in terms of the offset y. For offsets less than the depth of the target reflector, we can make an approximation using the Taylor expansion, giving

$$G_s \approx \frac{v_0}{v_{\text{rms}}^2 t(y)}\left[1 + \left(1 - \frac{v_{\text{rms}}^2}{v_0^2}\right)\left(\frac{v_0 y}{v_{\text{rms}}^2 t(y)}\right)^2 + \left(1 - \frac{v_{\text{rms}}^2}{v_0^2}\right)\left(\frac{v_0 y}{v_{\text{rms}}^2 t(y)}\right)^4\right]^{1/2} .$$ (A.37)

From equation (A.34), we have $v_0 y/[v_{\text{rms}}^2 t(y)] = (v_0/v_z)\sin\phi$, where ϕ is the angle of incidence at the reflector and v_z is the velocity at depth z. Equation (A.37) can then be expressed in terms of ϕ as follows:

$$G_s \approx \frac{v_0}{v_{\text{rms}}^2 t(y)}\left[1 + \left(1 - \frac{v_{\text{rms}}^2}{v_0^2}\right)\left(\frac{v_0}{v_z}\right)^2 \sin^2\phi + \left(1 - \frac{v_{\text{rms}}^2}{v_0^2}\right)\left(\frac{v_0}{v_z}\right)^4 \sin^4\phi\right]^{1/2} .$$ (A.38)

A.2.2   The effect of reflector curvature

The effect associated with the focusing and defocusing from reflector curvature can be defined as

$$G_c = \sqrt{\ell'/\ell} ,$$ (A.39)

where ℓ is the ray distance between the source and the incident point at the interface, and ℓ' is the virtual ray distance between the virtual source and the incident point. Because of the interface curvature, the virtual ray length ℓ' is given by

$$\ell' = \left\{\frac{1}{\ell} + \frac{2}{\cos\phi}\frac{d^2 z}{dx^2}\left[1 + \left(\frac{dz}{dx}\right)^2\right]^{-3/2}\right\}^{-1} .$$ (A.40)

This is deduced for the one-interface case from the formula derived in Appendix A.1. Given the length R of the normal-incidence ray (two-way), the ray distance ℓ can be estimated by

$$\ell = \frac{R}{2} \frac{\cos\theta}{\cos(\phi + \theta)} \, , \tag{A.41}$$

where $\theta = \tan^{-1}(dz/dx)$ is the local slope angle of the interface. The effect G_c can then be expressed as

$$G_c = \left[1 + \frac{R\cos^4\theta}{R_c\cos\phi\cos(\phi + \theta)}\right]^{-1/2} , \tag{A.42}$$

where $R_c = (d^2 z/dx^2)^{-1}$ is the radius of curvature of the reflector.

Making a small-dip approximation $\theta \approx 0$, we have

$$G_c \approx \left(1 + \frac{\Psi}{\cos^2\phi}\right)^{-1/2} \approx \left(1 - \frac{\Psi}{\cos^2\phi} + \frac{\Psi^2}{\cos^4\phi}\right)^{1/2} , \tag{A.43}$$

where $\Psi = R/R_c$, and here we assume $|\Psi| \ll 1$. Using the approximation

$$\frac{1}{\cos^2\phi} \approx 1 + \sin^2\phi + \sin^4\phi \, ,$$

we have G_c expressed in terms of $\sin^2\phi$ as

$$G_c \approx [1 - \Psi + \Psi^2 + \Psi(1 - 2\Psi)\sin^2\phi - \Psi(1 - 3\Psi)\sin^4\phi]^{1/2} . \tag{A.44}$$

If the interval velocity of the section can be approximated in the form $v = v_0 + kz$, the ray paths are arcs of circles whose centres are $k^{-1}v_0$ above the ground surface. Then the angle of incidence can be evaluated in terms of the offset y by (Ostrander, 1984)

$$\phi = \tan^{-1}\left[\frac{(v_0 + kz)y}{2v_0 z + kz^2 - \frac{1}{4}ky^2}\right] . \tag{A.45}$$

Substituting the relation $\sin\phi = v_z p$ and equation (A.34) into equation (A.44), we obtain the following expression,

$$G_c \approx \left[1 - \Psi + \Psi^2 + \Psi(1 - 2\Psi)\left(\frac{v_z}{v_{\text{rms}}^2 t(y)}\right)^2 y^2 - \Psi(1 - 3\Psi)\left(\frac{v_z}{v_{\text{rms}}^2 t(y)}\right)^4 y^4\right]^{1/2} , \tag{A.46}$$

expressed in terms of the offset y.

References

Aki K. and Richards P., 1980. *Quantitative Seismology: Theory and Method.* W.H. Freeman.

Almoghrabi H. and Lange J., 1986. Layers and bright spots. *Geophysics*, **51**, 699–709.

Amundsen L. and Ursin B., 1991. Frequency-wavenumber inversion of acoustic data. *Geophysics*, **56**, 1027–1039.

Beylkin G., 1985. Imaging of discontinuities in the inverse scattering problem by inversion of the causal generalized Radon transform. *J. Math. Phys.*, **26**, 99–108.

Bickel S.H., 1990. Velocity-depth ambiguity of reflection traveltimes. *Geophysics*, **55**, 266–276.

Birkhoff G., 1937. On product integration. *J. Math. Phys.*, **16**, 104–132.

Bishop T.N., Bube K.P., Cutler R.T., Langan R.T., Love P.L., Resnick J.R., Shuey R.T., Spindler D.A. and Wyld, H.W., 1985. Tomographic determination of velocity and depth in laterally varying media. *Geophysics*, **50**, 903–923.

Bleistein N., Cohen J.K. and Stockwell J.W. Jr., 2001. *Mathematics of Multidimensional Seismic Imaging, Migration and Inversion.* Springer-Verlag New York, Inc.

Blundell C.A., 1992. Illustrating the trade-off between velocity and reflector position in traveltime inversion by using a two-dimensional subspace. *Expl. Geophys.*, **23**, 27–32.

Bording R.P., Gersztenkorn A., Lines L.R., Scales J.A. and Treitel S., 1987. Applications of seismic traveltime tomography. *Geophys. J. R. Astr. Soc.*, **90**, 285–303.

Bortfeld R., 1961. Approximation to the reflection and transmission coefficients of plane longitudinal and transverse waves. *Geophys. Prosp.*, **9**, 485–502.

Bregman N.D., Bailey R.C. and Chapman C.H., 1989. Crosshole seismic tomography. *Geophysics*, **54**, 200–215.

Bregman N.D., Bailey R.C. and Chapman C.H., 1989. Ghosts in tomography: the effects of poor angular coverage in 2-D seismic traveltime inversion. *Can. J. Expl. Geophys.*, **25**, 7–27.

Broyden C.G., 1969. A new method of solving non-linear simultaneous equations. *Comput. J.*, **12**, 94–99.

Brzostowski M.A. and McMechan G.A., 1992. 3-D tomographic imaging of near-surface seismic velocity and attenuation. *Geophysics*, **57**, 396–403.

Bube K.P., Langan R.T. and Resnick J.R., 1995. Theoretical and numerical issues in the determination of reflector depths in seismic reflection tomography. *J. Geophys. Res.*, **100**, 12449–12458.

Bube K.P. and Meadows M.A., 1999. The null space of a generally anisotropic medium in linearized surface reflection tomography. *Geophys. J. Int.*, **139**, 9–50.

Buland A., Landrø M., Andersen M. and Dahl T., 1996. AVO inversion of Troll Field data. *Geophysics*, **61**, 1589–1602.

Bunks C., Saleck F.M., Zaleski S. and Chavent G., 1995. Multi-scale seismic waveform inversion. *Geophysics*, **60**, 1457–1473.

Burridge R., 1976. *Some Mathematical Topics in Seismology.* Courant Institute of Mathematical Sciences, New York University.

Castagna J.P., 1993. AVO analysis: tutorial and review. In *Offset-Dependent Reflectivity: Theory and Practice of AVO Analysis,* Castagna J.P. and Backus M.M. (eds.), Society of Exploration Geophysicists, pp. 3–36.

Castagna J.P., Batzle M.L. and Eastwood R.L., 1985. Relationship between compressional wave and shear wave velocities in clastic silicate rocks. *Geophysics,* **50**, 571–581.

Castagna J.P., Batzle M.L. and Kan T.K., 1993. Rock physics: the line between properties and AVO response. In *Offset-Dependent Reflectivity: Theory and Practice of AVO Analysis,* Castagna J.P. and Backus M.M. (eds.), Society of Exploration Geophysicists, pp. 135–171.

Castagna J.P. and Smith S.W., 1994. Comparison of AVO indicators: a modelling study. *Geophysics,* **59**, 1849–1855.

Červený V., 1985. The application of ray tracing to numerical modelling of seismic wave fields in complex structures. In *Seismic Shear Waves, Part A: Theory,* Dohr G.P. (ed.), Geophysical Press, pp. 1–124.

Červený V., 1989. Ray tracing in factorized anisotropic inhomogeneous media. *Geophys. J. Int.,* **99**, 91–100.

Červený V., 2001. *Seismic Ray Theory.* Cambridge University Press.

Červený V. and Ravindra R., 1971. *Theory of Seismic Head Waves.* University of Toronto Press.

Chapman C.H., 1976. Exact and approximate generalized ray theory in vertically inhomogeneous media. *Geophys. J. R. Astr. Soc.,* **46**, 201–233.

Chapman C.H., 1978. A new method for computing seismograms. *Geophys. J. R. Astr. Soc.,* **54**, 481–518.

Chapman C.H., 1985. Ray theory and its extensions: WKBJ and Maslov seismograms. *J. Geophys.,* **58**, 27–43.

Chapman C.H. and Drummond R., 1982. Body-wave seismograms in inhomogeneous media using Maslov asymptotic theory. *Bull. Seism. Soc. Am.,* **72**, S277–S317.

Chapman C.H. and Orcutt J.A., 1985. The computation of body wave synthetic seismograms in laterally homogeneous media. *Rev. Geophys.,* **23**, 105–163.

Chapman C.H., Chu J.-Y. and Lyness D.G., 1988. The WKBJ seismogram algorithm. In *Seismological Algorithms,* Doornbos D. (ed.), Academic Press, pp. 47–74.

Clenshaw C.W., 1962. Chebyshev series for mathematical functions. In *Mathematical Tables (vol. 5),* National Physical Laboratory, H.M. Stationery, London.

Cleveland W.S., 1979. Robust locally weighted regression and smoothing scatterplots. *J. Am. Statis. Assoc.,* **74**, 829–836.

Cleveland W.S. and Grosse E., 1991. Computational methods for local regression. *Statistics Comput.,* **1**, 47–62.

Connolly P., 1999. Elastic impedance. *The Leading Edge,* **18**, 438–452.

Constable S.C., Parker R.L. and Constable C.G., 1987. Occam's inversion: a practical algorithm for generating smooth models from electromagnetic sounding data. *Geophysics,* **52**, 289–300.

Deal M.M., Matteucci G. and Kim Y.C., 2002. Turning ray amplitude inversion: mitigating amplitude attenuation due to shallow gas. *Expanded Abstracts,* 72nd Annual Meeting of Society of Exploration Geophysicists, Salt Lake City, 2078–2081.

de Hoop M.V. and Bleistein N., 1997. Generalized Radon transform inversions for reflectivity in anisotropic elastic media. *Inverse Problems,* **13**, 669–690.

de Hoop M.V., Spencer C. and Burridge R., 1999. The resolving power of seismic amplitude data: an anisotropic inversion/migration approach. *Geophysics,* **64**, 852–873.

Delprat-Jannaud F. and Lailly P., 1992. What information on the Earth model do reflection traveltimes provide? *J. Geophys. Res.,* **97**, 19827–19844.

Delprat-Jannaud F. and Lailly P., 1993. Ill-posed and well-posed formulations of the reflection travel time tomography problem. *J. Geophys. Res.,* **98**, 6589–6605.

de Nicolao A., Drufuca G. and Rocca F., 1993. Eigenvectors and eigenvalues of linearized elastic inversion. *Geophysics*, **58**, 670–679.

Deregowski S.M., 1990. Common-offset migrations and velocity analysis. *First Break*, **8 (6)**, 225–234.

Dong W., 1996. A sensitive combination of AVO slope and intercept for hydrocarbon indication. *Extended Abstracts*, 58th Conference of European Association of Geoscientists & Engineers, Amsterdam, M044.

Drufuca G. and Mazzotti A., 1995. Ambiguities in AVO inversion of reflections from a gas-sand. *Geophysics*, **60**, 134-141.

Dyer B.C. and Worthington M.H., 1988. Seismic reflection tomography: a case study. *First Break*, **6 (11)**, 354–366.

Farra V., 1989. Ray perturbation theory for heterogeneous hexagonal anisotropic medium. *Geophys. J. Int.*, **99**, 723–738.

Farra V., 1990. Amplitude computation in heterogeneous media by ray perturbation theory: a finite element approach. *Geophys. J. Int.*, **103**, 341–354.

Farra V., 1992. Bending method revisited: a Hamiltonian approach. *Geophys. J. Int.*, **109**, 138–150.

Farra V. and Le Bégat S., 1995. Sensitivity of qP-wave traveltimes and polarization vectors to heterogeneity, anisotropy and interfaces. *Geophys. J. Int.*, **121**, 377–390.

Farra V. and Madariaga R., 1987. Seismic waveform modelling in heterogeneous media by ray perturbation theory. *J. Geophys. Res.*, **92**, 2697–2712.

Farra V. and Madariaga R., 1988. Non-linear reflection tomography. *Geophys. J.*, **95**, 135–147.

Farra V., Virieux J. and Madariaga R., 1989. Ray perturbation theory for interfaces. *Geophys. J. Int.*, **99**, 377–390.

Fatti J.L., Smith G.C., Vail P.J., Strauss P.J. and Levitt P.R., 1994. Detection of gas in sandstone reservoirs using AVO analysis: a 3-D case history using the Geostack technique. *Geophysics*, **59**, 1362–1376.

Fawcett J.A. and Clayton R.W., 1984. Tomographic reconstruction of velocity anomalies. *Bull. Seism. Soc. Am.*, **74**, 2201–2219.

Franklin J.N., 1970. Well-posed stochastic extensions of ill-posed linear problems. *J. Math. Anal. Appl.*, **31**, 682–716.

Frazer L.N. and Phinney R.A., 1980. The theory of finite frequency body wave synthetic seismograms in inhomogeneous elastic media. *Geophys. J. R. Astr. Soc.*, **63**, 691–713.

Futterman W.I., 1962. Dispersive body waves. *J. Geophys. Res.*, **67**, 5279–5291.

Gajewski D. and Pšenčík I., 1990. Vertical seismic profile synthetics by dynamic ray tracing in laterally varying layered anisotropic structures. *J. Geophys. Res.*, **95**, 11301–11315.

Gardner G.H.F., Gardner L.W. and Gregory A.R., 1974. Formation velocity and density: the diagnostic basics for stratigraphic traps. *Geophysics*, **39**, 770–780.

Gilbert F. and Backus G.E., 1966. Propagator matrices in elastic wave and vibration problems. *Geophysics*, **31**, 326–333.

Grau G. and Lailly P., 1993. Sequential migration-aided reflection tomography: an approach to imaging complex structures. *J. Appl. Geophys.*, **30**, 75–87.

Hampson D., 1991. AVO inversion, theory, and practice. *The Leading Edge*, **10 (6)**, 39–42.

Hargreaves N.D. and Calvert A.J., 1991. Inverse Q filtering by Fourier transform. *Geophysics*, **56**, 519–527.

Hatton L., Worthington M.H. and Makin J., 1986. *Seismic Data Processing: Theory and Practice*. Blackwell Scientific Publications.

Herman G.T., 1980. *Reconstructions from Projections: The Fundamentals of Computerized Tomography*. Academic Press.

Hilterman F., 1990. Is AVO the seismic signature of lithology? a case history of Ship Shoal-South addition. *The Leading Edge*, **9 (6)**, 15–22.

Hilterman F., 2001. *Seismic Amplitude Interpretation*. Distinguished instructor short course series No. 4, Society of Exploration Geophysicists and European Association of Geoscientists & Engineers.

Hron F., May B.T., Covey J.D. and Daley P.F., 1986. Synthetic seismic sections for acoustic, elastic, anisotropic, and vertically inhomogeneous media. *Geophysics*, **51**, 710–735.

Hu G. and Menke W., 1992. Formal inversion of laterally heterogeneous velocity structure from *P*-wave polarization data. *Geophys. J. Int.*, **110**, 63–69.

Hu T. and White R.E., 1998. Robust multiple suppression using adaptive beamforming. *Geophys. Prosp.*, **46**, 227–248.

Hubral P., 1977. Time migration: some ray theoretical aspects. *Geophys. Prosp.*, **25**, 738–745.

Ivansson S., 1986. Some remarks concerning seismic reflection tomography and velocity analysis. *Geophys. J. R. Astr. Soc.*, **87**, 539–557.

Jackson D.D., 1979. The use of *a priori* data to resolve nonuniqueness in linear inversion. *Geophys. J. R. Astr. Soc.*, **57**, 137–157.

Jannaud L.R., 1995. Reliability of traveltime data computed from interpreted migrated events. *J. Geophys. Res.*, **100**, 2135–2149.

Johansen T.A., Bruland L. and Lutro J., 1995. Tracking the amplitude versus offset (AVO) by using orthogonal polynomials. *Geophys. Prosp.*, **43**, 245–261.

Juhlin C. and Young R., 1993. Implications of thin layers for amplitude variation with offset (AVO) studies. *Geophysics*, **58**, 1200–1204.

Kendall J.-M. and Thomson C.J., 1989. A comment on the form of the geometrical spreading equations, with some numerical examples of seismic ray tracing in inhomogeneous, anisotropic media. *Geophys. J. Int.*, **99**, 401–413.

Kendall J.-M., Guest W.S. and Thomson C.J., 1992. Ray-theory Green's function reciprocity and ray-centred coordinates in anisotropic media. *Geophys. J. Int.*, **108**, 364–371.

Kendall J.-M. and Thomson C.J., 1993. Seismic modelling of subduction zones with inhomogeneity and anisotropy, I: teleseismic *P*-wavefront tracking. *Geophys. J. Int.*, **112**, 39–66.

Kennett B.L.N., Sambridge M.S. and Williamson P.R., 1988. Subspace methods for large inverse problems with multiple parameter classes. *Geophys. J.*, **94**, 237–247.

Kennett B.L.N. and Williamson P.R., 1988. Subspace methods for large-scale nonlinear inversion. In *Mathematical Geophysics: A Survey of Recent Developments in Seismology and Geodynamics*, Vlaar N.J., Nolet G., Wortel M.J.R. and Cloetingth S. (eds.), D. Reidel Publishing, pp. 139–154.

Kim K.Y., Wrolstad K.H. and Aminzadeh F., 1993. Effects of transverse isotropy on *P*-wave AVO for gas sands. *Geophysics*, **58**, 883–888.

Kjartansson E., 1979. Constant Q wave propagation and attenuation. *J. Geophys. Res.*, **84**, 4737–4748.

Kline M. and Kay I.W., 1965. *Electromagnetic Theory and Geometrical Optics*. John Wiley & Sons.

Knott C.G., 1899. Reflection and refraction of seismic waves with seismological applications. *Phil. Mag.*, **48**, 64–97.

Koefoed O., 1955. On the effect of Poisson's ratios of rock strata on the reflection coefficients of plane waves. *Geophys. Prosp.*, **3**, 381–387.

Kosloff D., Sherwood J., Koren Z., MacHet E. and Falkovitz Y., 1996. Velocity and interface depth determined by tomography of depth migrated gathers. *Geophysics*, **61**, 1511–1523.

Kosloff D.D. and Sudman Y., 2002. Uncertainty in determining interval velocities from surface reflection seismic data. *Geophysics*, **67**, 952–963.

Lailly P. and Sinoquet D., 1996. Smooth velocity models in reflection tomography for imaging complex geological structures. *Geophys. J. Int.*, **124**, 349–362.

Liner C.L., 1991. Theory of a 2.5-D acoustic wave equation for constant density media. *Geophysics*, **56**, 2114–2117.

Lines L., 1993. Ambiguity in analysis of velocity and depth. *Geophysics*, **58**, 596–597.

Liu X.-F. and Tromp J., 1996. Uniformly valid body-wave ray theory. *Geophys. J. Int.*, **127**, 461–491.

Lutter W.J. and Nowack R.L., 1990. Inversion for crustal structure using reflections from the PASSCAL Ouachita experiment. *J. Geophys. Res.*, **95**, 4633–4646.

Ma J.-F., 2002. Generalized elastic impedance forward modelling and inversion method in seismic exploration. *J. Chinese Geophys.*, **45**, issue 6.

MacDonald C., Davis P.M. and Jackson D.D., 1987. Inversion of reflection travel times and amplitudes. *Geophysics*, **52**, 606–617.

Mallick S., 1993. A simple approximation to the *P*-wave reflection coefficient and its implication in the inversion of amplitude variation with offset data. *Geophysics*, **58**, 544–552.

Mallick S., 1995. Model-based inversion of amplitude-variations-with-offset data using a genetic algorithm. *Geophysics*, **60**, 939–954.

Martinez R.D., 1993. Wave propagation effects on amplitude variation with offset measurements: a modelling study. *Geophysics*, **58**, 534–543.

Maslov V.P., 1972. *Théorie des Perturbations et Méthodes Asymptotiques*. Dunod, Paris.

Maslov V.P. and Fedoriuk M.V., 1981. *Semi-classical Approximation in Quantum Mechanics*. D. Reidel Publishing.

Matheney M.P. and Nowack R.L., 1995. Seismic attenuation values obtained from instantaneous frequency matching and spectral ratios. *Geophys. J. Int.*, **123**, 1–15.

Menke W., 1984. The resolving power of cross-borehole tomography. *Geophys. Res. Lett.*, **11**, 105–108.

Moser T.J., Nolet G. and Snieder R., 1992. Ray bending revisited. *Bull. Seism. Soc. Am.*, **82**, 259–288.

Mosher C.C., Keho T.H., Weglein A.B. and Foster D.J., 1996. The impact of migration on AVO. *Geophysics*, **61**, 1603–1615.

Müller G., 1985. The reflectivity method: a tutorial. *J. Geophys.*, **58**, 153–174.

Neele F., VanDecar J.C. and Snieder R., 1993a. A formalism for including amplitude data in tomographic inversions. *Geophys. J. Int.*, **115**, 482–496.

Neele F., VanDecar J.C. and Snieder R., 1993b. The use of *P*-wave amplitude data in a joint tomographic inversion with traveltimes for upper-mantle velocity structure. *J. Geophys. Res.*, **98**, 12033–12054.

Nercessian A., Hirn A. and Tarantola A., 1984. Three-dimensional seismic prospecting of the Mont Dore volcano, France. *Geophys. J. R. Astr. Soc.*, **76**, 307–315.

Neumann G., 1981. Determination of lateral inhomogeneities in reflection seismics by inversion of traveltime residuals. *Geophys. Prosp.*, **29**, 161–177.

Nichols D.E., 1996. Maximum energy traveltimes calculated in the seismic frequency band. *Geophysics*, **61**, 253–263.

Nolet G., 1987. Waveform tomography. In *Seismic Tomography*, Nolet G. (ed.), D. Reidel Publishing, pp. 301–322.

Nolet G., van Trier J. and Huisman R., 1986. A formalism for nonlinear inversion of seismic surface waves. *Geophys. Res. Lett.*, **13**, 26–29.

Nowack R.L. and Lutter W.J., 1988. Linearized rays, amplitude and inversion. *Pure Appl. Geophys.*, **128**, 401–421.

Nowack R.L. and Lyslo J.A., 1989. Fréchet derivatives for curved interfaces in the ray approximation. *Geophys. J.*, **97**, 497–509.

Nowack R.L. and Matheney M.P., 1997. Inversion of seismic attributes for velocity and attenuation structure. *Geophys. J. Int.*, **128**, 689–700.

O'Doherty R.F. and Anstey N.A., 1971. Reflections on amplitudes. *Geophys. Prosp.*, **19**, 430–458.

Oldenburg D.W., McGillivray P.R. and Ellis R.G., 1993. Generalized subspace methods for large-scale inverse problems. *Geophys. J. Int.*, **114**, 12–20.

Ory J. and Pratt R.G., 1995. Are our parameter estimators biased? the significance of finite-difference regularization operators. *Inverse Problems*, **11**, 397–424.

Ostrander W.J., 1984. Plane-wave reflection coefficients for gas sands at nonnormal angles of incidence. *Geophysics*, **49**, 1637–1648.

Pereyra V., 1996. Modelling, ray tracing and block nonlinear traveltime inversion. *Pure Appl. Geophys.*, **148**, 345–386.

Pratt R.G. and Chapman C.H., 1992. Traveltime tomography in anisotropic media, II: application. *Geophys. J. Int.*, **109**, 20–37.

Press W.H., Flannery B.P., Teukolsky S.A. and Vetterling W.T., 1989. *Numerical Recipes*. Cambridge University Press.

Ramos A.C.B. and Davis T.L., 1997. 3-D AVO analysis and modelling applied to fracture detection in coalbed methane reservoirs. *Geophysics*, **62**, 1683–1695.

Richards P.G. and Frasier C.W., 1976. Scattering of elastic waves from depth-dependent inhomogeneities. *Geophysics*, **41**, 441–458.

Richards P.G. and Menke W., 1983. The apparent attenuation of a scattering medium. *Bull. Seism. Soc. Am.*, **73**, 1005–1021.

Rutherford S.R. and Williams R.H., 1989. Amplitude-versus-offset variations in gas sands. *Geophysics*, **54**, 680–688.

Sambridge M.S., 1990. Non-linear arrival time inversion: constraining velocity anomalies by seeking smooth models in 3-D. *Geophys. J. Int.*, **102**, 653–677.

Santos L.T. and Tygel M., 2002. Reflection impedance. *Expanded Abstracts*, 72nd Annual Meeting of Society of Exploration Geophysicists, Salt Lake City, 225–228.

Scales J.A., Gersztenkorn A. and Treitel S., 1988. Fast l_p solution of large, sparse, linear systems: application to seismic traveltime tomography. *J. Comput. Phys.*, **75**, 314–333.

Shapiro S.A. and Hubral P., 1996. Elastic waves in finely layered sediments: the equivalent medium and generalized O'Doherty-Anstey formulas. *Geophysics*, **61**, 1282–1300.

Shapiro S.A., Zien H. and Hubral P., 1994. A generalized O'Doherty-Anstey formula for waves in finely layered media. *Geophysics*, **59**, 1750–1762.

Sheriff R.E., 1991. *Encyclopaedic Dictionary of Exploration Geophysics (3rd Edition)*. Society of Exploration Geophysicists.

Sheriff R.E. and Geldart L.P., 1995. *Exploration Seismology*. Cambridge University Press.

Shuey R.T., 1985. A simplification of the Zöppritz equations. *Geophysics*, **50**, 609–614.

Simmons J.L. Jr. and Backus M.M., 1994. AVO modelling and the locally converted shear wave. *Geophysics*, **59**, 1237–1248.

Simmons J.L. Jr. and Backus M.M., 1996. Waveform-based AVO inversion and AVO prediction-error. *Geophysics*, **61**, 1575–1588.

Smith G.C. and Gidlow P.M., 1987. Weighted stacking for rock property estimation and detection of gas. *Geophys. Prosp.*, **35**, 993–1014.

Smith G.C. and Sutherland R.A., 1996. The fluid factor as an AVO indicator. *Geophysics*, **61**, 1425–1428.

Snieder R. and Lomax A., 1996. Wavefield smoothing and the effect of rough velocity perturbations on arrival times and amplitudes. *Geophys. J. Int.*, **125**, 796–812.

Snieder R. and Spencer C., 1993. A unified approach to ray bending, ray perturbation and paraxial ray theories. *Geophys. J. Int.*, **115**, 456–470.

Spratt R.S., Goins N.R. and Fitch T.J., 1993. Pseudo-shear: the analysis of AVO. In *Offset-Dependent Reflectivity: Theory and Practice of AVO Analysis*, Castagna J.P. and Backus M.M. (eds.), Society of Exploration Geophysicists, pp. 37–56.

Stefani J.P., 1995. Turning-ray tomography. *Geophysics*, **60**, 1917–1929.

Stolt R.H., 1978. Migration by Fourier transform. *Geophysics*, **43**, 23–48.

Stolt R.H. and Weglein A.B., 1985. Migration and inversion of seismic data. *Geophysics*, **50**, 2458–2472.

Stork C., 1992a. Reflection tomography in the postmigrated domain. *Geophysics*, **57**, 680–692.

Stork C., 1992b. Singular value decomposition of the velocity-reflector depth tradeoff, part 1: introduction using a two-parameter model; part 2: high-resolution analysis of a generic model. *Geophysics*, **57**, 927–943.

Stork C. and Clayton R.W., 1992. Using constraints to address the instabilities of automated prestack velocity analysis. *Geophysics*, **57**, 404–419.

Swan H.W., 1991. Amplitude versus offset measurement errors in a finely layered medium. *Geophysics*, **56**, 41–49.

Tarantola A., 1987a. Inversion of travel times and seismic waveforms. In *Seismic Tomography*, Nolet G. (ed.), D. Reidel Publishing, pp. 135–157.

Tarantola A., 1987b. *Inverse Problem Theory: Methods for Data Fitting and Model Parameter Estimation*. Elsevier Science.

Tarantola A. and Vallette B., 1982. Generalized nonlinear inverse problems solved using the least squares criterion. *Rev. Geophys. Space Phys.*, **20**, 219–232.

Thomson C.J., 1983. Ray-theoretical amplitude inversion for laterally varying velocity structure below NORSAR. *Geophys. J. R. Astr. Soc.*, **74**, 525–558.

Thomson C.J. and Chapman C.H., 1985. An introduction to Maslov's asymptotic method. *Geophys. J. R. Astr. Soc.*, **83**, 143–168.

Tieman H.J., 1994. Investigating the velocity-depth ambiguity of reflection traveltimes. *Geophysics*, **59**, 1763–1773.

Tygel M., Schleicher J. and Hubral P., 1992. Geometrical spreading correction of offset reflections in a laterally inhomogeneous earth. *Geophysics*, **57**, 1054–1063.

Ursin B., 1990. Offset-dependent geometrical spreading in a layered medium. *Geophysics*, **55**, 492–496.

Ursin B. and Dahl T., 1992. Seismic reflection amplitudes. *Geophys. Prosp.*, **40**, 483–512.

Ursin B. and Ekren B.O., 1995. Robust AVO analysis. *Geophysics*, **60**, 317–326.

Ursin B., Ekren B.O. and Tjåland E., 1996. Linearized elastic parameter sections. *Geophys. Prosp.*, **44**, 427–455.

Ursin B. and Tjåland E., 1993. The accuracy of linearized elastic parameter estimation. *J. Seismic Expl.*, **2**, 349–363.

Ursin B. and Tjåland E., 1996. The information content of the elastic reflection matrix. *Geophys. J. Int.*, **125**, 214–228.

VerWest B., Masters R. and Sena A., 2000. Elastic impedance inversion. *Expanded Abstracts*, 70th Annual Meeting of Society of Exploration Geophysicists, Calgary, 1580–1582.

Virieux J., 1991. Fast and accurate ray tracing by Hamiltonian perturbation. *J. Geophys. Res.*, **96**, 579-594.

Virieux J. and Farra V., 1991. Ray tracing in 3-D complex isotropic media: an analysis of the problem. *Geophysics*, **56**, 2057–2069.

Virieux J., Farra V. and Madariaga R., 1988. Ray tracing in laterally heterogeneous media for earthquake location. *J. Geophys. Res.*, **93**, 6585–6599.

Walden A.T. and White R.E., 1984. On errors of fit and accuracy in matching synthetic seismograms and seismic traces. *Geophys. Prosp.*, **32**, 871–891.

Wang Y., 1999a. Simultaneous inversion for model geometry and elastic parameters. *Geophysics*, **64**, 182–190.

Wang Y., 1999b. Approximations to the Zöppritz equations and their use in AVO analysis. *Geophysics*, **64**, 1920–1927.

Wang Y., 2002. A stable and efficient approach to inverse Q filtering. *Geophysics*, **67**, 657–663.

Wang Y., 2003. Quantifying the effectiveness of stabilized inverse Q filtering. *Geophysics*, **68**, issue 1.

Wang Y. and Houseman G.A., 1994. Inversion of reflection seismic amplitude data for interface geometry. *Geophys. J. Int.*, **117**, 92–110.

Wang Y. and Houseman G.A., 1995. Tomographic inversion of reflection seismic amplitude data for velocity variation. *Geophys. J. Int.*, **123**, 355–372.

Wang Y. and Houseman G.A., 1997. Point-source τ-p: a review and comparison of computational methods. *Geophysics*, **62**, 325–334.

Wang Y. and Pratt R.G., 1997. Sensitivities of seismic traveltimes and amplitudes in reflection tomography. *Geophys. J. Int.*, **131**, 618–642.

Wang Y. and Pratt R.G., 2000. Seismic amplitude inversion for interface geometry of multi-layered structures. *Pure Appl. Geophys.*, **157**, 1601–1620.

Wang Y., White R.E. and Pratt R.G., 2000. Seismic amplitude inversion for interface geometry: practical approach for application. *Geophys. J. Int.*, **142**, 162–172.

Wang Y., Worthington M.H. and Pratt R.G., 2002. Decomposition of structural amplitude effect and AVO attributes: application to a gas-water contact. *Pure Appl. Geophys.*, **159**, 1305–1320.

Wapenaar C.P.A. and Herrmann F.J., 1996. True-amplitude migration taking fine layering into account. *Geophysics*, **61**, 795–803.

Wapenaar C.P.A., Verschuur D.J. and Herrmann P., 1992. Amplitude preprocessing of single and multicomponent seismic data. *Geophysics*, **57**, 1178–1188.

Whitcombe D.N., 1994. Fast model building using demigration and single-step ray migration. *Geophysics*, **59**, 439–449.

Whitcombe D.N and Carroll R.J., 1994. The application of map migration to 2-D migrated data. *Geophysics*, **59**, 1121–1132.

Whitcombe D. N., Connolly P. A., Reagan R. L. and Redshaw T. C., 2002. Extended elastic impedance for fluid and lithology prediction. *Geophysics*, **67**, 63–67.

White D.J., 1989. Two-dimensional seismic refraction tomography. *Geophys. J.*, **97**, 223–245.

White R.E., 1980. Partial coherence matching of synthetic seismograms with seismic traces. *Geophys. Prosp.*, **28**, 333–358.

Wiggins R.A., 1976. Body wave amplitude calculations II. *Geophys. J. R. Astr. Soc.*, **46**, 1–10.

Wiggins R.A. and Madrid J.A., 1974. Body wave amplitude calculations. *Geophys. J. R. Astr. Soc.*, **37**, 423–433.

Williamson P.R., 1990. Tomographic inversion in reflection seismology. *Geophys. J. Int.*, **100**, 255–274.

Williamson P.R. and Pratt R.G., 1995. A critical review of acoustic wave modelling procedures in 2.5 dimensions. *Geophysics*, **60**, 591–595.

Wong J., Bregman N.D., West G.F. and Hurley P., 1987. Cross-hole seismic scanning and tomography. *The Leading Edge*, **6 (1)**, 36–41.

Worthington M.H., 1984. An introduction to geophysical tomography. *First Break*, **2 (11)**, 20–26.

Zelt C.A. and Ellis R.M., 1990. Crust and upper mantle Q from seismic refraction data: Peace River region. *Can. J. Earth Sci.*, **27**, 1040–1047.

Zhu X., Sixta D.P. and Angstman B.G., 1992. Tomostatics: turning-ray tomography + statics correction. *The Leading Edge*, **11 (12)**, 15–23.

Zöppritz K., 1919. Über Erdbebenwellen, VIIb. *Göttingen Nachrichten*, **1**, 66–84.

Author Index

Topic Index